Basic Math for Game Development with Unity 3D

A Beginner's Guide to Mathematical Foundations

Second Edition

Kelvin Sung
Gregory Smith

Figures and illustrations: Clover Wai

Apress®

Basic Math for Game Development with Unity 3D: A Beginner's Guide to Mathematical Foundations, Second Edition

Kelvin Sung
Bothell, WA, USA

Gregory Smith
Caldwell, ID, USA

ISBN-13 (pbk): 978-1-4842-9884-8
ISBN-13 (electronic): 978-1-4842-9885-5
https://doi.org/10.1007/978-1-4842-9885-5

Managing Director, Apress Media LLC: Welmoed Spahr
Acquisitions Editor: Spandana Chatterjee
Development Editor: James Markham
Coordinating Editor: Mark Powers

Cover designed by eStudioCalamar

Distributed to the book trade worldwide by Apress Media, LLC, 1 New York Plaza, New York, NY 10004, U.S.A. Phone 1-800-SPRINGER, fax (201) 348-4505, e-mail orders-ny@springer-sbm.com, or visit www.springeronline.com. Apress Media, LLC is a California LLC and the sole member (owner) is Springer Science + Business Media Finance Inc (SSBM Finance Inc). SSBM Finance Inc is a **Delaware** corporation.

For information on translations, please e-mail booktranslations@springernature.com; for reprint, paperback, or audio rights, please e-mail bookpermissions@springernature.com.

Apress titles may be purchased in bulk for academic, corporate, or promotional use. eBook versions and licenses are also available for most titles. For more information, reference our Print and eBook Bulk Sales web page at http://www.apress.com/bulk-sales.

Any source code or other supplementary material referenced by the author in this book is available to readers on GitHub (https://github.com/Apress). For more detailed information, please visit https://www.apress.com/gp/services/source-code.

Paper in this product is recyclable

To my wife, Clover, and our girls, Jean and Ruth, for completing my life.

—Kelvin Sung

To my wife and our little one, thank you for making my life better each and every day.

—Gregory Smith

Table of Contents

About the Authors

Kelvin Sung is Professor with the Computing and Software Systems Division at the University of Washington Bothell (UWB). He received his Ph.D. in Computer Science from the University of Illinois at Urbana-Champaign. Kelvin's background is in computer graphics, hardware, and machine architecture. He came to UWB from Alias|Wavefront (now part of Autodesk), where he played a key role in designing and implementing the Maya Renderer, an Academy Award–winning image generation system. At UWB, funded by Microsoft Research and the National Science Foundation, Kelvin's work focuses on the intersection of video game mechanics, solutions to real-world problems, and mobile technologies. Together with his students and colleagues, Kelvin has co-authored six books: one in computer graphics and the others in 2D game engines with Apress.

Gregory Smith is a software engineer at Virtual Heroes, a company that focuses on creating training and simulation software in Unreal Engine. He received his undergraduate degree in Computer Science from Northwest Nazarene University in 2018 and earned a Master of Computer Science and Software Engineering degree from the University of Washington Bothell in 2020. Gregory also owns his own game company, Plus 2 Studios, which he works on in his spare time.

Acknowledgments

This book and the projects it relates to stem from the results of the authors' attempts to understand how to engage learners in exploring knowledge related to interactive computer graphics, introductory programming, and video games. Past funding for related projects includes support from the National Science Foundation for the projects "Essential Concepts for Building Interactive Computer Graphics Applications" (Award Number, CCLI-EMD, NSF, DUE-0442420) and "Game-Themed CS1/2: Empowering the Faculty" (Award Number DUE-1140410). Projects supported by Microsoft Research and Microsoft Research Connections include "XNA Based Game-Themed Programming Assignments for CS1/2" (Award Number 15871) and "A Traditional Game-Themed CS1 Class" (Award Number 16531). All of these past projects have laid the foundation for our perspectives and presentation of the materials in this book. We would also like to thank NSF officers Suzanne Westbrook, Jane Prey, Valerie Bar, and Paul Tymann for their invaluable discussions and encouragements, as well as Donald Brinkman and Kent Foster as they continue to be our best advocate and supporters at Microsoft. Lastly, we remember and continue to miss Steve Cunningham, John Nordlinger, and Lee Dirks for their early recognition of our vision and ideas.

A thank you must also go out to our students, whose honest, even when brutal, feedbacks and suggestions from CSS385: Introduction to Game Development, CSS451: 3D Computer Graphics, CSS452: Game Engine Development, and CSS551: Advanced 3D Computer Graphics at the University of Washington Bothell inspired us to explore the approach to present these materials based on an accessible game engine. They have tested, retested, contributed to, and assisted in the formation and organization of the contents of this book. The second author of this book is an alumnus of CSS551.

It must also be mentioned that the teaching brown bag hosted by Yusuf Pisan offered the opportunity for the discussions with Yusuf, Johnny Lin, Lesley Kalmin, and Mike Stiber on the topics of linear algebra applications which sparked the initial idea for this book. A sincere thank you goes to Yusuf for his enthusiasm and energy in organizing us and, of course, for the delicious-looking Tim Tam; one day, I will try them.

We also want to thank Spandana Chatterjee for believing in our ideas, her patience, and continual efficient and effective support. Nirmal Selvaraj organized everything and ensured proper progress was ongoing.

Lastly, a thank you must go to Peter Shirley, our technical reviewer, whose frank and precise comments made this a much easier to understand book.

The vehicle models used are free assets, UAA - City Props - Vehicles, downloaded from the Unity Asset Store under the Unity-EULA. The cone shape that represents the arrow heads for the axis frames and vectors in all examples is created based on the utilities developed and shared by Wolfram Kresse available at `https://wiki.unity3d.com/index.php/CreateCone`. The cosine function plot from Figure 5-5 is based on a screenshot taken from `www.desmos.com/calculator/nqfu5lxaij`.

Introduction

Welcome to *Basic Math for Game Development with Unity 3D*. Because you have picked up this book, you are probably interested in finding out more about the mathematics involved in game development or, maybe, in the details of fascinating applications like Unity. This can be the perfect book to begin with your exploration.

This book uses interactive examples in Unity to present each mathematical concept discussed, taking you on a hands-on journey of learning. The coverage of each topic always follows a pattern. First, the concept and its relevancy in video game functionality are described. Second, the mathematics, with a focus on applicability in game development and interactive computer graphics, are derived. Finally, an implementation of the concept and derived mathematics are demonstrated as an example in Unity.

Through interacting with these examples, you will have the opportunity to explore the implications and limitations of each concept. Additionally, you can examine the effects of manipulating the various related parameters. Lastly, and very importantly, you can study the accompanied source code and understand the details of the implementations.

In Chapter 2, you will begin by reviewing simple number intervals in the Cartesian Coordinate System. Chapters 3 and 4 let you examine and learn about vectors and the rules of their operations to formally relate positions in 3D space. Chapters 5 and 6 study the vector dot and cross products to relate vectors and the space that defines them. Chapter 7 leads you to work in multiple coordinate spaces simultaneously to address compound issues such as describing motions inside a navigating spaceship. Chapter 8 introduces quaternions and the rotation operator and Chapter 9 concludes with the basic math involved in game development. Throughout this book, you will learn the mathematical and implementation details of bounding boxes; bounding spheres; motion controls; ray castings; projecting points to lines and planes; computing intersections between fast-traveling objects; projecting objects onto 2D planes to create shadows; computing reflections; working in multiple coordinate spaces; rotations to align vectors; and much more!

Who Should Read This Book

This book is targeted toward video game enthusiasts and hobbyists who have some background in basic object-oriented programming. For example, if you are a student who has taken an introductory programming course, or are a self-taught programming enthusiast, you will be able to follow the concepts and code presented in this book with little trouble. If you do not have any programming background in general, it is suggested that you first become comfortable with the C# programming language before tackling the content provided in this book.

Besides a basic understanding of object-oriented programming, you will also need to be familiar with the Cartesian Coordinate System, basic algebra, and knowledge in trigonometry. Experience and working knowledge with Unity are not required.

Code Samples

Every chapter in this book includes examples that let you interactively experiment with and learn the new materials. You can download the source code for all the projects from the following page: `www.apress.com/`.

CHAPTER 1

Introduction and Learning Environment

After completing this chapter, you will be able to

- Know the details of what this book is about
- Understand the style that this book uses to present concepts
- Install Unity and an Integrated Development Environment (IDE) for developing programming code
- Access the accompanying source code and run the example projects
- Understand the Unity terminology used throughout this book
- Begin to appreciate the intricate details of math for game development

Introduction

When you think of math in a video game, you may picture health bars, attack stats, experience points, and other game mechanics. You may not consider the underlying math that enables the in-game physics world, such as calculating gravity, movements, or enemy chasing behaviors. Additionally, you may not consider physical interaction in a mathematical manner, such as collisions between different objects and the reflections of these objects after they collide. These underlying mathematical computations are critical to implementing a successful video game. When creating a game, whether you intend on using a game engine or you intend on performing the computations yourself, understanding the details and knowing how the underlying mathematics work and when to use them to create what you want, where you want, is vital.

K. Sung and G. Smith, *Basic Math for Game Development with Unity 3D*,
https://doi.org/10.1007/978-1-4842-9885-5_1

Traditionally, math is taught without any application contexts. Typically, theories are developed based on abstract symbols, formulas are derived to support these theories, and then numbers are used to verify the formulas. You are tested on whether you can generate the correct solution based on how the formulas are applied. It is believed that learning math in this manner has the benefit of granting learners the ability to understand the concepts being taught at the pure abstraction level. Then, once understood, the application of these concepts to different disciplinary contexts becomes straightforward. For many learners, this assumption is certainly true. However, for other types of learners, it can be difficult to appreciate the intricate details in the abstract without concrete examples or applications to build off. This fact is recognized by educators and often story problems are introduced after a basic understanding is established to help learners gain insights and appreciate the formulas. This learning approach is taken on and exploited in the context of linear algebra and video games.

This book takes you on the journey of learning linear algebra, a branch of mathematics that is the foundation of interactive graphical applications, like video games. While the underlying theories can be abstract and complicated, the application of these theories in graphical object interactions is relatively straightforward. For this reason, this book approaches linear algebra topics in a concrete manner, based around game-like examples that you can interact with. Through this book, you will learn a flavor of linear algebra that is directly applicable to video games and interactive computer graphics as a whole.

Every math concept presented in this book is accompanied with concrete examples that you can interact with and are relevant to video game development. It is the intent of this book that you will learn and know how to apply the concepts in solving the problems you are likely to encounter during game development. A direct consequence of this focused approach is that readers may find it challenging to apply the knowledge gained throughout this book to other disciplines, like machine learning or computer vision. For example, the dot product, which will be covered in Chapter 5, can be used to calculate intersection positions, and it can also be used in machine learning algorithms as a data reduction tool; however, this book will only focus on the video game applications of the dot product. If you are looking for general knowledge in linear algebra, you should consider a more traditional textbook. Such a book is likely to cover concepts at levels that are suitable for applications for multiple problem spaces. If you are interested in solving problems specific to interacting graphical objects, especially for game development, then this is the perfect book for you.

After the introduction to the game engine and terminologies in this chapter, Chapter 2 reviews the Cartesian Coordinate System and number intervals leading to the exploration of one of the most widely used tools in game development—bounding boxes. Chapter 3 continues bounding volume exploration by examining bounding spheres while also beginning the investigation of relationships between positions. Chapter 4 introduces vectors to formalize the relationships between positions in 3D space and applies vector concepts in controlling and manipulating object motions under external effects like wind or current flow. Chapter 5 presents the vector dot products to relate vectors, represents line segments based on vectors, and demonstrates the application of these concepts in computing distances between objects and motion paths when approximating potential collisions. Chapter 6 discusses the vector cross product, derives the space that defines vectors, defines vector plane equation, and illustrates the application of these concepts in computing intersections and reflections of moving objects and 2D planes. Chapter 7 examines the axis frame, or the derived space that contains vectors, analyzes the representation of vectors in different axis frames, and explains how to work with movements in axis frames that are dynamically changing, such as object motions in a navigating spaceship. Chapter 8 introduces the quaternion as a tool for rotating vectors, analyzes the relevant properties of quaternions, and demonstrates the alignments of 3D spaces based on quaternions. Finally, Chapter 9 summarizes all of the concepts presented in an aggregated example.

Choice of Unity Engine

Unity is the choice of platform for presenting the mathematical concepts covered in this book for three reasons. First, Unity provides elaborate utilities and efficient support for its user to implement and visualize solutions based on mathematical formulas. Its application programming interface (API) implements the basic and many advanced linear algebra functionalities, while the Entity-Component-System (ECS) game object architecture allows straightforward user scripting. These qualities give Unity a close pairing of math concepts to your programming code, assisting in the visualization of the mathematical solution that you are trying to understand. This close pairing cannot be understated and is the backbone of this book.

The second reason for choosing Unity is that, being a game engine, the system allows for a high degree of intractability with the solution as well as the ability to visualize that solution. For example, in addition to being able to examine the results of a ray and 2D plane intersection computation in real time, you will also be able to manipulate the ray and the 2D plane to observe the effects on the intersection. The ability to interact, manipulate, and examine the application of mathematical concepts in real time will give you a greater understanding and appreciation for that concept. Third and finally, Unity is chosen because there is no better way to learn math concepts for video games than through a popular game engine!

While this book is meant for readers who may be interested in building a video game in Unity, the focus of this book is on the math concepts and their implementations and not on how to use Unity. This book teaches the basic mathematical concepts that are relevant to video game development using Unity as a teaching instrument. This book does not teach how to use the math provided by Unity in building video games. You should focus on understanding the math rather than the Unity-specific functionality. For example, a position in 3D space in Unity is located at `transform.localPosition`; you should focus on working with that position and not be concerned about the `Unity.Transform` class. Ultimately, you should be able to take what you have learned in this book and apply to developing games in any game engine.

Note Unity Technologies is the name of the company; the game engine is most often referred to as Unity, though it is sometimes called Unity 3D. For simplicity, this book refers to the entire game engine system as Unity.

Setting Up Your Development Environment

There are two main applications that you will work with when using Unity. The first is the game engine editor, which will be referred to as Unity or Unity Editor throughout this book. The Unity Editor can be thought of as the graphical interface to the Unity game engine. The second application you will need is a script editing Integrated Development Environment (IDE). Microsoft's Visual Studio Community 2019 is the IDE of choice for developing the C# script examples in this book. This software will be referred to as the Script Editor, or the IDE, throughout the rest of this book.

To begin your download and installation of Unity and Visual Studio Community 2019, go to `https://store.unity.com/download?ref=personal`, accept the terms, and then download Unity Hub.

Note If you ever find yourself stuck at a certain point in this book, whether on installing Unity or just using it, there is a plethora of tutorials online, many of which were referenced in the development of this book and will be listed at the end of this chapter.

Notes on Installing Unity

This book is based on Unity in its most basic form. Unless you know what to specify when installing features or desire extra features, it is suggested you follow the default settings. Please begin downloading, installing, and launching the Unity Hub if you haven't already. When Unity Hub is up and running, navigate to the **Installs** tab on the left side, and select the **Install Editor** button in the top right. From here, you will be prompted with a list of different Unity versions. The version that this book uses is 2021.3.25f1. If you do not see this version in the selected list, you can go to this link `https://unity3d.com/get-unity/download/archive` and find it there to download. It should be noted that while this book is based on Unity 2021.3.25f1, any version at or newer than this version should suffice but is not guaranteed.

After selecting your Unity version, you will be prompted with options to install extra features. As mentioned previously, this textbook only requires the default options. These options, if you are running on Windows 10 or 11, should only be the suggested IDE, "Microsoft Visual Studio Community 2019." If you already have Visual Studio 2019 installed, then you may uncheck that option. Once you have selected all the features you want, begin the install process and then move onto the next section to begin familiarizing yourself with the source code used throughout this book.

Unity Editor Environment

It should be noted, again, that in this book Unity is used as a tool for learning math concepts for game development and not as a game building editor. This means many Unity-specific and game building–related information that do not pertain to the concept at hand will simply be skipped. For example, this book does not discuss how to create or save Scenes or how to build a final executable game. If these are subjects of interests, you should consider research through the many online tutorials or for example refer to the **Learn** tab of the Unity Hub. It should also be noted that all examples throughout this book will be run and interacted with through the editor and not as games. This will become clearer as the first example is discussed.

Now that you have Unity and the IDE installed and ready to go, you can refer to the GitHub repository located at `https://github.com/Apress/Basic-Math-for-Game-Development-with-Unity-3D`. After downloading the repository, open Unity Hub and add the `Chapter-1Introduction` project. Directions on how to do this can be seen in Figure 1-1.

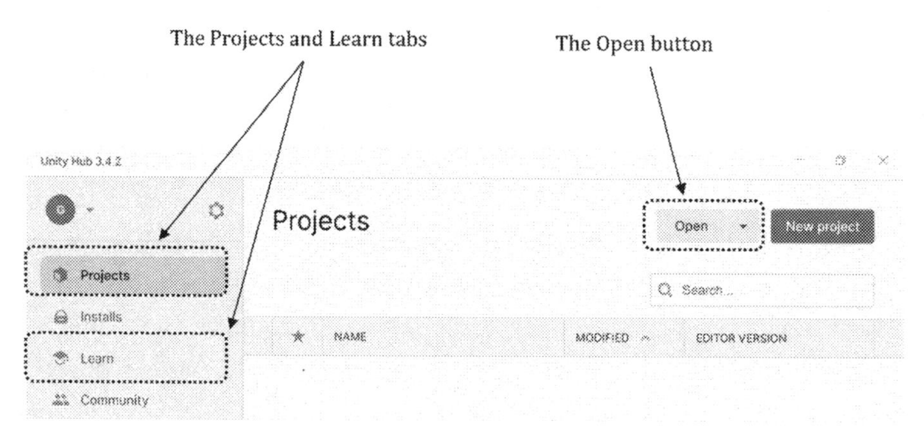

***Figure 1-1.** Opening `Chapter-1-Introduction` (the Intro to Unity Project) from Unity Hub*

As Figure 1-1 shows, to add a project, navigate to the **Projects** tab and then select the **Open** button. From here, navigate to where you downloaded the source code to this book. You will notice that the file structure is organized according to chapters. The first example you should open using the **Open** button is `Chapter-1-Introduction`. Note that after a project is opened, you need to click the newly opened project to launch it.

Figure 1-1 also establishes where the **Learn** tab is located. Here you can view and select Unity sponsored tutorials. The "Foundational Tutorial" category contains tutorials that will be very helpful to those who have never used Unity before as it contains tutorials such as "Welcome to Unity Essentials" and "Explore the Unity Editor." At the end of this chapter, there are some additional suggestions as to which tutorials to follow if you are new to Unity or just need a refresher.

Opening the Intro to Unity Project

To open a project from Unity Hub, simply click it. The first time you try to open any projects from this book, you will encounter the following two steps:

- Unity will invite you to select the version to use; you can simply select the version you just installed.
- Unity will display an information dialog box titled, "`Opening Project in Non-Matching Editor Installation`," you can simply click the `Continue` button.

The first time opening a project will take a while for Unity to copy the support library and perform system configuration. Once you open `Chapter1-Introduction`, you should be confronted with a window similar to the screenshot in Figure 1-2. If you do not see a screen similar to that of Figure 1-2, make sure the `IntroToUnity` scene is open and not an `Untitled` scene. To open the `IntroToUnity` scene, find it in Asset folder under the Project Tab and double-click to open it.

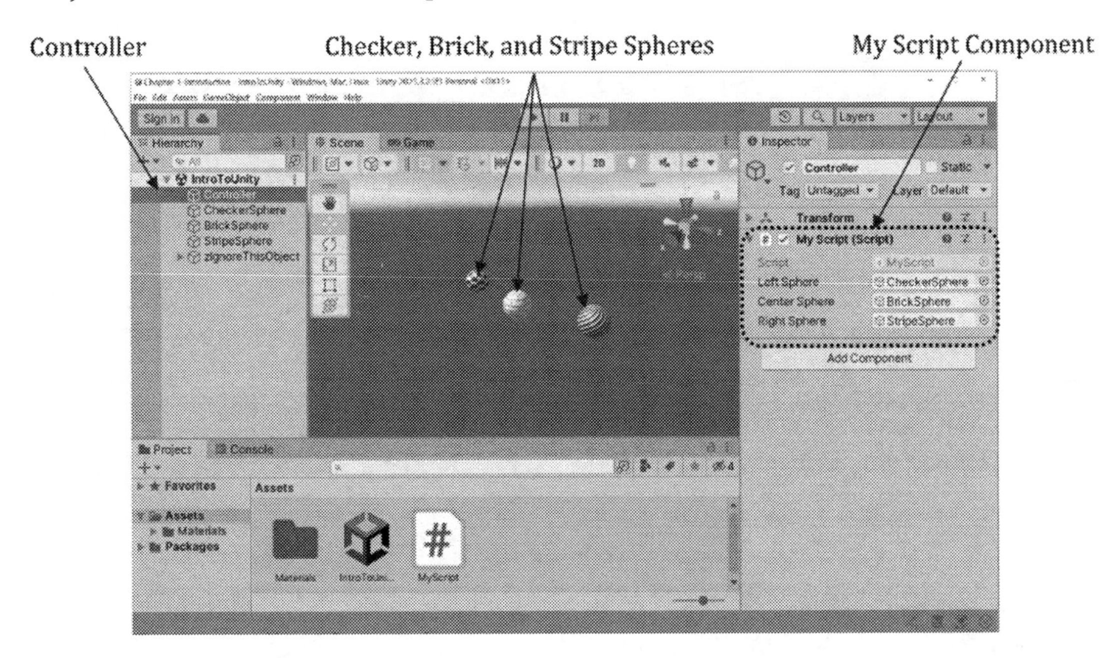

Figure 1-2. *Running the IntroToUnity scene in the Chapter-1Introduction project*

Figure 1-2 shows a very simple scene. There is the Controller game object and three different spheres. Each sphere is named after the design pattern placed upon it: CheckerSphere, BrickSphere, and StripeSphere. In this screenshot, the Controller object is selected so you can observe the MyScript component on the right. The Controller object and the MyScript component are present in every example in this book and will be described in detail. The purpose of this example is to familiarize you with how examples are organized and to establish terminologies that will be used throughout the book.

Working with the Unity Editor

Figure 1-2 is an example of what the Unity Editor looks like and is one of the two editors you will be working in. The other editor, the Script Editor, or IDE, will be discussed later. Figure 1-3 illustrates the various functionalities of the Unity Editor.

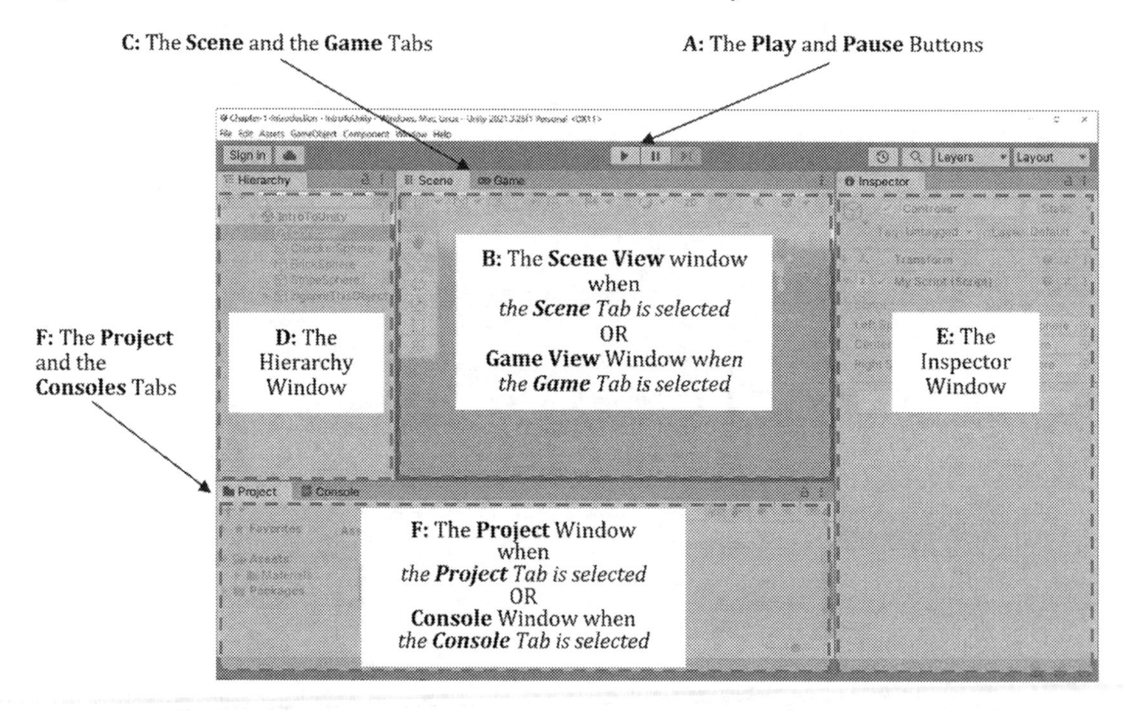

Figure 1-3. *The Unity Editor Environment*

Figure 1-3 overlays the editor in Figure 1-2 with labels identifying the different windows presented by the Unity Editor and establishes the terminologies that will be used from here on:

- A: The Play and Pause buttons: In the top-center area, you can see the Play and Pause buttons. These buttons control the running (or playing) of the game. Feel free to click the Play button, give the system a few seconds to load, and then observe the movements of the spheres in the scene. If you click the Play button again, the game will stop running. You will learn more about and work with these buttons later.
- B: The Scene View window: The main 3D window in the top-left region of the Unity Editor is the main area for performing interactive editing. In Figure 1-2, this window is displaying the Scene View of the game.
- C: The Scene and the Game View tabs: Above the Editor Window (B), you can spot the Scene and Game tabs. If you select the Game tab, then Unity will switch to the Game View which is what a player will see in an actual game. An example of the Scene View next to the Game View can be seen in Figure 1-4.

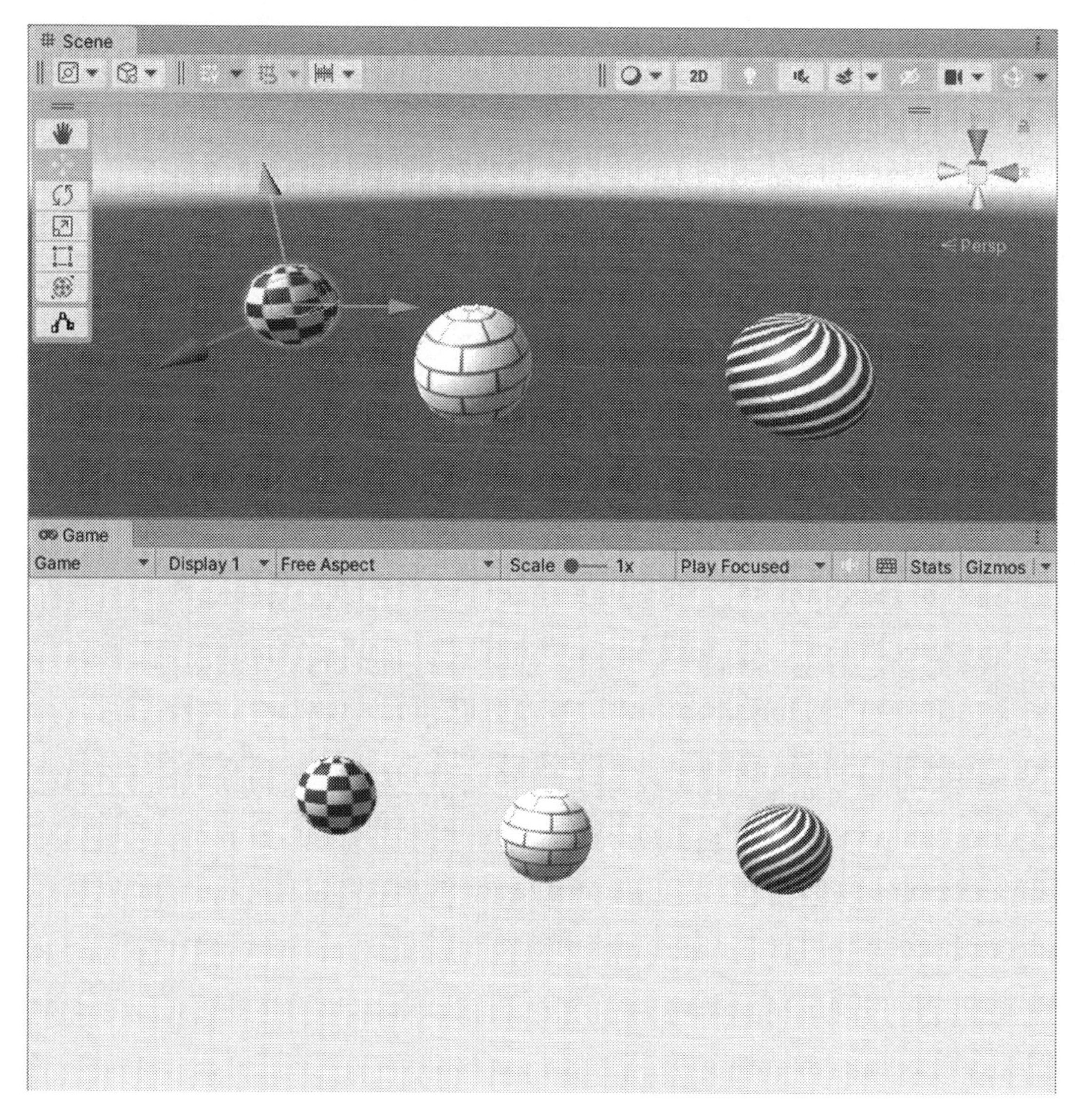

Figure 1-4. *The Scene View (top) and the Game View (bottom)*

Note Please pay attention to the differences between the Scene and Game Views. The Scene View is meant for the game designer to set up a game scene, while the Game View is what a player of the game would observe while playing the game. While both views can be invaluable tools for examining the intricate details of the mathematical concepts, you will be working exclusively with the Scene View.

Note To help distinguish between the Scene and the Game Views, as depicted in Figure 1-4, in all the examples for this book, the Scene View has a skybox-like background, while the Game View window has a constant, light blue backdrop. Once again, you will be working exclusively with the Scene View, the view with the skybox-like background.

EXERCISE

Working with the Scene View Window

Left-click and drag the Scene View tab to see that you can configure and place the Scene View window at different configuration locations throughout the Unity Editor or even outside as an independent window. This is the case for most of the Unity tabs, including the Game View window. Figure 1-4 shows the Scene View and Game View windows as two separate windows that can be examined simultaneously.

Figure 1-5 is a close-up view of the Hierarchy Window, which is labeled as D in Figure 1-3.

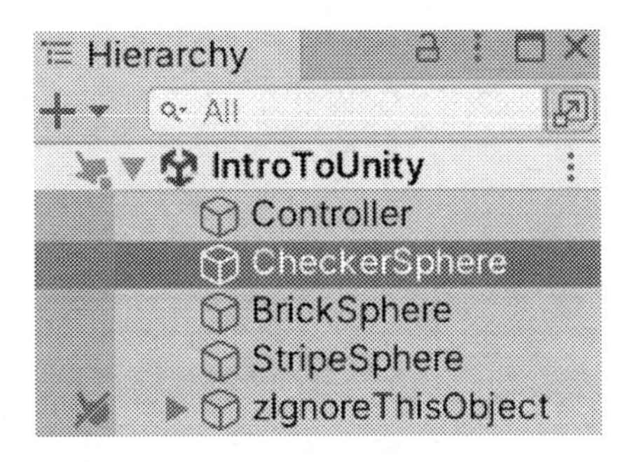

Figure 1-5. *The Hierarchy Window*

Note The crossed-out finger icon next to the last object, `zIgnoreThisObject`, disables click-select functionality in the editor window. In all examples, objects that are not meant to be interacted will have the crossed-out icon next to them.

- D: The Hierarchy window: In the Unity Editor, this window (Figure 1-5) is typically anchored to the left of the Scene View and above the Project/Console Windows (F). The Hierarchy Window displays every object and its parental relationship to other objects in the scene. Just like the Scene View and Game View, the Hierarchy Window can be moved and placed wherever you desire. You should observe the different objects within the Hierarchy Window. There is the `Controller`, which will be discussed later, but for now know that it contains the script that supports your interaction with the scene; the `CheckerSphere`, which is the checkered sphere; as well as the `BrickSphere` and `StripeSphere`, which also correspond to their object's descriptions. Finally, there is the `zIgnoreThisObject` object; this last object supports the setup of the game environment for the learning of math concepts specific to each example. You will never need to interact with this object, and therefore this book will ignore this object as its details can be distracting. You are, of course, more than welcome to examine and explore this object, and any others, at your leisure.

Note Try clicking the different objects in the Hierarchy Window and observe how the Scene View highlights the object you have selected while the Game View does not. This simple feature underscores how the Scene View is meant for scene edits while the Game View is not.

EXERCISE

Observe Differences Between the Scene View and Game View

Select different spheres in the Hierarchy Window and switch between the Scene and Game Views to observe the differences between these two views. You should notice that the selected sphere is highlighted in the Scene View and not in the Game View. It is essential to differentiate between these two views when you manipulate the scene in examining concepts. Once again, and very importantly, all examples in this book work exclusively with the Scene View.

Figure 1-6 is a close-up view of the Inspector Window, which is labeled as E in Figure 1-3.

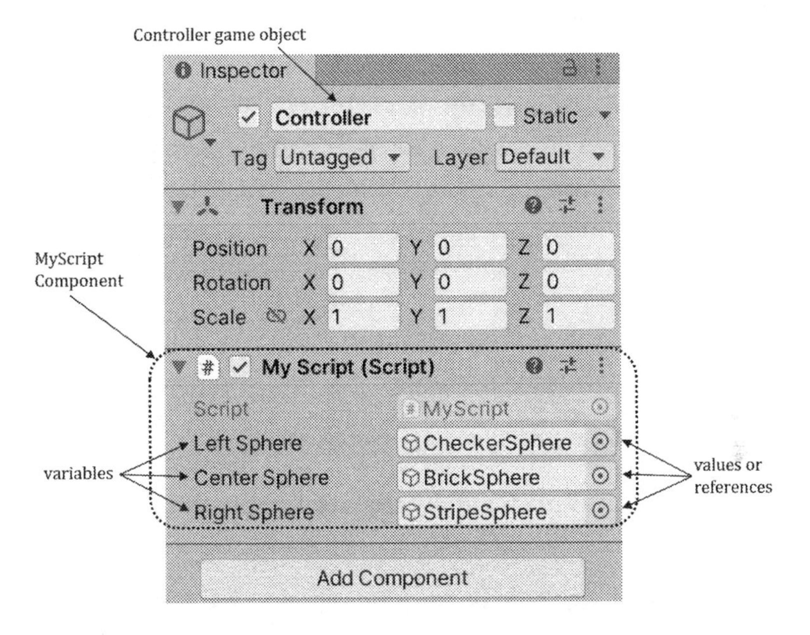

***Figure 1-6.** Inspector Window with the `Controller` object selected in the Hierarchy Window*

- E: The Inspector Window: The Inspector Window (Figure 1-6) displays the details of the selected object for the user to inspect and manipulate. The Inspector Window is typically located on the right of the Scene View. Just like all other windows described, it can be placed wherever you want. The selected object being displayed in Figure 1-6 is the `Controller`. Notice that there are two components attached to this object: `Transform` and `MyScript`. Figure 1-6 shows that you can expand and compress each of the components to examine or hide their details. In this case, the `Transform` and `MyScript` components are expanded. The `MyScript` component is the custom script developed for this book. Note that on the left side of the `MyScript` component are the names of the public variables defined in the script: `Left Sphere`, `Center Sphere`, and `Right Sphere`. Directly across from these variable names, you can see their values or the objects that the corresponding variables reference: `CheckerSphere`, `BrickSphere`, and `StripeSphere`. These aspects of the `MyScript` component will be explained in more detail in the next section.

- F: The Project and the Console windows and tabs: The Project Window displays the file structure of your project. This is where scripts, prefabs, materials, and everything else that will be loaded into your game are located. The Console Window is where Unity will output debug messages, warnings, and errors, all of which can be very helpful in debugging your code if something goes wrong. The Project Tab and Console Tab allow you to switch between these two windows just like the Game View and Scene View tabs do. These windows can also be moved around and placed wherever you decide.

Figure 1-3 shows the default layout used by this book. In the rest of this book, the corresponding windows will be referenced by their name as depicted in Figure 1-3. If you accidentally close one of these windows, they can be reopened by going to the Window drop-down menu at the top of the Unity Editor and then selecting the General option. There you will see a list of all of the windows that have been discussed.

Note In later chapters, there will be folders added to the Project Window such as Editor, Resources, and so on. These folders will include utilities that the book uses to create the examples. You are more than welcome to explore these. However, please keep in mind that the content in these folders will not be relevant to learning the mathematical concepts presented. For example, the Resources folder is a special folder that Unity searches for object blueprints known as prefabs. Knowing about these prefabs is irrelevant to learning the math concepts and therefore will not be covered.

Working with MyScript

In general, a Unity script is a component with code that can be attached to any game object. This script can then modify the behavior of that object or the entire game. All scripts presented in this book are written in C#.

Throughout this book, in each example you will only have to work with one script. This script will have `MyScript` be part of its name, for example, `EX_2_1_MyScript`, and will always be attached to the `Controller` object. It is important to note that the `Controller` object in all of the examples is empty (it does not contain anything visible) and does not perform any function other than to present the `MyScript` script for your interactions. The `MyScript` script always implements and demonstrates the concept being studied.

Figure 1-7 shows how you can open and edit `MyScript`.

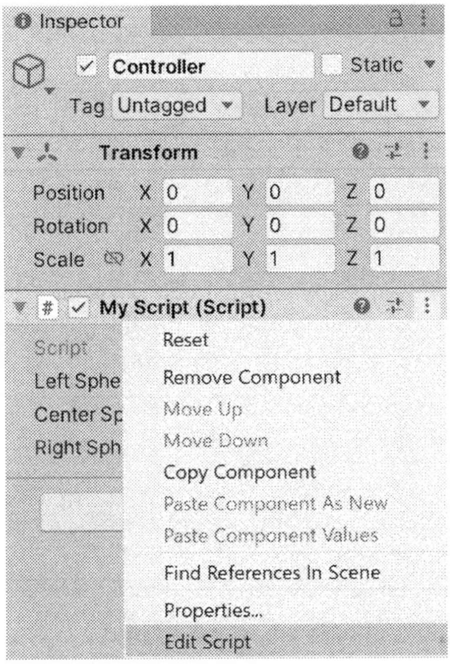

Figure 1-7. *Invoking the Script Editor*

There are two ways to open and edit scripts in Unity. The first method is depicted in Figure 1-7. To open and examine the source code of `MyScript`, select `Controller` in the Hierarchy Window, and then in the Inspector Window with the mouse pointer over the `MyScript` component, left-click the Settings button (the three-dots icon in the top right of the `MyScript` component) or right-click the name of the `MyScript` component ("`My Script (Script)`"). Both of these actions will trigger the pop-up menu as depicted in Figure 1-7. From there, select the "Edit Script" option at the very bottom. The second way to open and edit a script is by double-clicking the script icon in the Project Window. In all of the examples, `MyScript` is located in the `Assets/` folder. Once you open `MyScript`, you should see a pop-up window showing the progress of Unity invoking the IDE.

After your Script Editor has loaded, you should see a screen similar to that of Figure 1-8, which shows the `MyScript`'s code using Visual Studio under the light theme.

```
5    public class MyScript : MonoBehaviour
6    {
7        public GameObject LeftSphere = null;        // Sphere to the left in the init Editor View
8        public GameObject CenterSphere = null;      // Sphere in the center in the init Editor View
9        public GameObject RightSphere = null;       // Sphere to the right in the init Editor View
10
11       private readonly float kSmallDelta = 0.01f; // amount to translate
12
13       // Start is called before the first frame update
14       void Start()[...]
20
21       // Update is called once per frame
22       void Update()[...]
37   }
```

Figure 1-8. *Overview of the code in `MyScript`*

Figure 1-8 is a screenshot of the IDE with `MyScript` opened. Notice that the name of the script, `MyScript`, is also the name of the C# class and is a subclass of the Unity class `MonoBehavior`. Once again, the name of the script in each example will always contain the `MyScript` substring. In each example, with each script, you will only need to pay attention to the following three items:

- Variables: Make sure you take note of the variables in each script. A public variable will show as a variable that can be edited through the Inspector Window, which was seen previously in Figure 1-6. A private variable is one that can only be accessed in the code. In Figure 1-8, you can see the three public variables, `LeftSphere`, `CenterSphere`, and `RightSphere`. Notice how these are the same variables from Figure 1-6, demonstrating that public variables are indeed accessible from the Inspector Window when the corresponding game object (in this case, `Controller`'s `MyScript` component) is selected. In this example, each of the variables is of the `GameObject` type. This means each variable can hold a reference to one of the game objects in the scene. The other variable, `kSmallDelta`, is the only private variable. Notice how this variable does not appear in the Inspector Window in Figure 1-6. The `k` in front of the variable name is a convention that indicates the variable is a constant (read-only) variable.
- Start() function: This function will be called once the Play button from Figure 1-3 (A) is clicked. In this book, the `Start()` function always initializes the scene.

- Update() function: This function is called after the `Start()` function is executed and continues to be called at a real-time rate, or about 60 times per second while the Play button is active. In this book, the `Update()` function continuously computes the corresponding math concepts and supports interaction.

The `Start()` function of `MyScript` is listed as follows:

```
void Start(){
    Debug.Assert(LeftSphere != null);     // Make sure proper
                                             editor setup
    Debug.Assert(CenterSphere != null);   // Assume properly
                                             initialized to
    Debug.Assert(RightSphere != null);    // Checker, Brick,
                                             and, Stripe
}
```

In this example, the `Start()` function ensures that all of the public variables are properly initialized. Note that the `Start()` function does not attempt to make assignments to these variables; instead, it prints out an error message to the Console Window if the variables have not been assigned values by the time the user hits the Play button. In Figure 1-8 lines 7 through 9, these three public variables are set to null references. However, if you launch the game, you'll notice that these three spheres in the scene are moving. These observations indicate that the public variables must have been properly initialized somewhere such that no debug errors are printed by the `Start()` function. As will be explained, in this scenario, the user has assigned proper references to these variables through the Inspector Window.

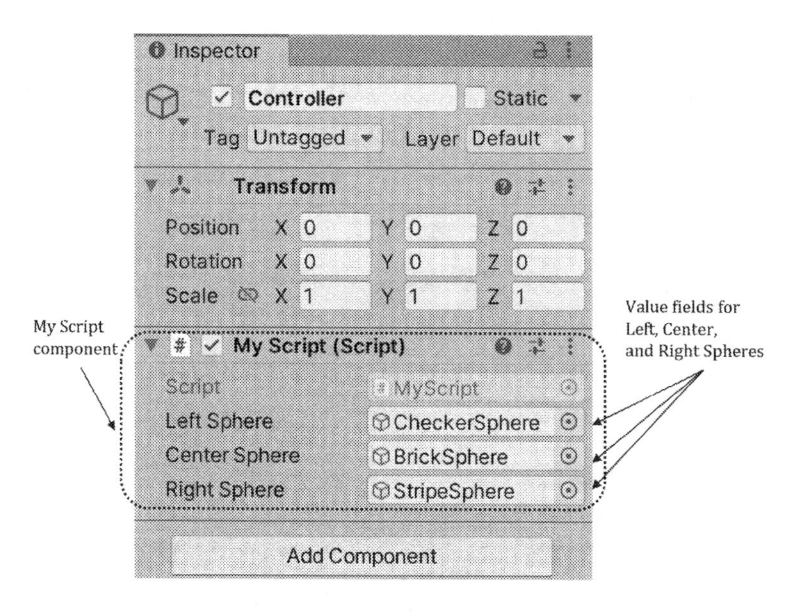

Figure 1-9. *Accessing public variables of* `MyScript` *in the Inspector Window*

Unity allows you to drag and drop game objects from the Hierarchy Window into the value fields of matching variable types in the Inspector Window to establish variable to object references. In this case, as depicted in Figure 1-9, one way to establish initial values for the three public variables is by selecting the `Controller` object in the Hierarchy Window and then dragging the `CheckerSphere` game object and dropping it in the value field of `LeftSphere` variable and the `BrickSphere` and `StripeSphere`, respectively, in the `CenterSphere` and `RightSphere` value fields. With these initial values assigned, when the script begins to run, any reference to the `LeftSphere`, `CenterSphere`, or `RightSphere` variables will result in accessing the `CheckerSphere`, `BrickSphere`, or `StripeSphere` game object in the scene. This functionality of assigning values to variables through the Inspector Window is not unique to `MyScript` and is supported for any public variable in any script.

Note For readability, Unity adopts the strategy of labeling an identifier by dividing the name at the capital letter positions. For example, the variable identifier, or name, "LeftSphere" is labeled as "Left Sphere" in the Inspector Window. For convenience and consistency, as you have already seen, this book will refer to all game objects and variables by their identifier, that is, LeftSphere.

The Update() function of MyScript is listed as follows:

```
void Update(){
    // This prints the argument string to the Console Window
    Debug.Log("Printing to Console:
               Convenient way to examine state.");
    // Update the sphere positions
    //      Left moves in the positive X-direction
    LeftSphere.transform.localPosition
        += new Vector3(kSmallDelta, 0.0f, 0.0f);

   // Center moves in the positive Y-direction
   CenterSphere.transform.localPosition
        += new Vector3(0.0f, kSmallDelta, 0.0f);

    // Right moves in the positive Z-direction
    RightSphere.transform.localPosition
        += new Vector3(0.0f, 0.0f, kSmallDelta);
}
```

The very first line of code, Debug.Log(), prints the string argument to the Console Window. Debug.Log() statements and other debug statements such as Debug.Assert(), Debug.LogWarning(), and Debug.LogError() are excellent ways of verifying the state of your application and will be used throughout this book. These debug statements will be examined more closely in an exercise at the end of this chapter.

The next three lines of code in the Update() function increment the position of each of the left, center, and right spheres by kSmallDelta in the X-, Y-, and Z-axes correspondingly. The value for this variable, as shown in Figure 1-8, is 0.01. The "f" after 0.01 indicates that this number is a floating-point data type and not a double data type.

Recall from Figure 1-9 that the LeftSphere references the CheckerSphere object, CenterSphere references the BrickSphere, and the RightSphere references the StripeSphere. Now if you click the Play button again, you should observe that the LeftSphere moves along the X-axis, the CenterSphere moves along the Y-axis, and the RightSphere moves along the Z-axis, just as the script programs. In this way, these objects' positions are controlled by MyScript. Note that the script is in control only when the Play button is depressed. Lastly, and very importantly, please ensure that you are examining the game in the Scene View and not the Game View.

Note `transform.localPosition` is how Unity accesses an object's position in 3D space. You can also access an object's position from the Inspector Window via the `Transform` component.

EXERCISES

Investigate the Manipulators

The goal of this exercise is to manipulate a selected object. As the case when working with any example in this book, make sure you are in the Scene View for this exercise. Now, click to select the `CheckerSphere` and then click the different object manipulation tools as illustrated in Figure 1-10. These object manipulation tools are located in the top left of the Unity Editor. You should experiment with each tool, especially the first four. With the first tool selected, the Hand Tool, perform the following actions:

- Move (or track) the camera: Left-click drag
- Rotate (or tumble) the camera: Right-click drag
- Zoom (or dolly) the camera: Middle mouse button wheel scroll or Alt-right-click drag

The second, third, and fourth icons activate the translate, rotate, and scale manipulators for the selected object in your scene. Try clicking the `CheckerSphere` object and then the multi-direction arrow icon to translate the `CheckerSphere's` position. You will use these object manipulation tools repeatedly when examining relevant math concepts, so make sure you are familiar with them now.

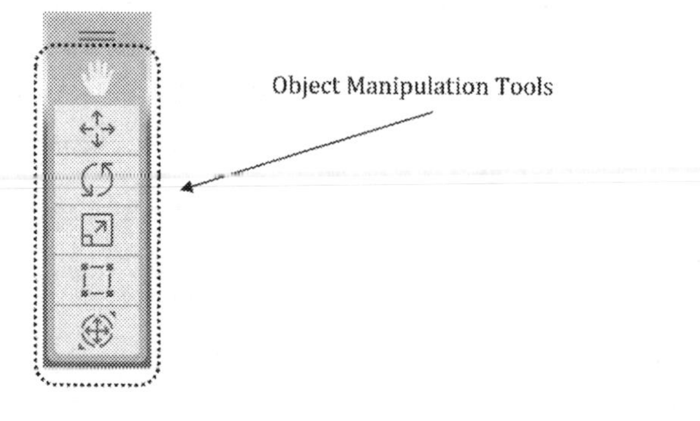

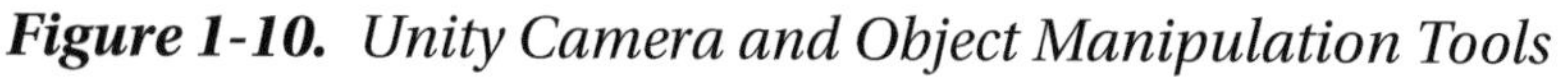

Figure 1-10. *Unity Camera and Object Manipulation Tools*

Notice that as you translate, rotate, or scale the CheckerSphere, or any object for that matter, the corresponding values in the Transform component in that object's Inspector Window also update accordingly.

Use the Implicit Sliders to Adjust an Object's Transform Values

Look closely at the Transform component for a selected object. Place your mouse pointer in between the label and the corresponding value, as shown in Figure 1-11. Notice the mouse pointer switching to a small left-right arrow icon. At this point, you can left-click and drag the mouse to the left or right to update the corresponding floating-point value as though you were adjusting a slider bar. This shortcut is referred to as the **Implicit Slider** in this book and works for any floating-point value in the Inspector Window. You will be using the Implicit Slider to control parameters in almost every example.

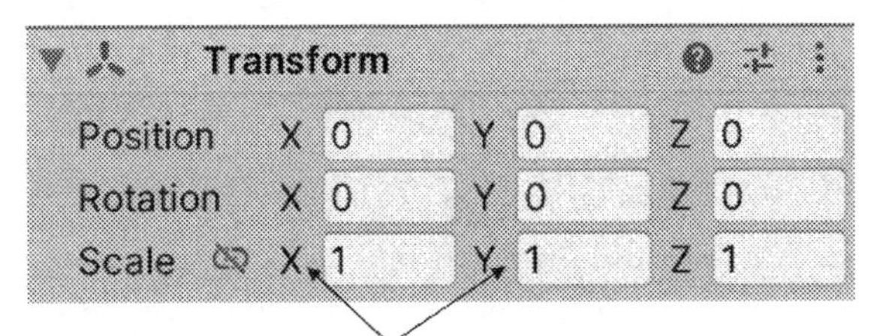

Figure 1-11. *How to find the Implicit Slider to manipulate float values in the Inspector Window*

Initialize Public Variable of MyScript in the Inspector Window

With the Controller selected, left-click the CheckerSphere in the Hierarchy Window, and drag, without releasing your original left-click, to the value location of CenterSphere under the MyScript component. By doing so, you have changed CenterSphere to reference CheckerSphere.

Now, click the Play button and observe that the BrickSphere is not traveling anymore, but the CheckerSphere is now moving simultaneously in both the X and Y directions. You now have the experience to initialize any future game objects via the Inspector Window.

Delete Initial Values of Public Variables and Observe Errors

With the Controller selected, click the CenterSphere value location and then hit the delete key to remove the initial reference. You should observe the following message in the value location of CenterSphere: None (Game Object).

Next, click the Play button and observe that none of the spheres are moving. Navigate to the Console Window (Figure 1-3 (F)) to observe the error messages. Recall that the `Start()` function in `MyScript` asserts that all three public variables must be properly initialized. In this case, the `CenterSphere` is a null reference which results in an assertion failure. These errors can be observed in Figure 1-12.

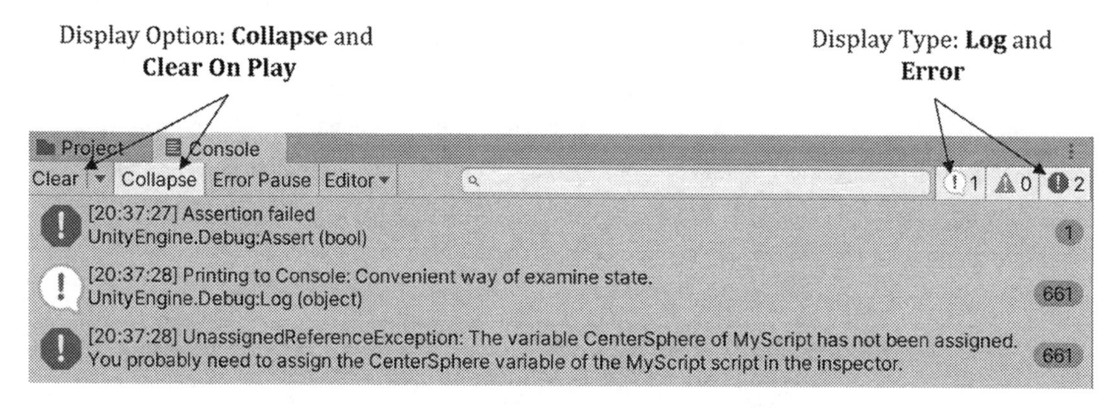

Figure 1-12. *Console Window displaying options and message types*

As indicated in Figure 1-12, the Console Window supports different display options and message types. Try enabling different options and observe that the Collapse option allows you to collapse identical messages into one. You can also show or hide log, warning, and error messages. We have found it convenient to show all message types and to enable the Collapse and Clear on Play options.

Edit Operations During Play Mode Are Ignored After the Play Mode

Now re-initialize `CenterSphere` to refer to `BrickSphere`. Remember, this can be done by selecting the `Controller` and then dragging `BrickSphere` from the Hierarchy Window to the value field of `CenterSphere` in the Inspector Window.

Click the Play button to begin the game. You should notice all three spheres are moving once again. Next, remove the `CenterSphere` reference by clicking the `CenterSphere` value and hitting the delete key. You will notice that the `BrickSphere` has stopped moving and error messages show up in the Console Window.

Next, stop the game by clicking the Play button again and notice that the value field of `CenterSphere` is no longer empty, but restored to its previous value of `BrickSphere`.

In general, and with few exceptions, edit operations performed when the game is running are undone when the game is stopped. This can be both invaluable and frustrating. Invaluable because you are free to perform editing operations while playing to examine the effects and verify mathematical concepts. Frustrating because you will likely forget that you are in play mode and perform a series of editing operations only to have those edits be undone once play mode is terminated.

Note Always be aware of the current game play mode when performing edit operations.

To Learn More About Working with Unity

We have covered only operations in Unity that are relevant to learning the math concepts for this book. It is very important to note that what you have learned about Unity in this chapter is focused on preparing you to work with and learn from examples in this book. This knowledge may or may not be relevant in being a competent game designer. If you are interested in learning more about Unity, you can find useful resources under the **Learn** tab in the Unity Hub as mentioned previously. Here are some additional tutorials that can be helpful:

- All of the Foundational Tutorials
- The Create with Code tutorial under Beginner Course

If you are new to C#, we suggest that you follow this link `https://learn.unity.com/learn/search?k=%5B%22q%3AScripting%22%5D` and examine the Beginner and Intermediate Scripting tutorials.

How to Use This Book

This book begins with the most fundamental mathematical concept that is relevant to game development, working with a single position, and then covers number intervals, introduces vectors, and advances to the powerful and regularly applied concepts in vectors: the dot and cross products, vector spaces, and rotation of vectors. For each topic, an introduction is followed by a simple example that demonstrates the associated applications that are relevant to interactive graphical or video game development. The examples are simple, always a single scripting file, featuring the details of solutions implemented based on the topics being discussed. The scripting file and the associated C# class will always be with the same name containing the substring `MyScript`, for example, `EX2_1_MyScript`. This script, as mentioned previously, will always be attached to the `Controller` object in each example. It is important to note that the implementation of the scripts, setup of the game objects, and structure of the examples are designed to feature the math concepts being discussed. This organization allows you to analyze the concepts, examine the implementation, and experiment and interact with the game objects such that you can gain understanding and insights into the topics discussed. The contents of this book do not attempt to address any specific issues in game design or development as no such issues were considered.

The best way to read this book is by downloading the Unity projects, reading the book section that describes the concepts, running the corresponding examples while paying attention to the highlighted topics, examining the source code, and, finally, tinkering and experimenting with the implementation accordingly.

As a final reminder, this book does not explain and will not explain how the game objects were created, how to customize their behaviors, or how the examples and scenes were built. Those features deal with how to work with Unity in general and are outside the scope of this book.

Summary

Through this chapter, you have learned how to install Unity and an IDE for script editing, as well as how to open, run, and tinker with the examples that accompanied this book. You have also learned about the Unity Editor and the terminologies that will be used throughout the book to work with the examples. You were given some extra resources to investigate if you want to learn more about how to work with Unity and C#, as well as what this book will be covering along with a friendly reminder of the goals and scope of this book. In the next chapter, you will begin with the math concept of intervals and learn about bounding boxes.

References

To learn more about linear algebra based on a traditional approach, free from specific application context, there are a number of textbooks. For example:

- Gilbert Strang, *Introduction to Linear Algebra, Fifth Edition*, Wellesley-Cambridge Press, 2016. ISBN: 978-0980232776
- Online, Linear Algebra section of the Khan Academy: `www.khanacademy.org/math/linear-algebra`

To learn more about how mathematics is relevant to video games without explicit implementation examples:

- Fletcher Dunn and Ian Parberry, *3D Math Primer for Graphics and Game Development*, 2nd Ed, CRC Press, 2011. ISBN: 978-1482250923
- James M. Van Verth and Lars M. Bishop, *Essential Mathematics for Games and Interactive Applications*, 3rd Ed, CRC Press, 2016. ISBN: 978-1568817231

CHAPTER 2

Intervals and Bounding Boxes

After completing this chapter, you will be able to

- Use Unity to work with positions in the 3D Cartesian Coordinate System
- Program with intervals along the three major axes
- Define axis-aligned bounding areas in 2D and axis-aligned bounding boxes (AABB) in 3D
- Determine if a position is inside of an axis-aligned bounding area or box
- Approximate inter-object collision using AABBs
- Compute the intersection of two AABBs
- Appreciate the strengths and weaknesses of AABBs

Introduction

This chapter begins by reviewing the Cartesian Coordinate System, continues with the exploration of 3D positions and number intervals, and wraps up with how you can apply these simple comparisons to approximate object boundaries and collisions between objects. While comparing numbers is rather trivial, the generalization and application of these simple concepts lead directly to one of the most powerful and widely used tools in video games: the axis-aligned bounding box, or AABB. AABB is an important topic in video games because it allows for simple and efficient approximation of object

K. Sung and G. Smith, *Basic Math for Game Development with Unity 3D*,
https://doi.org/10.1007/978-1-4842-9885-5_2

proximity. In other words, AABB is intuitive to comprehend and can quickly compute how close two objects are to each other, including if the objects are currently colliding.

Generally speaking, it is difficult and time-consuming to determine if geometrically complex objects are physically close to each other or if they are currently colliding. AABBs can be used to address this issue. Imagine, with your eyes closed, someone put a pizza in front of you with several slices removed. In this situation, without opening your eyes, how would you determine if your extended hand is about to touch the pizza? Now, if the pizza was placed in a pizza box, or a bounding box, then the solution can be approximated by answering the question of whether your hand has come into contact with the pizza box. Notice that with slices removed, touching the box can only warn you that you are about to touch the pizza. It does not tell you if you are going to actually touch the pizza. AABB, or bounding box, related computations involve simple number comparisons, trading accuracy for simplicity, and are thus efficient. Unless your AABB exactly matches your object's shape (i.e., your shape is a box), your proximity calculations will only be approximated; however, in many cases this is sufficient to deliver satisfactory game play.

Mathematically, this chapter should be a relatively straightforward review as it will cover concepts that are generally taught in the late middle school to early high school years in the United States. In addition to refamiliarization with these concepts, this review process can also serve as an excellent opportunity to learn more about and to become more comfortable with the Unity environment, the involved utilities, custom tools, and the approach that this book takes in discussing topics. In this book, after each concept is described, you will be introduced to a new Unity scene which presents that concept, led through interactions with the scene to explore and experience aspects of that concept, and then instructed to examine the implementation associated with that concept. At the end of this process, the book summarizes a list of key points for you to verify your learning.

Review of Cartesian Coordinate System

Recall that the 3D Cartesian Coordinate System defines an origin position (0, 0, 0) and three perpendicular axes, X, Y, and Z, known as the major axes. Each axis begins from the origin and extends in both its positive and negative directions. This can be seen in Figure 2-1 where the checkered sphere in the middle is intersected by all three arrows and is the origin. Each arrow represents a major axis; the direction of the arrow represents the positive direction along that axis.

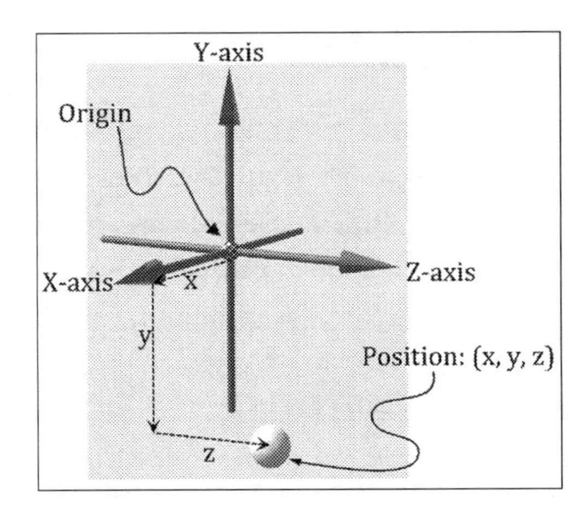

Figure 2-1. *The 3D Cartesian Coordinate System*

In the Cartesian Coordinate System, the position of a point is defined by a three-variable tuple (x, y, z), the point's distance as measured along the three major axes. For example, in Figure 2-1, the sphere's center position is x-value along the X-axis, y-value along the Y-axis, and z-value along the Z-axis. In this case, since the sphere is below the origin and the Y-axis has upward as its positive direction, the y-value will be negative. If the x-, y-, or z-values are altered, you can expect the corresponding object to be relocated in the coordinate system accordingly.

It is important to remember that the major axes are always perpendicular to each other and with a unit that is convenient for the specific application. For example, when applying the Cartesian Coordinate System in describing positions in a room, you may define the origin to be at one corner of the room, the X- and Z-axis to be along the floor edges, and the Y-axis to be along the wall pointing upward toward the ceiling. In this case, a convenient unit may be in meters. With such a coordinate system definition, all positions in the room will be of values (x, y, z) measured in meters from the corner that was identified as the origin. Note that there can be infinite number of Cartesian Coordinate Systems defined for the room, for example, choosing a different corner to be the origin or identifying the center of the room to be the origin with inches as the unit.

What is important to remember is that a Cartesian Coordinate System always has perpendicular major axes with an arbitrary unit that is convenient for the specific application. The coordinate values are measurements from the origin along the major axes in the defined units.

Intervals: Min-Max Range

The Cartesian Coordinate System allows for straightforward comparison between positions along its major axes. For example, Figure 2-2 shows a transparent region along the Y-axis where this region is defined by two values, a min (minimum) value and a max (maximum) value. The Y-axis direction, noted by the arrow, indicates the direction of increasing coordinate value. In this case, the minimum value is always below the maximum value, both literally and pictorially. A region defined by min and max values along a major axis is referred to as an **interval**.

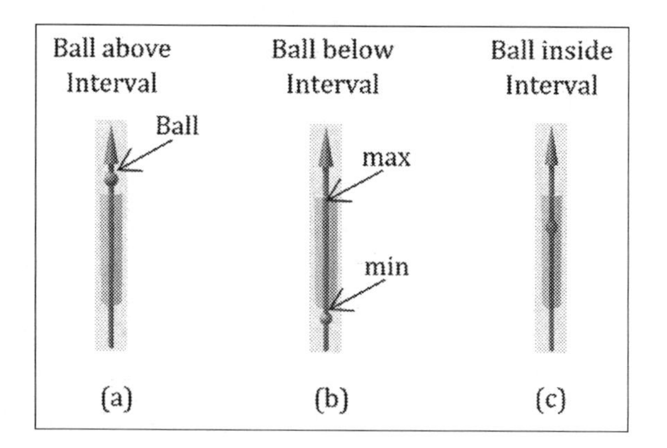

Figure 2-2. *A simple min-max interval along the Y-axis*

An interval is convenient for determining if a given position is within a specific range. For example, the Ball in Figure 2-2(a) is above the interval, and thus you know the y-value of the center of the Ball is greater than the maximum value of the interval. Figure 2-2(b) shows that the reverse is true as well: if the Ball is below the interval, then the y-value of the center of the Ball is less than the minimum value of the interval. Figure 2-2(c) on the other hand, shows that the Ball is inside the interval when the y-value of its center is in between the given max and min values. The determination of these conditions can be simplified as follows and is referred to as the inside-outside test:

```
if ((Ball.y >= Interval.Min) && (Ball.y <= Interval.Max))
    // Ball is inside the Interval
else
    // Ball is outside the Interval
```

Note that the comparison symbol is greater or less than *and* equal to. This means if the Ball is right on the boundary, it will be considered as being inside the interval. Now that you have reviewed the Cartesian Coordinate System and how to program intervals, you are ready to explore the different examples and concepts presented in this chapter. However, before you do that, you'll need to understand how to work with the Unity examples given in this book.

Working with Examples in Unity

Before you dive into any examples, you'll first have to know how the examples are organized within each chapter. Figure 2-3 shows you the different scenes and their corresponding `MyScript` for this chapter and how future chapters will be laid out.

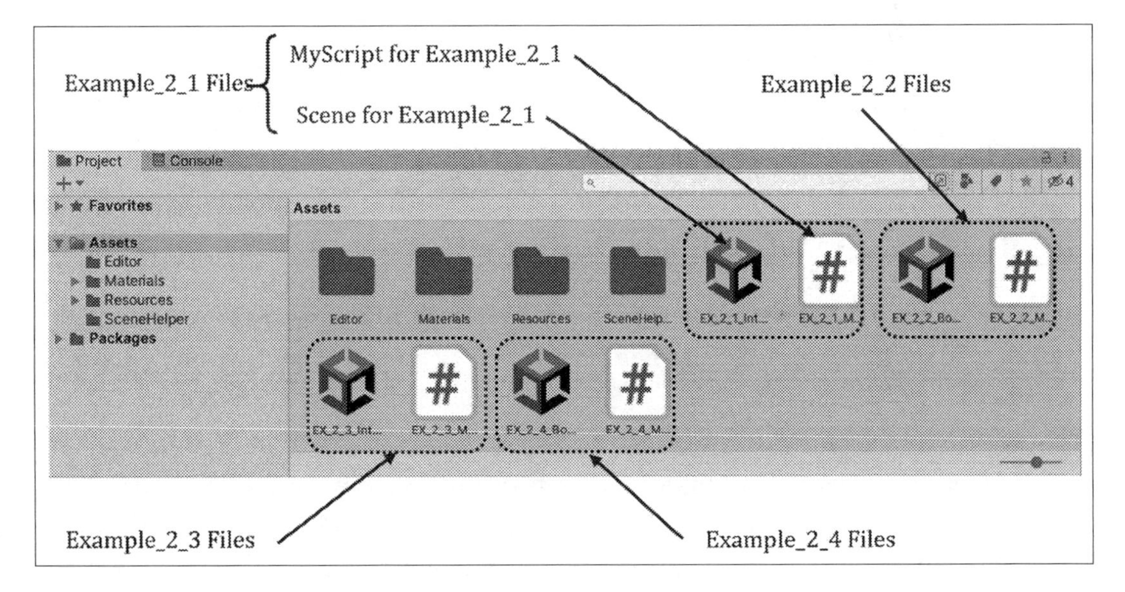

Figure 2-3. *The Project Window of Chapter-2-Intervals+AABB project*

One Unity project is defined for each chapter, and each example for the associated concepts in that chapter is organized as a separate Unity **Scene** in that project. As mentioned in Chapter 1, each example or Scene has only one script with a name that includes the string, `MyScript`. For example, all examples in Chapter 2 are defined in the Unity project that is in the `Chapter-2-Intervals+AABB/` folder. Figure 2-3 illustrates that after you open the project and navigate to the `Assets/` folder of the Project Window, you will observe two files for each example. The first is the Scene file named EX_2_x_title, and the second is a corresponding MyScript file named EX_2_x_MyScript. EX stands for example, the 2 stands for this chapter's number, the x is the sequence of the example in its chapter (e.g., EX_4_3 would translate to Chapter 4's third example), and finally, title refers to the title of that example. For simplicity, the term MyScript will be used to refer to the EX_MyScript associated with the current example.

When you are ready to examine an example, simply double-click the corresponding scene file. This will load the scene into the Unity Editor. The Controller of that scene will already have the corresponding MyScript component attached to it, and therefore no further setup is required. Remember, to open a script in the IDE, you can simply double-click its icon in the Assets/ folder of the Project Window.

Now open the Chapter-2-Interval+AABB project and double-click the EX_2_1_IntervalBoundsIn1D scene file in the Assets/ folder of the Project Window to load it into the Unity Editor. You can tell what scene is currently open in your project by looking at the Hierarchy Window; the very first item is always the name of the scene you have open.

The Interval Bounds in 1D Example

This example reviews the Cartesian Coordinate System, introduces you to working with a customized script (MyScript), and demonstrates how to work with the Unity Vector3 data type. In a nutshell, this example defines a 1D bound along the Y-axis, allows you to interactively adjust the max and min values of the interval, as well as examines an implementation of the interval inside-outside test as depicted in Figure 2-2. Figure 2-4 shows a screenshot of running the Interval Bounds in 1D scene from the Chapter-2-Intervals+AABB project. As discussed in the previous section, this scene can be opened by double-clicking the EX_2_1_IntervalBoundsIn1D scene file in the Assets/ folder of the Project Window.

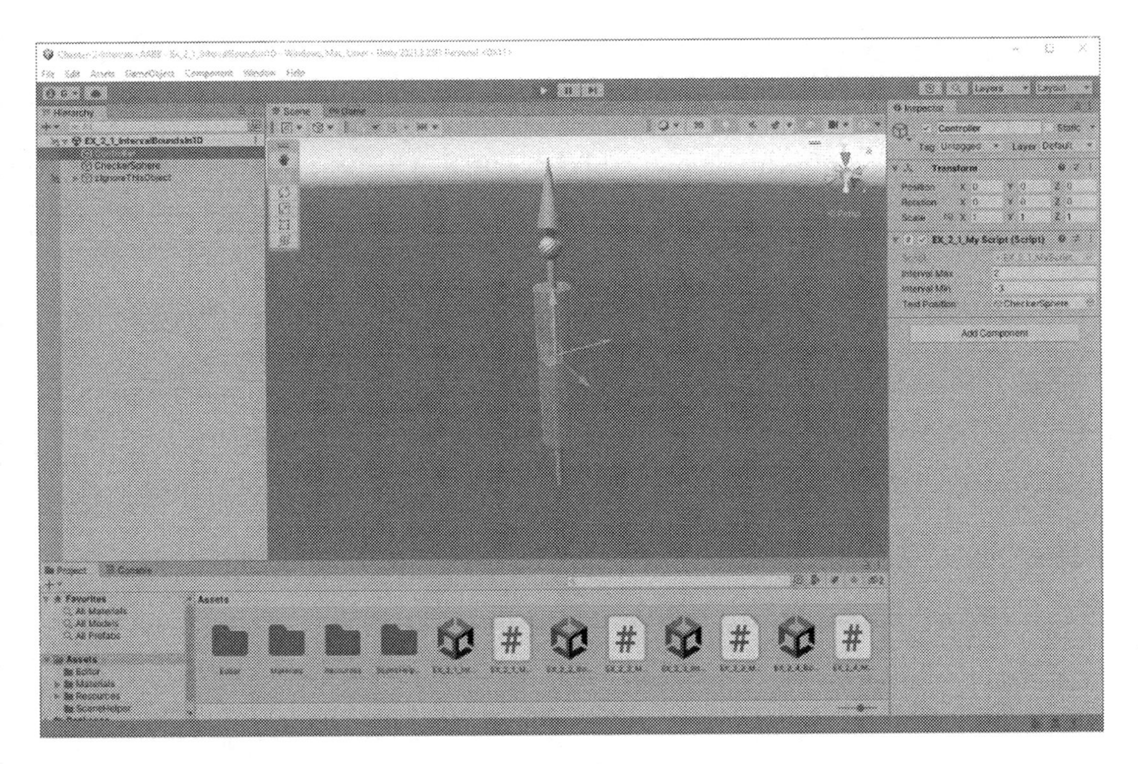

Figure 2-4. *Running the Interval Bounds in 1D example*

Note Please take note of the separated Scene and Game Views. Make sure to differentiate between these two views. All object manipulations must be carried out through the Scene View and not the Game View.

The goals of this example are for you to

- Review the Cartesian Coordinate System
- Experience adjusting positions of game objects in Unity
- Begin familiarizing yourself with the `Vector3` class
- Understand and interact with intervals along an axis
- Examine the implementation of an interval inside-outside test

Examine the Scene

Take a look at the `Example_2_1_IntervalBoundsIn1D` scene and observe the predefined game objects in the Hierarchy Window. This is a very simple scene where, besides `Controller`, there is only one other defined object, the `CheckerSphere`. In this example, you will manipulate the position of the `CheckerSphere` object to examine the results of the interval inside-outside test along the Y-axis.

Note Please continue to ignore the `zIgnoreThisObject` in the Hierarchy Window. Once again, this game `object` hides miscellaneous and distracting scene supporting objects that do not pertain to the math you are learning.

Analyze Controller MyScript Component

Select `Controller` in the Hierarchy Window. Please refer to Figure 2-5 and make sure your Inspector Window looks the same by locating the `EX_2_1_MyScript` component and ensuring it is expanded so you can examine its variables and the corresponding values and references. There are three variables that you can access from the Inspector Window with this script:

- `IntervalMax`: The maximum value of the interval
- `IntervalMin`: The minimum value of the interval
- `TestPosition`: Holds a reference to the `CheckerSphere` such that `MyScript` can access the position of the `CheckerSphere` game object in the scene

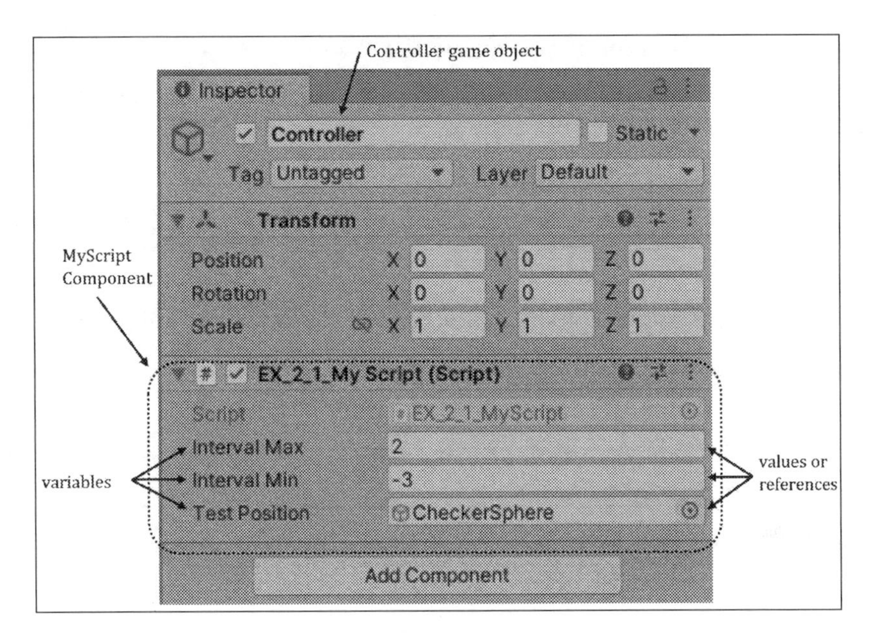

Figure 2-5. The MyScript Component of Controller

Note Once again, make sure to differentiate between the Scene View and the Game View windows. Remember, the Scene View has a skybox-like background, and the Game View has a constant light blue background. The Scene View is the window where you can select and manipulate game objects. If you try to click an object in the Game View, nothing will happen.

The name of the script you will work with is actually EX_2_1_MyScript; once again, for simplicity and brevity, MyScript is used in the text. This will be the case for all examples in the rest of this book.

Interact with the Example

Click the Play button to run the example. While running, select the `CheckerSphere` either through the Hierarchy Window or by clicking the `CheckerSphere` in the Scene View window. Once selected, change the position of the `CheckerSphere` by invoking the Implicit Slider (refer to Figure 1-11 if you forgot how to do this). You can also change the position of the `CheckerSphere` by simply typing into the value fields of the corresponding variables in its Transform component. Try increasing and decreasing the x-, y-, and z-values of the CheckerSphere and observe the corresponding movement. Notice that the CheckerSphere does indeed move along the major axes of the Cartesian Coordinate System, obeying the positive and negative directions as expected.

Note You can also manipulate the Transform component of a game object by selecting the corresponding object in the Hierarchy Window and using the different Object Manipulation Tools as shown in Figure 1-10.

Now observe the transparent cylinder along the Y-axis. This is the interval defined by the IntervalMax and IntervalMin values. Notice how the color of the cylinder changes as you change the y-value of the CheckerSphere position to be either above or below the interval. Also, note that changing the x- or z-position of the CheckerSphere has no effect on the color of the interval.

You can adjust the IntervalMax and IntervalMin values by selecting the Controller object in the Hierarchy Window and modifying the values of the corresponding variables in the MyScript component. Notice how the transparent cylinder or interval object responds to your adjustments while maintaining its proper behavior of adjusting its color depending on if the CheckerSphere is inside or outside of it.

Details of MyScript

Open the MyScript for this example (EX_2_1_MyScript) and examine the implementation source code in your IDE. Once again, to open a script, you can either right-click over the MyScript component's name ("EX_2_1_My Script (Script)") in the Inspector Window when Controller is selected (refer back to Figure 1-7 if you need a refresher) or double-click the MyScript ("EX_2_1_MyScript" for this example) icon in the Assets/ folder of the Project Window. In the future, you will not be given these reminders and will simply be told to open MyScript. The following variable definitions can be observed:

```
private MyIntervalBoundInY AnInterval = null;
public float IntervalMax = 1.0f;
public float IntervalMin = 1.0f;
// Use sphere to represent a position
public GameObject TestPosition = null;
```

Notice the one-to-one correspondence between the public variables and those accessible via the Inspector Window, as illustrated in Figure 2-5. Recall that `TestPosition` is set up to reference the `CheckerSphere` game object, and thus your changes to the `CheckerSphere` game object can be accessed via the `TestPosition` variable. The private variable of data type `MyIntervalBoundInY` is defined and used to visualize the interval defined by the `IntervalMax` and `IntervalMin` values.

Note The code in `MyScript` is only executed when the Play button is active.

Figure 2-6 shows that, besides the drawing support (e.g., `DrawInterval` and `IntervalColor`), the `MyIntervalBound` class only defines and uses the `MinValue` and `MaxValue` variables, which is the definition of an interval. The `MyIntervalBoundInY` class is a simple subclass that overrides the `PositionToDraw()` function. The `PositionToDraw()` function is used to visualize intervals along a major axis. The `MyIntervalBound` class and its subclasses and functions can be found in the `Assets/SceneHelper/` folder in the Project Window. Please do feel free to explore its implementation. To avoid distraction from learning the mathematics, the details of the `MyIntervalBound` class and all other classes for supporting visualization (other scripts in the `Assets/SceneHelper/` folder) will not be discussed in this book.

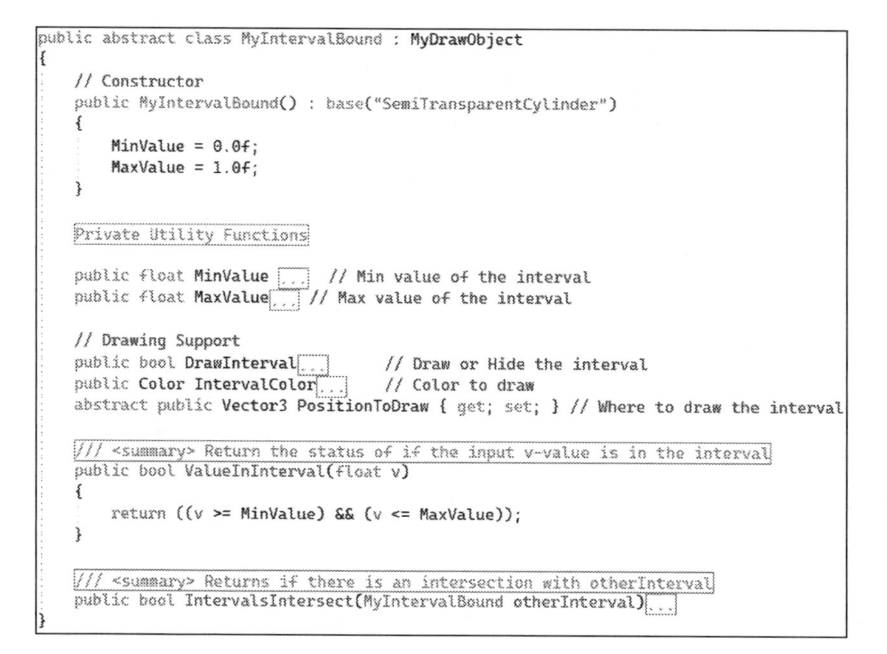

public abstract class MyIntervalBound : MyDrawObject
{
// Constructor
public MyIntervalBound() : base("SemiTransparentCylinder")
{
MinValue = 0.0f;
MaxValue = 1.0f;
}
Private Utility Functions
public float MinValue ... // Min value of the interval
public float MaxValue ... // Max value of the interval
// Drawing Support
public bool DrawInterval ... // Draw or Hide the interval
public Color IntervalColor ... // Color to draw
abstract public Vector3 PositionToDraw { get; set; } // Where to draw the interval
/// <summary> Return the status of if the input v-value is in the interval
public bool ValueInInterval(float v)
{
return ((v >= MinValue) && (v <= MaxValue));
}
/// <summary> Returns if there is an intersection with otherInterval
public bool IntervalsIntersect(MyIntervalBound otherInterval) ...
}

Figure 2-6. *The MyIntervalBound class for visualizing an interval along a major axis*

When the game first begins to run, the `Start()` function instantiates the visualization object, `AnInterval`, for displaying the semi-transparent interval. Details of the `Start()` function are as follows:

```
void Start() {
    Debug.Assert(TestPosition != null);  // Ensure proper setup
    AnInterval = new MyIntervalBoundInY();
}
```

Next, you will examine the `Update()` function. Recall that the `Update()` function is invoked at a real-time rate of about 60 times per second to update the state of the application, hence the name of the function. The details of the `Update()` function are as follows:

```
void Update() {
    // Updates AnInterval with values entered by the user
    AnInterval.MinValue = IntervalMin;
    AnInterval.MaxValue = IntervalMax;
```

```
    // Assume point is outside
    AnInterval.IntervalColor = MyDrawObject.NoCollisionColor;
    // computes inside-outside of the current TestPosition.y value
    Vector3 pos = TestPosition.transform.localPosition;
    bool isInside = (pos.y >= IntervalMin)
                && (pos.y <= IntervalMax);
    if (isInside) {
        Debug.Log("Position In Interval! ("
                + IntervalMin + ", " + IntervalMax + ")" );
        AnInterval.IntervalColor = MyDrawObject.CollisionColor;
        // MyYInterval supports the inside functionality
        Debug.Assert(AnInterval.ValueInInterval(pos.y));
    }
}
```

The first three code lines of Update() ensure that AnInterval is updated with the latest values entered by the user. Next, the inside interval test is performed based on comparing the y-value of the TestPosition object, which, as you may recall, was set to reference the CheckerSphere through the Inspector Window (see Figure 2-5 or look at your own project for confirmation).

Sixty times every second, AnInterval is updated with user input, and the y-value of the CheckerSphere position is compared against the user-specified IntervalMin and IntervalMax, changing the color of the interval as necessary. This fast update rate conveys a sense of instantaneous modifications to the user. An important detail is that the variable pos or the data type for TestPosition.transform.localPosition is Vector3. A Vector3 with x-, y-, and z-values is designed to represent a position and, as detailed in the later chapters, a vector. Click to view the Console Window (please refer to Figure 1-3, label F) and observe the text output generated by the Debug.Log() function. This is an excellent way to examine and debug the state of your game.

The very last line of the Update() function demonstrates that the MyIntervalBoundInY class has also implemented the inside-outside test and the Debug.Assert() verifies the consistency of the test results. The MyIntervalBound.ValueInInterval() is a convenient function that will be used in later examples.

Notice that the variable AnInterval only supports drawing and does not participate in any way in the logic and computation of the inside-outside test. For example, you can remove all occurrences of the AnInterval variable and the example will execute perfectly, only, without visual feedback. For this reason, the details of the MyIntervaBoundInY class are irrelevant to the understanding of the interval computation and can be distracting.

Takeaway from This Example

In this very simple example, you have experienced interacting with and moving objects in the 3D Cartesian Coordinate System while observing the results of mathematical computations. You have also learned how to establish a reference of a GameObject to a variable in MyScript in order to gain access to and manipulate the position of that game object. Additionally, you have begun to work with the Unity Vector3 class and reviewed floating-point number comparisons. Lastly, you have learned how to determine if a position is within the bounds of an interval along a major axis of the Cartesian Coordinate System.

Relevant mathematical concepts covered include

- Cartesian Coordinate System
- Position of an object in the 3D Cartesian Coordinate System
- Intervals along a major axis defined by minimum and maximum values
- Testing for being inside or outside of an interval along a major axis

Unity tools

- A GameObject's position is defined by its transform.localPosition
- Vector3 can be used to represent an object's position
- Debug.Assert() can be used for assertion of conditions
- Debug.Log() can be used for printing text messages to the Console Window
- MyIntervalBoundInY is a custom-defined class to support the visualization of intervals along the Y-axis

Interaction technique

- Use a sphere GameObject to represent and manipulate a position.

Limitation

- The idea of an interval is straightforward. However, the inside-outside test implementation is straightforward only for cases where the interval is defined along one of the three major axes. In later chapters, you will learn about vectors and vector dot products. Those concepts can help generalize interval testing and support the inside-outside tests along a non-major axis. Intervals and inside-outside test will be revisited later.

EXERCISES

Checking for Error

Note that it is possible to set the `IntervalMin` to be a value greater than that of `IntervalMax`. Please modify the `Update()` function to prevent this situation.

Drawing Location of MyIntervalBoundInY

Run the game. Open the `zIgnoreThisObject` (by clicking the small triangle beside this object in the Hierarchy Window to expand the object structure and observe its children objects) and select `SemiTransparentCylinder(Clone)`. Notice that the interval along the Y-axis in the Scene View is highlighted when this object is selected. This is the instance of `MyIntervalBoundInY` that was instantiated in `MyScript` for visualizing the interval. Now try to adjust the position of this object, for example, change the x-position value and observe the object shift in the x-direction. Notice that you can adjust both the x- and z-positions but not the y-position. This is because the y-position of the object is constantly being set and updated by the user-specified `IntervalMax` and `IntervalMin` values in the `MyScript` component on `Controller`. From this exercise, you have learned that it is possible to draw the y-interval at any x- and z-position.

Extending the Inside-Outside Test to Other Axes

Notice that the inside-outside test in the `Update()` function is specific to the Y-axis. Please define four additional variables. These variables will represent the minimum and maximum interval values for the other two axes, X and Z. With these new values, you can now detect if the `TestPosition` is within the specific interval bounds of the X-, Y-, and Z-axes. Although you may not be able to visualize all three intervals, you can still compute and echo the inside conditions using the Debug.Log() utility. The next example will examine the topic of interval inside-outside testing more closely to define the simple and yet powerful axis-aligned bounding box utility.

Axis-Aligned Bounding Boxes: Intervals in Three Dimensions

An interval along a major axis is simply a line segment where positions inside the interval are points of that line segment. When working with two intervals along two different major axes, for example, an interval along the X-axis and a second interval along the Z-axis, the combined result is a 2D rectangular region or an axis-aligned rectangular plane.

As illustrated in Figure 2-7, the rectangular region on the X-Z plane is defined by the horizontal interval along the X-axis with xMin and xMax values and by the vertical interval along the Z-axis with zMin and zMax values. Figure 2-7 (a) shows that to determine if the given ball position is inside the rectangular region, the position must satisfy the inside-outside tests of both intervals. Figures 2-7 (b) and (c) depict the conditions when a position is only inside one of the intervals but not both. In (b) the ball's position is within the horizontal interval but outside of the vertical. In (c) the ball's position is inside the vertical, but not within the horizontal interval.

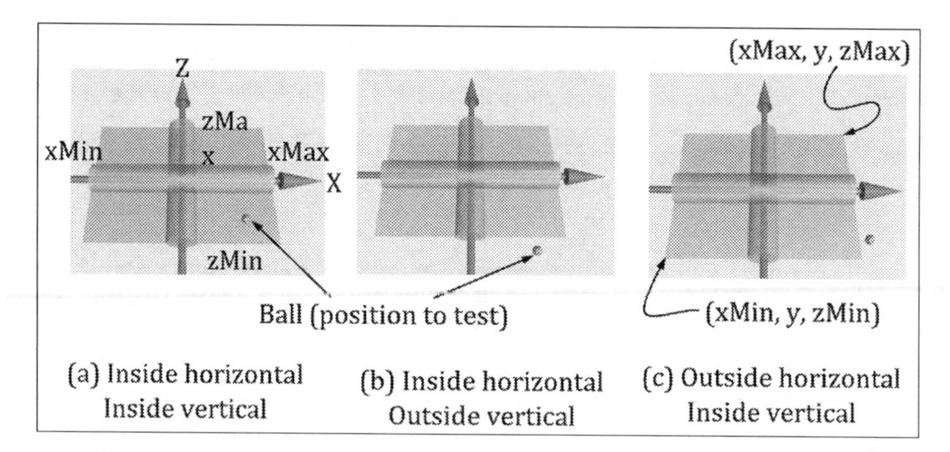

Figure 2-7. *Inside-outside test of two intervals along the X- and Z-axis*

When working with three intervals along all three major axes, the rectangular region changes into a 3D axis-aligned rectangular box, known as an **axis-aligned bounding box (AABB)**. In this case, a position is inside a given AABB only when it satisfies the inside-outside tests for all three intervals. This condition testing can be implemented as follows:

```
// if in all intervals
if (
    (Ball.x >= xInterval.Min) && (Ball.x <= xInterval.Max)
    &&  // x-axis
    (Ball.y >= yInterval.Min) && (Ball.y <= yInterval.Max)
    &&  // y-axis
    (Ball.z >= zInterval.Min) && (Ball.z <= zInterval.Max)
    // z-axis
   )
     // Ball is inside the bounding box
else
     // Ball is outside the bounding box
```

As you can observe, the logic and computation involved in the AABB inside-outside test are straightforward and efficient. For this reason, AABBs are a widely used utility for approximating object proximity and collisions. AABBs are so important and useful that Unity defines its own class, `Bounds`, that implements the AABB functionality (`https://docs.unity3d.com/ScriptReference/Bounds.html`). At the end of this chapter, you will see what is in this class compared to what you will have implemented on your own throughout the examples in this chapter.

Note For brevity, the rest of this book refers to axis-aligned bounding boxes, or AABB, simply as bounding boxes. A bounding box that is not aligned with the major axes is referred to as a *general* bounding box.

The Box Bounds Intervals in 3D Example

This example demonstrates the functionality of bounding boxes, implements the point inside-outside test, and allows you to interact with and examine its implementation. Figure 2-8 shows a screenshot of running the `EX_2_2_BoxBounds_IntervalsIn3D` scene of the `Chapter-2-Intervals+AABB` project. This scene can be opened by double-clicking the `EX_2_2_BoxBounds_IntervalsIn3D` scene file in the `Assets/` folder of the Project Window.

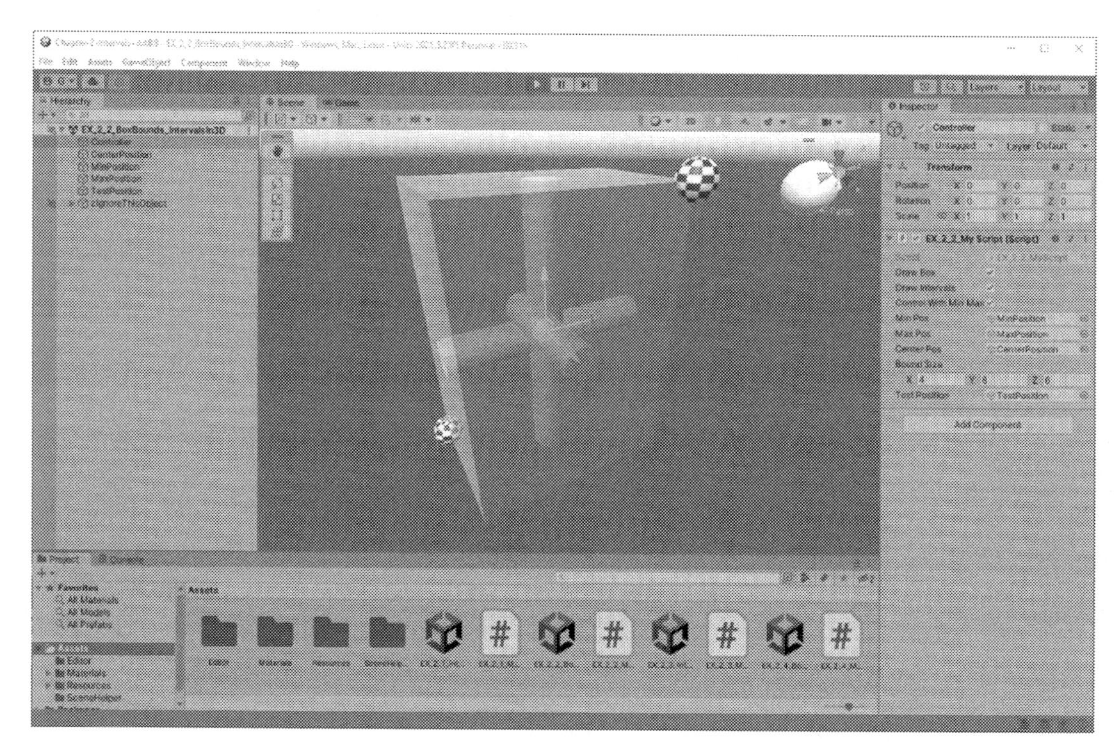

***Figure 2-8.** Running the Box Bounds Intervals in 3D example*

The goals of this example are for you to

- Understand and interact with a bounding box
- Examine the implementation of a bounding box inside-outside test

Examine the Scene

The Hierarchy Window shows that in addition to the `Controller`, there are four other game objects: `CenterPosition`, `MinPosition`, `MaxPosition`, and `TestPosition`. The center, min, and max position objects are three separate checkered spheres representing the corresponding positions of a bounding box. The `TestPosition` is the white sphere. Just like the `IntervalBoundsIn1D` example, you can manipulate the position of the white sphere to trigger the inside-outside test and examine results.

Analyze Controller MyScript Component

The `MyScript` component on the `Controller` object shows eight public variables supporting three general functionalities:

- Drawing support
    - `DrawBox`: Used to determine whether to draw or hide the bounding box
    - `DrawInterval`: Used to determine whether to draw the three axis-aligned intervals that implement the bounding box
- Box control support
    - `ControlWithMinMax`: Gives the user two options for manipulating the bounding box
    - Option 1: Manipulate the box by specifying the `MinPos` and `MaxPos` positions.
    - Option 2: Manipulate the box by specifying the `CenterPos` position and interval size (`BoundSize`) along each axis.
- Testing position support
    - `TestPosition`: References the white sphere and is used for testing the inside-outside implementation

Interact with the Example

Run the game and notice the transparent box that bounds the three intervals along the X-, Y-, and Z-axes. On the minimum and maximum corners of the transparent box are two checkered spheres. These two spheres are the `MinPosition` and `MaxPosition` game objects, which are referenced in `MyScript` by the `MinPos` and `MaxPos` variables, respectively.

Now select `Controller` and look at the `MyScript` component. Experiment with hiding and showing the box and intervals. The important observation to make here is that the box and the three intervals both define the same 3D volume. These two are complementary ways of perceiving and visualizing the volume defined by the intervals. With the intervals hidden, notice that there is a checkered sphere located at the center of the box. This is the `CenterPosition` game object; when the interval visualization objects are displayed, it is hidden inside the cylinders representing the intervals.

Next, manipulate the box with the min and max position game objects. Select the `MinPosition` or `MaxPosition` game object from the Hierarchy Window and translate their position to a new location. Notice how the bounding box continuously tracks and maintains these two positions as its min and max corners. Now, select the `CenterPosition` in the Hierarchy Window and try to manipulate it; you'll notice that its position is not changeable. When you adjust the min and max positions, the center position is computed based on your input values from these positions. This is the same case for the `BoundSize` variable in the `MyScript` component: the min and max positions determine the `BoundSize`.

In order to manipulate the bounding box by manipulating the `CenterPosition` and the `BoundSize`, you must disable the `ControlWithMinMax` check box on the `Controller's` `MyScript` component. Now, you can experience changing the entire box position by dragging the `CenterPosition` object and changing the box size by adjusting the `BoundSize` variable in the `MyScript` component. However, notice that the `MinPosition` and `MaxPosition` are no longer adjustable. These two positions are now defined by the user-specified `CenterPosition` and the `BoundSize`.

Lastly, select and drag the `TestPosition` game object to manipulate its position within the scene. Notice that the box changes colors when the `TestPosition` object moves from outside to inside its bounds. Before you look at the script, try entering and leaving the region from different sides of the box, for example, the left, right, top, bottom, and so on. Note that for each case, you will get the same inside-outside test results.

Details of MyScript

Open MyScript and examine the source code in the IDE. The instance variables are as follows:

```
private MyBoxBound MyBound = null;          // For visualizing AABB
public bool DrawBox = true;                 // Show/hide the 3D box
public bool DrawIntervals = true;           // Show/hide intervals
public bool ControlWithMinMax = true;       // min/max vs. center
public GameObject MinPos = null;            // Min position of the box
public GameObject MaxPos = null;            // Max position of the box
public GameObject CenterPos = null;         // Center of the box
public Vector3 BoundSize = Vector3.one;  // Interval size
public GameObject TestPosition = null;      // Position for testing
```

All the public variables have been discussed when analyzing the MyScript component. The private variable of the MyBoxBound data type is there to support the visualization of the bounding box. Figure 2-9 shows the public interface of the MyBoxBound class.

```
public class MyBoxBound : MyBoxDrawObject
{
    public MyIntervalBoundInX XInterval = new MyIntervalBoundInX();
    public MyIntervalBoundInY YInterval = new MyIntervalBoundInY();
    public MyIntervalBoundInZ ZInterval = new MyIntervalBoundInZ();
    private utility

    /// <summary> Position: (XInterval.MinValue, YInterval.MinValue, ZInterval.MinVa ...
    public new Vector3 MinPosition...

    /// <summary> Position: (XInterval.MaxValue, YInterval.MaxValue, ZInterval.MaxVa ...
    public new Vector3 MaxPosition...

    /// <summary> Center position = 0.5 * (MinPosition + MaxPosition)
    public new Vector3 Center...

    /// <summary> Size = MaxPosition - MinPosition
    public new Vector3 Size...

    // Drawing and Color Support
    public bool DrawIntervals ...          // Draw or Hide the intervals
    public bool DrawBoundingBox ...        // Draw or Hide the box
    public new void ResetBoxColor()...     // Reset box color to default (transparent white)
    public void SetBoxColor(Color c)       // Sets the color for the box...

    /// <summary> Return the status of if point is inside the box
    public bool PointInBox(Vector3 point)
    {
        return
            XInterval.ValueInInterval(point.x)   // in x interval
              &&                                 //  and
            YInterval.ValueInInterval(point.y)   // in y interval
              &&                                 //  and
            ZInterval.ValueInInterval(point.z);  // in z interval
    }

    /// <summary> Return the status of two boxes intersect
    public bool BoxesIntersect(MyBoxBound otherBound)...
}
```

Figure 2-9. *The MyBoxBound class*

Figure 2-9 shows the definition and public properties and functions of the `MyBoxBound` class. Note that this class is indeed built with three interval objects: `XInterval`, `YInterval`, and `ZInterval`, which are instances of the same `MyIntervalBound` class from the `IntervalBoundsIn1D` example. As in all previous classes defined for visualization, this file can be found in the `Assets/SceneHelper/` folder. As usual, to avoid distracting from the mathematical concepts discussion, the implementation details of this class are left for you to explore independently. The `Start()` function for `MyScript` is listed as follows:

```
void Start() {
    // Ensure proper setup in the Hierarchy Window
    Debug.Assert(CenterPos != null);
    Debug.Assert(MinPos != null);
```

```
    Debug.Assert(MaxPos!= null);
    Debug.Assert(TestPosition != null);
    MyBound = new MyBoxBound();          // For visualization
}
```

The Debug.Assert() calls ensure proper setup of referencing the appropriate game objects via the Inspector Window, while the MyBound variable is instantiated in order to visualize the bounding box. The Update() function is listed as follows:

```
void Update() {
    // Step 1: update drawing options
    MyBound.DrawBoundingBox = DrawBox;
    MyBound.DrawIntervals = DrawIntervals;
    // Step 2: control the box
    if (ControlWithMinMax) {
        // User controls Min/Max Position
        MyBound.MinPosition = MinPos.transform.localPosition;
        MyBound.MaxPosition = MaxPos.transform.localPosition;
        // Show bound center and size
        BoundSize = MaxPos.transform.localPosition -
            MinPos.transform.localPosition;
        CenterPos.transform.localPosition = 0.5f *
            (MaxPos.transform.localPosition +
             MinPos.transform.localPosition);
    } else {
        // User control center position and the size
        MyBound.Center = CenterPos.transform.localPosition;
        MyBound.Size = BoundSize;
        // Show Min/Max Position in the MyScript component
        MinPos.transform.localPosition =
            CenterPos.transform.localPosition -
            (0.5f * BoundSize);
        MaxPos.transform.localPosition =
            CenterPos.transform.localPosition +
            (0.5f * BoundSize);
    }
```

```
    // Step 3: perform inside/outside test
    Vector3 pos = TestPosition.transform.localPosition;
    Vector3 min = MinPos.transform.localPosition;
    Vector3 max = MaxPos.transform.localPosition;

    if ((pos.x > min.x) && (pos.x < max.x) &&
             // point in x-interval   AND
        (pos.y > min.y) && (pos.y < max.y) &&
             // point in y-interval   AND
        (pos.z > min.z) && (pos.z < max.z) )
            // point in z-interval
    {
        Debug.Log("TestPosition Inside!");
        MyBound.SetBoxColor(MyDrawObject.CollisionColor);
    } else {
        MyBound.ResetBoxColor();
    }
}
```

The `Update()` function implements the interaction with the user in three main steps:

- Step 1: Drawing control: The first two lines of code set the box and interval drawing options according to the user input.
- Step 2: Bounding box manipulation: The bounding box is manipulated either via receiving the min and max position from the user and then computing and setting the center and size values or through receiving the center and size values from the user and then computing and setting the min and max positions. Note that the size of an interval is always max-min and is true for 3D bounding boxes as well. Additionally, the center position is always 0.5 * (max + min).
- Step 3: Inside-outside test: Compute the `TestPosition`'s position against the inside-outside condition and update the box color accordingly.

Take note that the MyBound variable does indeed only serve as a visualization tool. For example, you can delete all occurrences of the variable and still be able to run the example. Only, in that case, there will be no visual feedback of the bounding box or the results of the inside-outside tests. Lastly, an important observation to make is that the Vector3 "-" and "+" operators subtract and add the corresponding x-, y-, and z-component values of their operands.

Takeaway from This Example

This example expands on the very simple concept of an interval along a major axis to create three intervals along each major axis, resulting in a 3D bounding box.

Through interacting with this example, you have learned that there are two fundamental approaches in defining a bounding box, either with the min and max corner positions or with the center position and the size. This knowledge informs you that the internal representation of a bounding box class can either be min/max or center/size.

You have also learned that the inside-outside test for a bounding box is simply the inside-outside test for one interval, three times. Finally, you have observed that bounding boxes are simple to program with efficient runtime performance.

Relevant mathematical concepts covered include

- 3D bounding boxes
- Testing for being inside or outside of a bounding box
- The two alternative approaches to manipulate a bounding box: min/max or center/size

Unity tools

- MyBoundBox: A custom-defined class to support the visualization of a bounding box
- MyIntervalBoundInX, MyIntervalBoundInY, and MyIntervalBoundInZ: Custom-defined classes to support the visualization of intervals along the X-, Y-, and Z-axes, respectively

EXERCISE

Testing and Printing Position Status for Each of the Intervals Separately

Modify the `Update()` function to print out (with `Debug.Log()`) the status of the `TestPosition`'s position with respect to each of the X-, Y-, and Z-intervals of the bounding box. Through this exercise, you can practice implementing interval testing yourself, and you can verify that a given position can be inside one or two of the intervals of the bounding box and still be outside of the box.

Collision of Intervals

Now that it is possible to efficiently detect if the position of an object is inside a 3D bounding box, the next question to answer is how do you detect when two bounding boxes intersect? Answering this question is key for detecting a collision between two objects, for example, two vehicles in a video game. One approach to study this problem is by first examining how two axis-aligned intervals intersect. In the same manner as understanding and extending a 1D interval to a 3D bounding box, a 1D interval collision can be generalized to a 3D bounding box collision.

Figure 2-10 shows two intervals defined along the Y-axis, the Green (G) and the Blue (B) intervals. To ensure clear visualization of overlapping intervals, the two are drawn on different sides of the Y-axis, with the third interval representing the intersection drawn centered around the Y-axis (where the colors overlap each other). Figure 2-10 shows all the different combinations that the two intervals can intersect or overlap. These include

(a) No intersection.

(b) G.min is inside the B interval, while G.max is outside, which is equivalent to B.max being inside the G interval, but B.min being outside.

(c) The entire G interval is inside the B interval.

(d) The entire B interval is inside the G interval.

(e) G.max is inside the B interval, while G.min is outside, which is equivalent to B.min being inside the G interval, but B.max being outside.

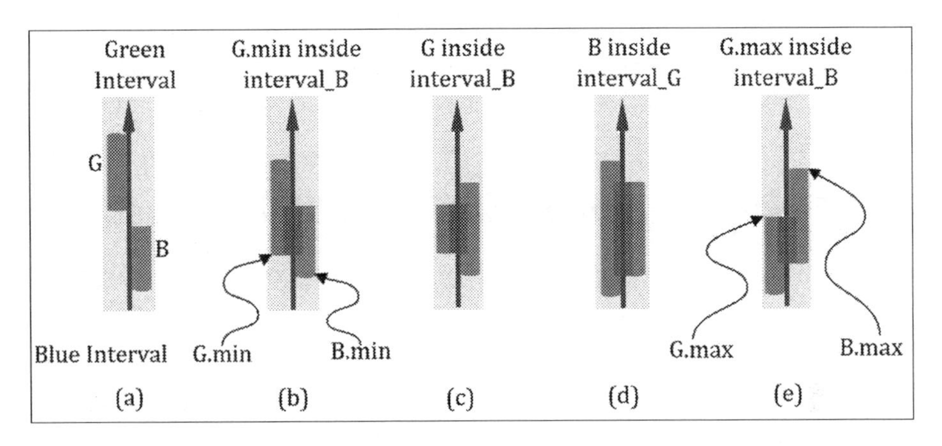

***Figure 2-10.** The different possible ways that two intervals can intersect*

Notice that when two intervals overlap (or intersect or collide), the result is always a new interval that is equal to or smaller than the original intervals. In fact, the overlapping interval is always the smaller of the two original max values and the larger of the two original min values. This fact is illustrated in Figure 2-11.

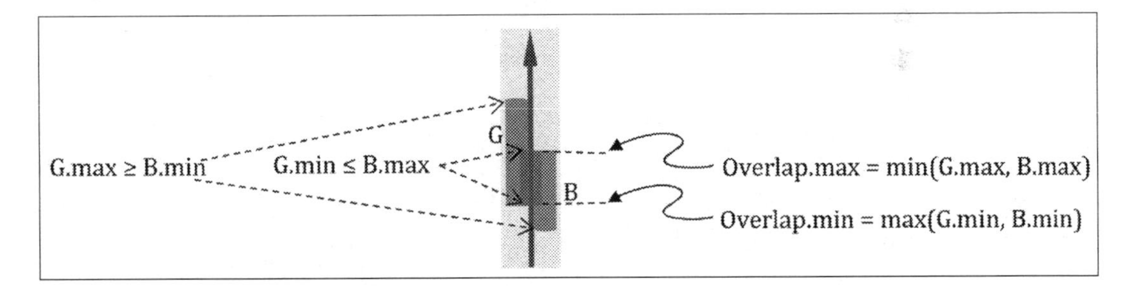

***Figure 2-11.** The condition and results of a two-interval intersection*

Figure 2-11 shows how the smaller of the two max values and the larger of the two min values define a valid interval. This condition is summarized as follows. Feel free to analyze this code against all conditions depicted in Figure 2-10 to verify that the `resultInterval` is indeed correct for all possible ways that the intervals can intersect:

```
if ( (G.max >= B.min) && (G.min <= B.max) ) {
        // Intervals G and B intersect
        resultInterval.max = min(G.max, B.max)
        resultInterval.min = max(G.min, B.min)
} else
        // No intersection
```

The Interval Bound Intersections Example

This example computes and visualizes the intersection of two intervals. It allows you to adjust and examine the different ways that two intervals can intersect. Figure 2-12 shows a screenshot of running the `EX_2_3_IntervalBoundIntersections` scene from the `Chapter-2-Intervals+AABB` project. This scene can be opened by double-clicking the `EX_2_3_IntervalBoundIntersections` scene file in the `Assets/` folder of the Project Window.

***Figure 2-12.** Running the Interval Bound Intersections example*

The goals of this example are for you to

- Examine and verify the different ways two intervals can intersect
- Understand the implementation of intersecting two intervals

Examine the Scene

The Hierarchy Window shows that the initial scene setup is extremely simple where the only predefined object is `Controller`. When this example runs, it will display a Green and a Blue interval along the Y-axis and will allow you to adjust these two intervals while examining the intersection results.

Analyze Controller MyScript Component

The `MyScript` component on the `Controller` shows six variables. These variables are three sets of min and max values, one set for each interval: `GreenInterval`, `BlueInterval`, and `OverlapInterval`. You can adjust the minimum and maximum values of the `GreenInterval` and the `BlueInterval` to create the `OverlapInterval` and thus its min and max values. The values of `OverlapInterval` are the computed intersection results and cannot be adjusted.

Interact with the Example

Run the game and observe the Green and Blue intervals along the Y-axis. Try adjusting the minimum and maximum values for each of the intervals by adjusting their corresponding min/max values on the `MyScript` component of `Controller`. Note that when the two intervals do not overlap, there is no overlap interval and the min/max values of `OverlapInterval` are both displayed as NaN (Not a Number).

Next, adjust the minimum and maximum values of the Green and Blue intervals to reproduce the different scenarios in Figure 2-10. Notice that when the two intervals intersect, the overlap region can be described by a new interval which is a cylinder centered on the Y-axis. This is the `OverlapInterval` with minimum and maximum values displayed in the `OverlapInterval` min and max variables. For each scenario, verify that the `OverlapIntervalMax` is indeed the smaller of the two maximum values from the Blue and Green intervals and that `OverlapIntervalMin` is the larger of the two minimum values.

Details of MyScript

Open MyScript and examine the source code in the IDE. The instance variables are as follows:

```
// For visualizing the Green Interval
private MyIntervalBoundInY GreenInterval = null;
// Max/Min values for Green interval
public float GreenIntervalMax = 1.0f;
public float GreenIntervalMin = 0.0f;
// For visualizing the Blue Interval
private MyIntervalBoundInY BlueInterval = null;
// Max/Min values of the Blue Interval
public float BlueIntervalMax = 1.0f;
public float BlueIntervalMin = 0.0f;
// For visualizing the overlap interval
private MyIntervalBoundInY OverlapInterval = null;
// Max/Min values of the overlap interval
public float OverlapIntervalMax = float.NaN;
public float OverlapIntervalMin = float.NaN;
```

Notice the three sets of intervals and their corresponding minimum and maximum values. The public variables, the min and max variables for each interval, were discussed earlier. The private variables are of the MyIntervalBoundInY data type which, as pointed out in the first example of this chapter, are designed for visualizing the Y-axis intervals. The Start() function is listed as follows:

```
void Start() {
    // Define the Green Interval
    GreenInterval = new MyIntervalBoundInY();
    GreenInterval.IntervalColor = GreenColor;
    // Draw slightly offset from the axis
    GreenInterval.PositionToDraw = new Vector3(0.6f, 0, 0);
    // Define the Blue Interval
    BlueInterval = new MyIntervalBoundInY();
    BlueInterval.IntervalColor = BlueColor;
```

```
    // Draw slightly offset from the axis
    BlueInterval.PositionToDraw = new Vector3(-0.6f, 0, 0);
    // The overlap interval
    OverlapInterval = new MyIntervalBoundInY();
    OverlapInterval.DrawInterval = false; // Initially hide
    // Draw on the axis
    OverlapInterval.PositionToDraw = new Vector3(0.0f, 0, 0);
    OverlapInterval.IntervalColor =  OverlapColor;
}
```

Once again, you can observe a pattern of three sets of similar functions: instantiating the variables, setting the corresponding color, and setting the interval's position. For the case of the `OverlapInterval`, it is initially set to be hidden because it is only displayed when an intersection between the Green and Blue intervals occurs. The `Update()` function is listed as follows:

```
void Update() {
    // Update Green Interval with user input
    GreenInterval.MinValue = GreenIntervalMin;
    GreenInterval.MaxValue = GreenIntervalMax;
    // Update Blue Interval with user input
    BlueInterval.MinValue = BlueIntervalMin;
    BlueInterval.MaxValue = BlueIntervalMax;
    // Intersect Green and Blue Intervals
    if (GreenIntervalMin <= BlueIntervalMax&&
        GreenIntervalMax >= BlueIntervalMin) {
       // overlap condition
       OverlapInterval.DrawInterval = true;
           // show the overlap interval
       // set the max/min values
       OverlapIntervalMax = Mathf.Min(GreenIntervalMax,
                                      BlueIntervalMax);
       OverlapIntervalMin = Mathf.Max(GreenIntervalMin,
                                      BlueIntervalMin);
        // display these values for the user
        OverlapInterval.MaxValue = OverlapIntervalMax;
        OverlapInterval.MinValue = OverlapIntervalMin;
```

```
        // Implemented in theMyIntervalBound class
        Debug.Assert(GreenInterval.IntervalsIntersect
                     (BlueInterval));
    } else {
        OverlapInterval.DrawInterval = false;
        OverlapIntervalMax = float.NaN;
        OverlapIntervalMin = float.NaN;
    }
}
```

The first four lines set the user entered min and max values into the Green and Blue interval min and max values, respectively, for visualization. The `if` condition tests for the intersection of two intervals and, when an overlap is detected, sets the min and max values of the `OverlapInterval`. The logic for setting the `OverlapInterval` follows exactly as depicted in Figure 2-11; the smaller of the two max values and the larger of the two min values define the intersecting intervals. Notice that `MyIntervalBound.IntervalsIntersect()` is defined and the `Debug.Assert()` function verifies that the function does indeed return the condition if two intervals have collided. This is a convenient utility function that will be used in later examples.

Takeaway from This Example

You have examined how two simple intervals can overlap, analyzed the conditions of this overlap, and verified the implementation that checks for an overlap between these two intervals. Although two intervals can intersect in many ways, the intersection detection logic is relatively straightforward.

Similar to the case of extending the inside-outside test for one interval to support 3D bounding boxes, the interval intersection knowledge can also be generalized to support 3D bounding box collisions and intersections as you will see in the next section.

Relevant mathematical concepts covered include

- All interval intersection conditions
- Testing for an intersection between two intervals
- Computing the minimum and maximum values of the intersecting or overlapping interval

Unity tools

- `MyIntervalBound`: Custom-defined abstract class to support the visualization of intervals along the X-, Y-, and Z-axes

Interaction technique

- The use of NaN to communicate invalid float values

EXERCISE

Point in Multiple Intervals

In this exercise, you will program the logic to perform the inside-outside test for a point that can be in any combination of the three intervals from this example. For example, inside the Green interval but outside of the Blue and Overlap intervals. Please derive the appropriate logic such that for any test position, you can print out the inside-outside test results for all three intervals. Note that the `OverlapInterval` is only defined when the user overlaps the Green and Blue Intervals and thus will not always be available for the inside-outside test.

Collision of Bounding Boxes

Recall that the volume in a bounding box is defined by the three corresponding intervals along the three major axes. This fact is reflected in the inside-outside test, where a given position is inside the bounding box if and only if it is inside all three major axes' intervals.

In exactly the same manner, based on exactly the same reasoning, two bounding boxes are colliding, if and only if each of the three intervals that define the two boxes collided with each other along their corresponding axis. Additionally, since a new interval is the result of each interval collision, bounding boxes' intersections always result in a new bounding box. The new bounding box's maximum and minimum points can be computed in exactly the same fashion that a new interval is calculated from the results of an interval collision. The maximum position of the colliding bounding box is the minimum of all the intervals' maximum values, and the minimum position is the maximum of all the intervals' minimum values. This condition is listed as follows:

```
if ((box1.XInterval.IntervalIntersects(box2.Xinterval) &&
     // intersects in X
    (box1.YInterval.IntervalIntersects(box2.Yinterval) &&
     // intersects in Y
    (box1.ZInterval.IntervalIntersects(box2.Zinterval)
     // intersects in Z
   ) {
     // The two boxes are colliding
     // result of the xInterval intersection
     overlapBox.Xinterval.min = max(box1.Xinterval.min,
                                    box2.XInterval.min)
     overlapBox.XInterval.max = min(box1.Xinterval.max,
                                    box2.XInterval.max)
     // result of the yInterval intersection
     overlapBox.Yinterval.min = max(box1.Yinterval.min,
                                    box2.YInterval.min)
     overlapBox.YInterval.max = min(box1.Yinterval.max,
                                    box2.YInterval.max)
     // result of the zInterval intersection
     overlapBox.Zinterval.min = max(box1.Zinterval.min,
                                    box2.ZInterval.min)
     overlapBox.ZInterval.max = min(box1.Zinterval.max,
                                    box2.ZInterval.max)
}
```

Note that when intersection occurs, the resulting `overlapBox` is a properly defined 3D bounding box with three intervals defined along the three major axes: `overlapBox.XInterval`, `overlapBox.YInterval`, and `overlapBox.ZInterval`.

The Box Bound Intersections Example

This example demonstrates the intersection of two bounding boxes. It allows you to interact with and examine the geometries creating the bounding boxes as well as manipulate the boxes to approximate where the geometries intersect with each other. Figure 2-13 shows a screenshot of running `EX_2_4_BoundingBoxIntersections` scene from the `Chapter-2-Intervals+AABB` project. This scene can be opened by double-clicking the `EX_2_4_BoundingBoxIntersections` scene file in the `Assets/` folder of the Project Window.

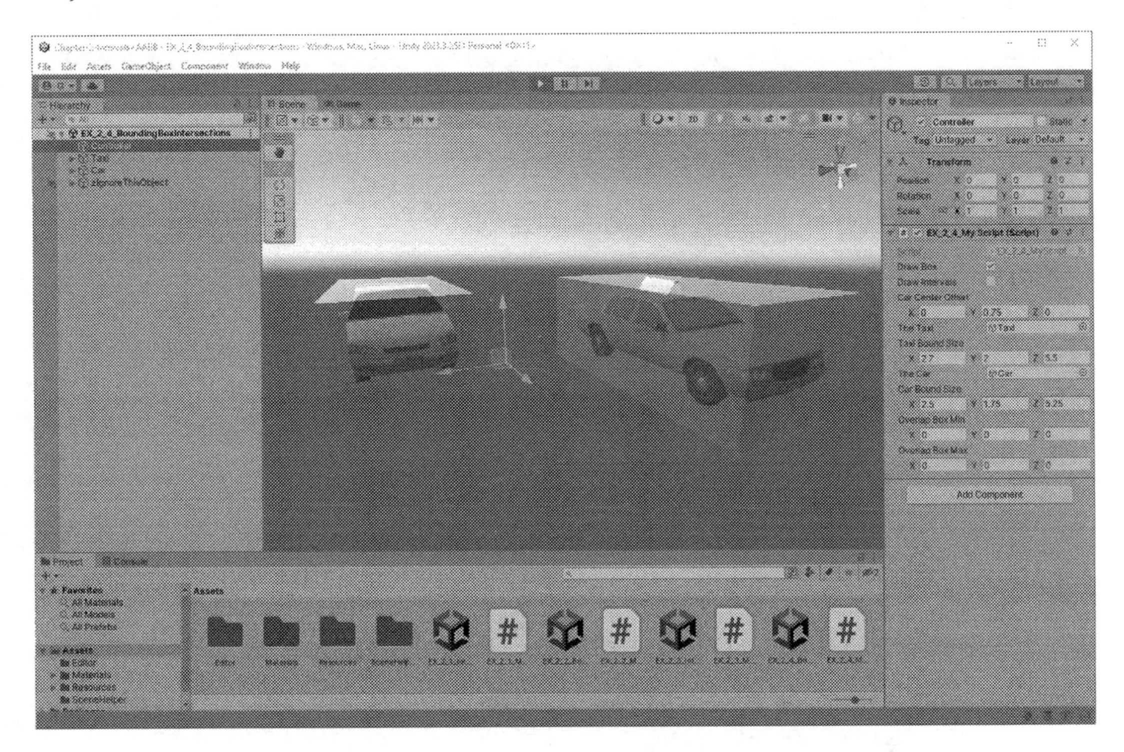

***Figure 2-13.** Running the Box Bound Intersections example*

The goals of this example are for you to

- Examine complex geometric objects bounded by their own bounding boxes
- Interact and adjust bounding boxes of objects
- Experiment with manipulating bounding boxes for collisions
- Understand and verify the bounding box intersection implementation

Examine the Scene

Besides Controller, there are only two other objects in the Hierarchy Window, Car and Taxi. These objects represent their corresponding vehicles in the scene. Running this example will build a bounding box around each of these two vehicles and allow you to manipulate these bounding boxes. You will also examine the details of bounding box intersection.

Analyze Controller MyScript Component

The MyScript component on the Controller shows nine variables that can be classified into approximately three different categories:

- Bounding box drawing support: Used to show and hide the drawing of the bounding boxes and the intervals that define the bounding boxes
  - DrawBox: Shows or hides the bounding boxes around both vehicles
  - DrawInterval: Shows or hides the three intervals that make up the bounding boxes around both vehicles
- Placement of bounding box
  - CarCenterOffset: Ensures the proper centering of the bounding box over the vehicles. More details will be provided when discussing the interaction with this example.
- Bounding box information
  - TheTaxi: Reference to the Taxi object
  - TheCar: Reference to the Car object
  - TaxiBoundSize: The size of the bounding box around the Taxi
  - CarBoundSize: The size of the bounding box around the Car
  - OverlapBoxMin: The minimum corner position of the overlapped box created when the taxi's and the car's bounding boxes collide
  - OverlapBoxMax: The maximum corner position of the overlapped box created when the taxi's and the car's bounding boxes collide

---

Note The information presented for the overlapped bounding box is its min and max values. This is contrasted to how you can control the other two bounding boxes—via center and size information. Using min and max values for the overlapped box allows you to verify the correctness of its computation.

Interact with the Example

Run the game and observe the two transparent boxes around each of the vehicles. These transparent boxes represent the corresponding bounding boxes of each vehicle. Try toggling the `DrawBox` and `DrawInterval` options under the `MyScript` component on `Controller`. Notice how toggling these options gives you control over displaying or hiding these boxes and their corresponding intervals. Additionally, take note that you can adjust the size of the two bounding boxes by changing the bound size variable for each vehicle (`CarBoundSize` and `TaxiBoundSize`).

Note The `Car` and `Taxi` game objects consist of corresponding children game objects. You can verify this by clicking the small triangle beside these objects in the Hierarchy Window to expand the object structure and observe their children objects. Take care that you are only manipulating these objects at the parent level, ensuring you don't change or manipulate any of their children.

Placement of the Bounding Box over the Vehicles

Try adjusting the values of `CarCenterOffset` and observe the relative position changes between the boxes and their corresponding vehicle. Recall that you have learned two ways to define a bounding box, either by specifying the maximum and minimum corner positions or by specifying its center and size. As you will see when examining the source code in `MyScript`, in this example the bounding boxes are defined according to their center and size information. The center position is defined by the position of the corresponding game object, that is, the values of that object's `transform.localPosition` variable. The size of the bounding box is specified by the user via the `MyScript` component on `Controller`. By changing `CarCenterOffset`, you are changing how much the bounding box's center differs from its corresponding vehicle's center.

CarCenterOffset has an initial offset in the Y-axis of about 0.75. This is because the center position of the vehicle model is not located in the middle of the vehicle, but at the height where the tires would meet the road. Thus, to ensure that the bounding box covers the entire car, its center position is raised to the approximate location of the vehicle's true center. You can verify this by setting CarCenterOffset's y-value to 0 and observing that the resulting bounding box does not cover the upper half of its corresponding vehicle.

Bounding Box Collisions

With CarCenterOffset set to (0, 0.75, 0), adjust the position of the Taxi object such that the vehicle overlaps with the Car object. Notice a new bounding box appearing in the overlapping region of the bounds. Examine the OverlapBoxMax and OverlapBoxMin values in the Inspector Window and verify that these are the smallest of the maximum corresponding interval values and the largest of the minimum corresponding interval values.

Void Space of a Bounding Box

When the CarCenterOffset is set to zero, the bottom half of the bounding box is outside of the vehicle and thus does not bound any useful information. This emptied bound volume is referred to as **void space**: the space where the bounding box can cause false collision detection. The potential of significant void space is the major drawback of the bounding box collision approximation. In general, all bounding boxes should be defined to minimize void space.

Details of MyScript

Open MyScript and examine the source code in the IDE. The instance variables are as follows:

```
// For visualizing the three bounding boxes
private MyBoxBound CarBound = null;
private MyBoxBound TaxiBound = null;
private MyBoxBound OverlapBox = null;
public bool DrawBox = true;          // Controls what to show/hide
public bool DrawIntervals = false;
```

```
// Offset between the geometry and box centers
public Vector3 CarCenterOffset = Vector3.zero;
// Note: The y-center of the car is at ground level
// Reference to the Taxi game object
public GameObject TheTaxi = null;
// User sets desirable taxi bounding box size
public Vector3 TaxiBoundSize = Vector3.one;
// Reference to the Car game object
public GameObject TheCar = null;
// User sets the desirable car bounding box size
public Vector3 CarBoundSize = Vector3.one;
// Min position of the overlapping bounding box
public Vector3 OverlapBoxMin = Vector3.zero;
// Max position of the overlapping bounding box
public Vector3 OverlapBoxMax = Vector3.zero;
```

As in all previous examples, the public variables listed have been analyzed. The private variables are once again for visualizing the bounding boxes. The `Start()` function for `MyScript` is listed as follows:

```
void Start() {
     // Ensure that proper reference setup in Inspector Window
     Debug.Assert(TheTaxi != null);
     Debug.Assert(TheCar != null);

     // Instantiate the visualization variables
     TaxiBound = new MyBoxBound();
     CarBound = new MyBoxBound();
     OverlapBox = new MyBoxBound();
     OverlapBox.SetBoxColor(new Color(0.4f, 0.9f, 0.9f, 0.6f));

     // hide the overlap box initially
     OverlapBox.DrawBoundingBox = false;
     // not showing this in this example
     OverlapBox.DrawIntervals = false;
}
```

As in all cases, the Start() function ensures proper game object reference setup in the Inspector Window and instantiates the private variables. Additionally, the Start() function also assumes no initial collision and hides the overlapped bounding box. The Update() function is listed as follows:

```
void Update() {
      // Step 1: Set the user specify drawing state
      TaxiBound.DrawBoundingBox = DrawBox;
      TaxiBound.DrawIntervals = DrawIntervals;
      CarBound.DrawBoundingBox = DrawBox;
      CarBound.DrawIntervals = DrawIntervals;
      // Step 2: Update the bounds (Taxi first, then Car)
      TaxiBound.Center = TheTaxi.transform.localPosition +
                         CarCenterOffset;
      TaxiBound.Size = TaxiBoundSize;
      CarBound.Center = TheCar.transform.localPosition +
                        CarCenterOffset;
      CarBound.Size = CarBoundSize;

      // Step 3: test for intersection ...
      // Two box bounds overlap when all three intervals overlap ...
      if (((TaxiBound.MinPosition.x <= CarBound.MaxPosition.x)
             && // X overlap
            (TaxiBound.MaxPosition.x >= CarBound.MinPosition.x))
           &&                          // AND
           ((TaxiBound.MinPosition.y <= CawrBound.MaxPosition.y)
             && // Y overlap
            (TaxiBound.MaxPosition.y >= CarBound.MinPosition.y))
           &&                         // AND
           ((TaxiBound.MinPosition.z <= CarBound.MaxPosition.z)
             && // Z overlap
            (TaxiBound.MaxPosition.z >= CarBound.MinPosition.z))) {
              // Min/Max of the overlap box bound
              Vector3 min = new Vector3(
                       // set with max of x, y, and z min values
```

```
            Mathf.Max(TaxiBound.MinPosition.x,
                      CarBound.MinPosition.x),
            Mathf.Max(TaxiBound.MinPosition.y,
                     CarBound.MinPosition.y),
            Mathf.Max(TaxiBound.MinPosition.z,
                     CarBound.MinPosition.z));

        Vector3 max = new Vector3(
            // set with min of x, y, and z max values
            Mathf.Min(TaxiBound.MaxPosition.x,
                     CarBound.MaxPosition.x),
            Mathf.Min(TaxiBound.MaxPosition.y,
                     CarBound.MaxPosition.y),
            Mathf.Min(TaxiBound.MaxPosition.z,
                     CarBound.MaxPosition.z));
        OverlapBox.DrawBox = TaxiBound.DrawBox;
        OverlapBox.DrawIntervals = TaxiBound.DrawIntervals;
        OverlapBox.MinPosition = min;
        OverlapBox.MaxPosition = max;

        // Update to show the overlap bound's min and max
        OverlapBoxMax = max;
        OverlapBoxMin = min;

         // functionality is implemented in the BoxBound
        Debug.Assert(TaxiBound.BoxesIntersect(CarBound));
    } else {
        OverlapBox.DrawBox = false;
        OverlapBox.DrawIntervals = false;
        OverlapBox.MinPosition = Vector3.zero;
        OverlapBox.MaxPosition = Vector3.zero;
    }
}
```

The Update() function sets the state of the application in three simple steps:

- Step 1: Set drawing state: Assign the user-specified drawing states of DrawBox and DrawInterval to TaxiBound and CarBound.
- Step 2: Update bound information: Update the Taxi bounding box (TaxiBound) and the Car bounding box (CarBound) with the user-specified values. Notice the use of CarCenterOffset to correct transform.localPosition, ensuring the bound is centered at the desired location.
- Step 3: Test for collision and create the overlapped bounding box: Test the bounds for an intersection, and when the condition is favorable, the min and max positions of the overlapped bounding box are computed and the new box is displayed for the user.

Note the very last line in the collision computation, the Debug.Assert() statement shows that the bounding box intersection functionality is also implemented in the MyBoxBound class. This line of code verifies the correctness of the bounding box collision test.

Takeaway from This Example

You have experienced bounding geometric objects with bounding boxes and learned that there might exist an offset between the center of the object and its bounding box. From working with TheTaxi and TheCar bounds, you have observed that when defining a bound, it is convenient to work with center and size information. This is in contrast with the case of the OverlapBound, where it is important to verify the computation results in the min and max positions.

For bounding boxes, just as in the case of the interval inside-outside test, the condition for intersection and the new bounding box resulting from that intersection are both straightforward and efficient to compute. Bounding boxes are one of the most widely used tools in interactive graphical applications because of their simplicity. The main shortcoming of bounding boxes is the potential for significant void space. However, the void space problem can be mitigated by defining multiple bounding boxes for one object, or a hierarchy of bounding boxes. You will work with hierarchy bounding boxes slightly in the exercises.

Relevant mathematical concepts covered include

- Testing for collisions between two bounding boxes
- Computing the minimum and maximum values of the intersecting (overlapping) bounding box

Unity tools

- `Vector3` addition operation that adds the corresponding operand x-, y-, and z-component values

EXERCISES

Manipulate CarBound with the Min/Max Values Implementation

Implement the functionality to replace the `CarBoundSize` with `CarBoundMin` and `CarBoundMax` variables. Notice that in this case, the min/max user interaction involves more complicated computations. In general, it is easier to define bounds of objects based on their center and size information than it is to use minimum and maximum corner positions.

Experiment with Void Space

Select and rotate `TheTaxi` by 45 degrees around the X-axis. Observe that a larger bounding box is now required to completely enclose the rotated vehicle. As a result, the void space has increased. This example illustrates a major limitation of bounding boxes: because of the axis-aligned requirement, they are ill-suited for bounding non-axis-aligned objects, for example, a rotated car or a human limb in motion. In the next chapter, you will learn about bounding spheres, another bounding volume with its own challenges, which can sometimes remedy the shortcomings of bounding boxes. If you were to rotate the bounding box with the taxi, then the bounding box would no longer be axis-aligned and the mathematics and algorithms developed in this chapter would not apply.

Experience with Hierarchical Bounding Boxes

One approach to remedy the potentially excessive void space for a bounding box is by defining a hierarchy of bounds. For example, define two *children* bounding boxes inside the given `CarBound` (or `TaxiBound`) and place them at the centers of the front and back wheels. Now, when a position is inside the parent bound (e.g., `CarBound`), you can perform the inside-outside tests with the two children bounds to better approximate if a collision has really occurred.

Final Words on Bounding Boxes

In general, there are other geometric volumes that can be used to bound complex geometries for proximity or collision determination. These approximation geometries are referred to as **bounding volumes** or **colliders**. As mentioned previously, Unity has defined its own bounding box class, `Bounds`. You will learn about bounding spheres and Unity's `BoundingSphere` class in the next chapter. These are both examples of bounding volumes for collision approximation. The general requirements for bounding volumes are as follows:

1. Representation: Their representation must be compact.
2. Efficiency: They must be algorithmically simple and computationally efficient.
3. Bound tightness: The void space must be *tolerable* for the target geometric shape.

In this chapter, you have learned that bounding boxes are easy to represent, either being two positions or a position and three floats, and are straightforward and efficient to compute collisions. Additionally, you have observed bounding boxes to be effective with relatively minimal void spaces when it comes to bounding rectangular shape geometries, for example, cars, still humans, or still animals. However, it is also true that when these objects rotate off-axis, for example, rotating a car about the X-axis by 45 degrees or a human leaning forward, the bounding box void space can increase significantly and thus greatly affect the accuracy of the collision approximation.

Unfortunately, this variability of approximation accuracy is true in general. All bounding volumes have variable efficiencies depending on the profile and orientation of the geometric shape that they bound. It is up to the game designer to determine the best types of bounding volumes to use for their purpose.

The Unity Bounds Class

Unity Application Programming Interface (API) describes the `Bounds` class as

> *An axis-aligned bounding box, or AABB for short, is a box aligned with coordinate axes and fully enclosing some object. Because the box is never rotated with respect to the axes, it can be defined by just its center and extents, or alternatively by min and max points.*

Unity `Bounds` defines the following properties and public functions (`https://docs.unity3d.com/ScriptReference/Bounds.html`):

- Properties
  - center: The center of the bounding box.
  - extents: The extents of the Bounding Box. This is always half of the size of the Bounds.
  - max: The maximal point of the box. This is always equal to center+extents.
  - min: The minimal point of the box. This is always equal to center–extents.
  - size: The total size of the box. This is always twice as large as the extents.
- Public methods
  - `ClosestPoint`: The closest point on the bounding box.
  - Contains: Is point contained in the bounding box?
  - `Encapsulate`: Grows the Bounds to include the point.
  - Expand: Expands the bounds by increasing its size by amount along each side.
  - `IntersectRay`: Does ray intersect this bounding box?
  - Intersects: Does another bounding box intersect with this bounding box?
  - SetMinMax: Sets the bounds to the min and max value of the box.
  - `SqrDistance`: The smallest squared distance between the point and this bounding box.

Through this chapter, you have learned the implementation details of all the functionality with bolded names, for example, the `size` property, or the `Contains` method. The mathematics behind the other functionality will be covered in different chapters in the rest of this book.

Summary

This chapter begins with covering the 3D Cartesian Coordinate System and follows by reviewing intervals along a major axis. These topics were used to build into the concept of an axis-aligned bounding box in 3D space that can be applied in determining the proximity of objects and approximating collisions. The chapter then reviews how to compute the intersection of intervals along a 1D axis before generalizing into the intersection of 3D bounding boxes. Besides learning the details of bounding boxes, it is important to recognize the merits of the foundational concepts that make up the bounding box. This chapter went from simple number comparisons to efficient collision approximation between complex geometries.

Through this chapter, you have also become familiar with this book's approach to presenting concepts. For each concept, the book always begins with explanations and presentations of pseudocode that is independent of Unity. This is then typically followed with a Unity project where you are guided to interact with and appreciate the effects and results of applying that concept. You are then led to analyze the parameters that control or implement the concepts being demonstrated via studying the variables on the `MyScript` component of the `Controller` game object. Lastly, you are walked through the examination of the actual source code. You can expect this rhythm to continue throughout the rest of this book.

CHAPTER 3

Distances and Bounding Spheres

After completing this chapter, you will be able to

- Compute the distance between any two positions
- Define bounding spheres for objects
- Perform inside-outside tests for bounding spheres
- Detect collisions between bounding spheres
- Appreciate the strengths and weaknesses of bounding spheres

Introduction

Now that you have more familiarity with the Unity environment and the learning tools that this book utilizes, it is time to review some slightly more advanced, yet still fundamental math concepts for video game creation. Similar to how Chapter 2 took simple number comparisons and used them to create bounding boxes, this chapter will develop the simple concepts of distances and the applications of the Pythagorean Theorem to create another powerful and widely used tool in video games: bounding spheres, which are also called sphere colliders.

From the previous chapter, you have learned that bounding boxes are created with and executed from simple logic statements and have an efficient runtime. However, you also learned that they are ill-suited for bounding objects that are not axis-aligned. Spheres can be represented simply by a point (the center) and a radius and are perfectly symmetrical with respect to its center. Their simple and compact representation and, as you will discover in this chapter, the efficient computations involved mean that

K. Sung and G. Smith, *Basic Math for Game Development with Unity 3D*,
https://doi.org/10.1007/978-1-4842-9885-5_3

spheres are prime candidates for serving as the geometry of bounding volumes. The elegant symmetrical property implies that the efficiency and effectiveness of bounding spheres are independent of object axis alignment or the rotations of objects. For these reasons, bounding spheres or sphere colliders are one of the most widely used tools in video games.

This chapter begins by reviewing distance computation, then follows by applying the results of this computation to sphere inside-outside tests, and finally wraps up with developing the bounding sphere functionality. Take note of the use of the `Vector3` data type in these discussions. Although this data type encapsulates three separate entities, the x-, y-, and z-values of a position, `Vector3` objects will be increasingly referenced and utilized as one unified entity with its own operators including addition, subtraction, magnitude, dot product, and so on. These observations will lead to the topic of vectors in the next chapter.

Distances Between Positions

Recall that, as depicted in Figure 2-1, the position of an object (x, y, z) is simply the distance measured from the origin along each corresponding major axis, for example, x-distance along the X-axis. Very conveniently and by design, the major axes of the Cartesian Coordinate System are perpendicular to each other. For this reason, the relationship between any position and the origin can be characterized by two right-angle triangles. This characterization is illustrated in Figure 3-1. Notice how the given position, D, is connected to the origin via two triangles, ABC and ACD, where both are right-angle triangles.

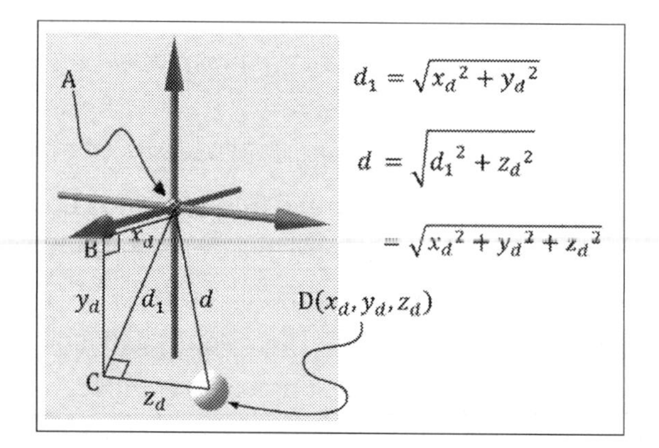

Figure 3-1. *The distance between the origin and a position, D (x_d, y_d, z_d)*

Triangle ABC is defined by vertices A (the origin), B, and C. The lengths of the edges of this triangle are as follows:

- Edge AB: The length along the X-axis, x_d
- Edge BC: The length along the Y-axis, y_d
- Edge AC: The length along the hypotenuse, computed via the Pythagorean Theorem,

$$d_1 = \sqrt{x_d^{\ 2} + y_d^{\ 2}}$$

Triangle ACD is defined by vertices A, C, and D (the position of interest). The lengths of the edges for this triangle are as follows:

- Edge AC: The length along the hypotenuse of the triangle ABC, d_1
- Edge CD: The length along the Z-axis, z_d
- Edge AD: The length along the hypotenuse, computed via the Pythagorean Theorem,

$$d = \sqrt{d_1^{\ 2} + z_d^{\ 2}} = \sqrt{x_d^{\ 2} + y_d^{\ 2} + z_d^{\ 2}}$$

Notice that the length of the edge AD, *d*, is simply the distance between the position (in this case, D) and the origin. The distance formula states that the distance between a position and the origin is the square root of the sum of the distances between that position and the origin measured along each major axis. In this case, those distances are x_d, y_d, and z_d. As illustrated in Figure 3-2, this concept can be generalized to compute distances between any two positions in the Cartesian Coordinate System.

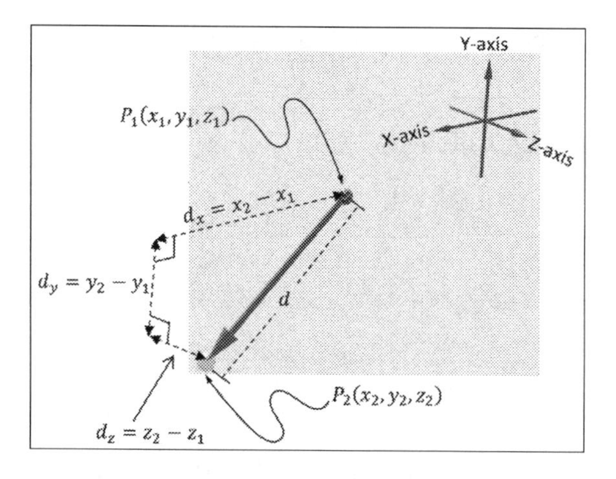

***Figure 3-2.** Calculating the distance between any two positions: P_1 and P_2*

Please refer to Figure 3-2 and consider the situation where the vertex A from Figure 3-1 has moved away from the origin to position $P_1(x_1, y_1, z_1)$. In this case, the distance between P_1 and any position $P_2(x_2, y_2, z_2)$ can still be determined by computing the distances along each of the major axes:

- Distance along the X-axis: $d_x = x_2 - x_1$
- Distance along the Y-axis: $d_y = y_2 - y_1$
- Distance along the Z-axis: $d_z = z_2 - z_1$
- Distance between P_1 and P_2:

$$d = \sqrt{d_x^2 + d_y^2 + d_z^2} = \sqrt{(x_2 - x_1)^2 + (y_2 - y_1)^2 + (z_2 - z_1)^2}$$

Note that since this equation squares the distances (the subtraction results) along each axis, the order of subtraction does not matter. This can be explained intuitively as the distance between P_1 and P_2 is the same as the distance between P_2 and P_1.

The Positions and Distances Example

The focus of this example is to demonstrate, allow you to interact with, and verify the distance computation between two positions. Figure 3-3 shows a screenshot of running the `EX_3_1_PositionsAndDistances` example from the `Chapter-3-Distances+BoundingSpheres` project. Recall that this scene can be opened by double-clicking the corresponding scene file in the Project Window.

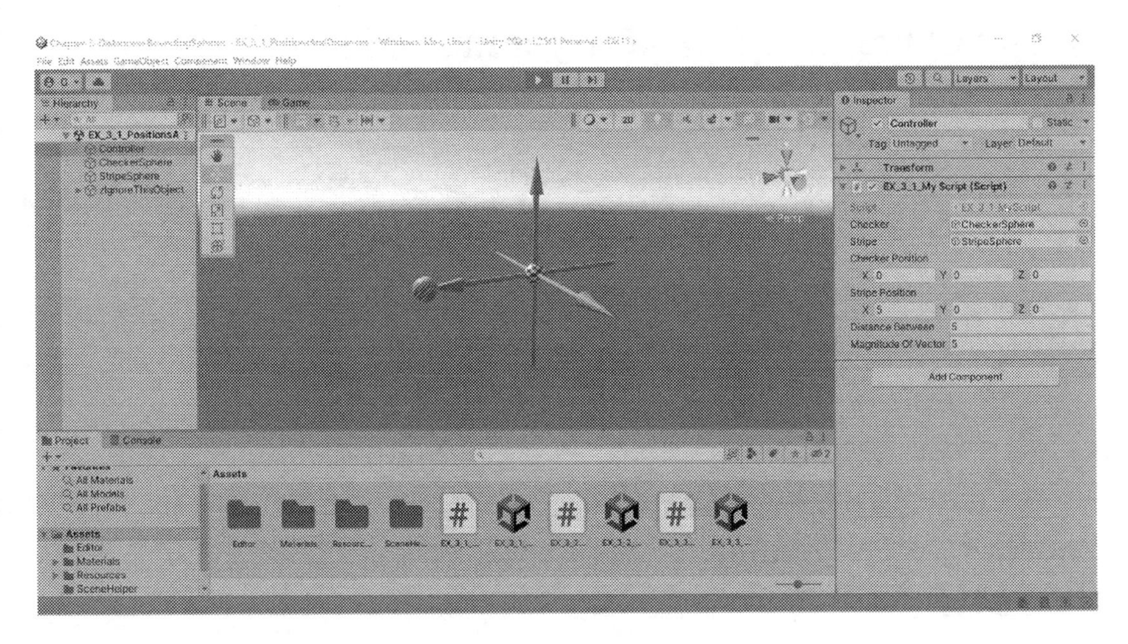

***Figure 3-3.** Running the Positions And Distances example*

The goals of this example are for you to

- Apply the Pythagorean Theorem for distance computation
- Manipulate positions and verify the results of distance computation
- Work with relevant, predefined functions of Unity's `Vector3` class

Examine the Scene

Examine the Hierarchy Window to observe that besides `Controller`, the two objects that you will interact with in this example are `CheckerSphere` and `StripeSphere`. This example allows you to manipulate the position of the `CheckerSphere` and the `StripeSphere` while it continuously computes and updates the distance between these spheres in two different ways, first by explicitly applying the Pythagorean Theorem and second by invoking a predefined `Vector3` function.

Note The three arrows, representing the major axes of the Cartesian Coordinate System, are defined in the `zIgnoreThisObject`. The axis frame is displayed as a reference for supporting your object manipulation exercise.

Analyze Controller MyScript Component

There are six defined variables in the `MyScript` component of `Controller`:

- `Checker`: A reference to the `CheckerSphere` game object
- `Stripe`: A reference to the `StripeSphere` game object
- `CheckerPosition`: The position of the `CheckerSphere`
- `StripePosition`: The position of the `StripeSphere`
- `DistanceBetween`: The distance between `CheckerSphere` and `StripeSphere`
- `MagnitudeOfVector`: The magnitude or length of a `Vector3` data type

The last two variables, `DistanceBetween` and `MagnitudeOfVector`, are the focus of this example.

Interact with the Example

Run the game and notice that as soon as the game begins, the values in `Controller`'s `MyScript` component have changed. While `CheckerSphere` and `StripeSphere` are still at their starting positions, the distance variables no longer have a value of 0. The `DistanceBetween` and `MagnitudeOfVector` variables are now both showing a value of 5. This value is the distance between the center of `CheckerSphere` and the center of `StripeSphere`. This can be easily verified by observing that the `CheckerSphere` is located at the origin and that the `StripeSphere` is located at (5, 0, 0), proving that the distance is indeed 5.

Set the position of the `CheckerSphere` to be (6.4, 0, 0) by manipulating the `CheckerPosition` variable. Verify that the `CheckerSphere` did move in the scene and is now located just beyond the tip of the red arrow representing the X-axis. More importantly, note that the `DistanceBetween` and `MagnitudeOfVector` variables are both showing the new correct distance value of 1.4. Try moving the `CheckerSphere` along the X-axis and verify that the computed distance for both variables is always correct.

After verifying the computed distances are correct along the X-axis, move the two spheres randomly off the X-axis. Observe the thin red, green, and blue lines that run parallel to the three major axes and connect the `CheckerSphere` to the `StripeSphere`. These three lines are used to help visualize the d_x, d_y, and d_z values between the center positions of the two spheres. If you do not see these lines, make sure you are looking at the Scene View window and that the example is running as the lines are not shown in the Game View window because they are meant for debugging in the Unity Editor.

With the two spheres located at random positions, examine the distances computed. Though it can be challenging to eyeball and verify that the computed distance is correct, rest assured, they are.

Details of MyScript

Open MyScript and examine the source code in the IDE. The instance variables are as follows:

```
public GameObject Checker = null;    // The spheres to work with
public GameObject Stripe = null;
public Vector3 CheckerPosition = Vector3.zero;
public Vector3 StripePosition = Vector3.zero;
public float DistanceBetween = 0.0f;
public float MagnitudeOfVector = 0.0f;
```

All of these variables have been discussed when analyzing the MyScript component. Next, you will examine the Start() function. It is similar to the Start() functions in other examples where assertion statements are used to verify game object references. In this case, it checks that the Checker and Stripe variables are indeed properly initialized in the Inspector Window.

```
void Start() {
    Debug.Assert(Checker!= null);  // Ensure proper init
    Debug.Assert(Stripe != null);
}
```

The Update() function is the essence of this example and is listed as follows:

```
void Update() {
    // Update the sphere positions
    Checker.transform.localPosition = CheckerPosition;
    Stripe.transform.localPosition = StripePosition;
    // Apply Pythagorean Theorem to compute distance
    float dx = StripePosition.x - CheckerPosition.x;
```

```
        float dy = StripePosition.y - CheckerPosition.y;
        float dz = StripePosition.z - CheckerPosition.z;
        DistanceBetween = Mathf.Sqrt(dx*dx + dy*dy + dz*dz);
        // Compute the magnitude of a Vector3
        Vector3 diff = StripePosition - CheckerPosition;
        MagnitudeOfVector = diff.magnitude;
        #region Display the dx, dy, and dz
}
```

The `Update()` function sets the state of this example in four simple steps:

- Step 1: Sets the `GameObject` positions with their corresponding position variables. This step allows you to change the location of the `StripeSphere` and the `CheckerSphere` via the `CheckerPosition` and `StripePosition` variables.
- Step 2: Applies the Pythagorean Theorem to compute distance. Based on the center of the two spheres, this step computes the distances along the X-, Y-, and Z-axes and then takes the square root of the sum of the squared axis distances.
- Step 3: Calculates the distance by working with the `Vector3` class. This step demonstrates that the distance between two positions is also calculated by the `magnitude` property of the `Vector3` class. You may recall from previous examples that the subtract operator, "-", of `Vector3` computes the differences of the corresponding x-, y-, and z-components. For this reason, the results in the variable, `diff`, are identical to the computed results, `dx`, `dy`, and `dz`. Interestingly, the `magnitude` property of `Vector3` class computes the square root of the sum of squares of the components, in this case, the distance between the two spheres. The next chapter will examine vectors and the `Vector3` class in detail. For now, simply take note of the convenience of working with the `Vector3` class.

- Step 4: Visualizes the distance along each axis with lines. The last step is hidden in the collapsed "`Display the` d_x, d_y, and d_z" region. This region hides the logic that visualizes the d_x, d_y, and d_z displacements along their corresponding axis. This code will be straightforward to follow after the coverage of vectors in the next chapter. For now, note that `Debug.DrawLine()` is a handy function for drawing debug lines in the Scene View window.

The `Vector3` subtraction operator and the `magnitude` property are convenient shortcuts for avoiding the tedious per-coordinate x-, y-, and z-component access required when computing the distance between positions. As you will see in the next chapter, Unity's `Vector3` class and its operators are not designed specifically to support distance computation. Instead, they are a part of a powerful set of operators that belong to an important topic, vectors, that will be the focus of study for most of the rest of this book.

Note With the Microsoft Visual Studio IDE, a `#region` can be hidden or expanded by clicking the "+" or "–" symbols to the left of the corresponding line of code.

Takeaway from This Example

You have verified the application of the Pythagorean Theorem in computing distances between positions, and you have begun to work with the `magnitude` property of the `Vector3` class.

Relevant mathematical concepts covered include

- Pythagorean Theorem for computing the distance between two positions

Unity tools

- `Vector3`
    - Subtraction operator for computing distances measured along the major axes between two positions
    - The `magnitude` property that computes the Pythagorean Theorem

- `Debug.DrawLine()` function for drawing debugging lines in the Scene View window

Interaction technique

- Use a sphere game object to represent and manipulate a position.

EXERCISES

Order of Subtraction

Recall that because the Pythagorean Theorem computation involves the sum of squared distances, the following two statements compute the same results:

```
float distance1 = (pointA - pointB).magnitude;
float distance2 = (pointB - pointA).magnitude;
```

Verify this statement by switching the subtraction order on the `CheckerSphere` and `StripeSphere` when computing the distance and confirm that the results are still correct.

Any Position

You have learned that the distance computation is applicable to compute the distance between any two positions. Modify the `Update()` function to include a third position, for example, `ThirdPosition`. Now compute and display the distance between the `ThirdPosition` and the `CheckerSphere` and the distance between the `ThirdPosition` and the `StripeSphere`. Now manipulate the `CheckerSphere` and the `StripeSphere` to observe the two computed distances to the `ThirdPosition`. Notice that these distances converge to the same value when you move the two spheres to be close to each other. This exercise demonstrates that the computation learned does indeed compute the distance between any two positions.

Sphere Colliders or Bounding Spheres

Recall that in 2D space, a compass sketches a circle by fixing one point and then tracing out all points that are at a fixed distance from that one point. The fixed position is the center and the fixed distance is the radius of the circle. A point is inside the circle when its distance to the center is smaller than the radius of the circle; otherwise, the position is outside of the circle. This simple observation can be generalized from 2D to 3D space. A point is inside a sphere when its distance to the center is less than the radius of the sphere; otherwise, it is outside of the sphere.

Based on this simple observation, it is possible to use the Pythagorean Theorem to determine if a point is within the bounds of a sphere. In this way, it becomes straightforward to determine if an object bounded by a sphere is colliding or within a certain proximity of a given position. This concept is illustrated in Figure 3-4, where a car is bounded by a sphere.

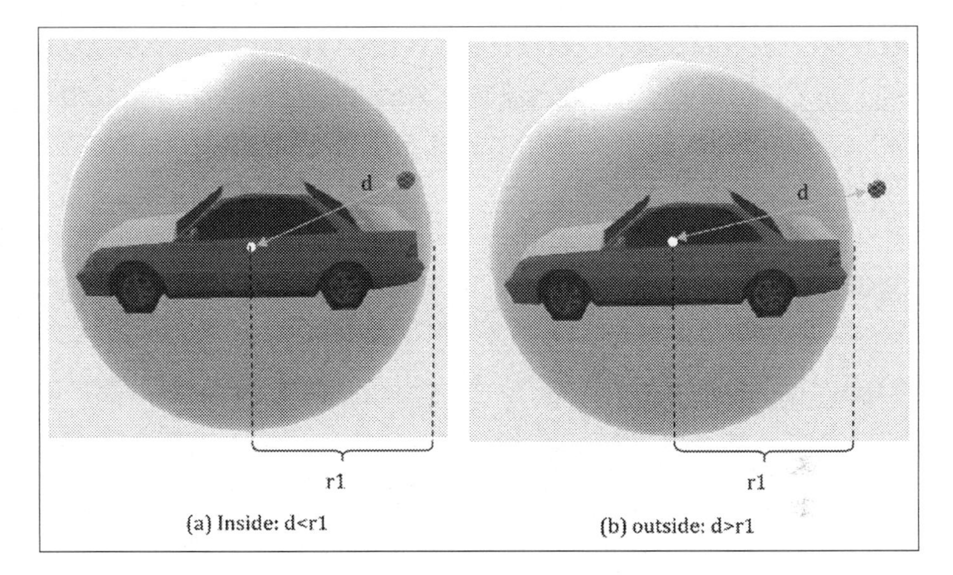

***Figure 3-4.** Determining if a position is inside or outside a sphere*

With the spherical bound shown in Figure 3-4, it becomes possible to determine if a position is inside (close enough to the car) or outside (not close enough to the car) of the sphere. These conditions can be determined by comparing the distance between the position (the checkered sphere) and the center of the sphere, d, to the radius of the sphere, r1. This is the inside-outside test of the bounding sphere; the logic for this test is listed as follows:

```
float d = (Position - Sphere.Center).magnitude;
if (d <= Sphere.Radius)
     // Position is inside the sphere: Figure 3-4(a)
else
     // Position is outside the sphere: Figure 3-4(b)
```

The less-than-or-equal test for the inside condition says that when positions are located on the circumference of the sphere, they are considered as inside the sphere. The spherical bound is referred to as a `SphereBound` or `SphereCollider` or `BoundingSphere`. Similar to the case of bounding boxes, this type of bound is widely used and important enough that Unity has defined its own `BoundingSphere` class, `https://docs.unity3d.com/ScriptReference/BoundingSphere.html`, that implements the associated functionality. At the end of this chapter, after you have learned some of the involved algorithms and implementations, you will examine this Unity class in more detail.

The Sphere Bounds Example

This example implements and demonstrates the strengths and weaknesses of the bounding sphere functionality. Figure 3-5 shows a screenshot of running the `EX_3_2_SphereBounds` example.

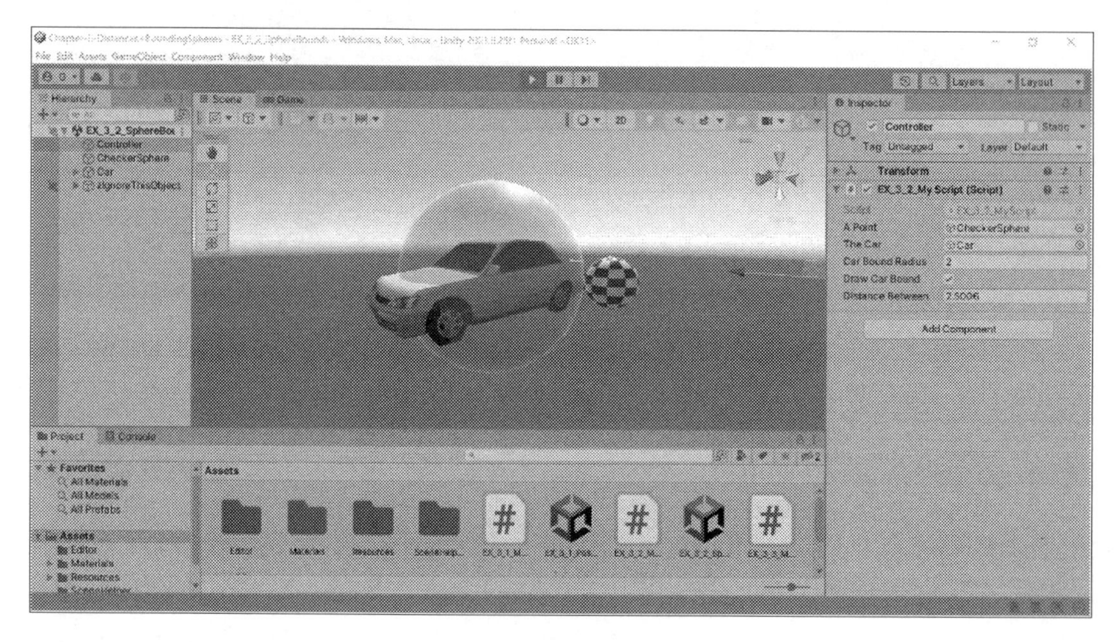

Figure 3-5. *Running the Sphere Bounds example*

The goals of this project are for you to

- Review the application of the Pythagorean Theorem
- Examine the details of the bounding sphere inside-outside test implementation

Examine the Scene

Look at the Hierarchy Window in the EX_3_2_SphereBounds scene and observe that besides Controller, the two objects that you will interact with in this example are CheckeredSphere and Car. This example defines a sphere bound around the Car and demonstrates the approximation of collision between the Car and the CheckeredSphere.

Analyze Controller MyScript Component

Select Controller and examine the MyScript component in the Inspector Window. You will see five variables:

- APoint: The reference to the CheckerSphere game object
- TheCar: The reference to the Car game object
- CarBoundRadius: The radius of the sphere bound around the Car object
- DrawCarBound: A toggle determining if the car bound should be drawn
- DistanceBetween: The computed distance between the center of TheCar and APoint

Interact with the Example

Run the game to observe a transparent white sphere covering the Car object. This transparent sphere represents the SphereBound of the Car. Select and manipulate the position of the CheckerSphere. Notice the color of the car bound changes when the center of the CheckerSphere is within its bounds. This same behavior can be observed by manipulating the position of the Car.

By design, the car bound sphere does not completely cover the Car. For example, the front and rear bumpers are outside of the bounding sphere. This means that the system is not able to detect when the CheckerSphere is colliding with the front or the rear of the car. You can change the size of the car sphere bound by adjusting the CarBoundRadius variable. Finally, notice the large amount of void space in between the Car and its spherical bound. In general, spherical bounds are not suitable for bounding rectangular objects.

Details of MyScript

Open MyScript in your IDE and observe the instance variables. Once again, you can observe and verify the one-to-one correspondence between the public variables defined in the script source code and the user manipulatable variables of the MyScript component in the Inspector Window. These variables are as follows:

```
public  GameObject APoint = null;          // CheckerSphere position
private MySphereBound SphereBound = null; // The car sphere bound
public  GameObject TheCar = null;          // Reference to the car
public  float CarBoundRadius = 2.0f;       // Sphere bound radius
public  bool DrawCarBound = true;          // To draw/hide bound
public float DistanceBetween = 0.0f;       // Car to APoint distance
```

The SphereBound variable is the only private variable and is defined for visualizing the car bounding sphere. In Figure 3-6, you can see the public fields and functions of the MySphereBound class. This class is used to help visualize and create the bounding sphere. Notice that besides the two fields for supporting drawing, DrawBound and BoundColor, the class only defines a Center and a Radius—the definition of a sphere.

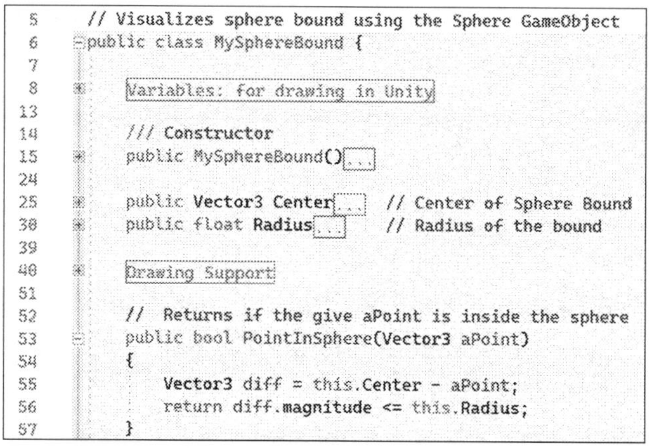

```
5     // Visualizes sphere bound using the Sphere GameObject
6     public class MySphereBound {
7
8         Variables: for drawing in Unity
13
14        /// Constructor
15        public MySphereBound()...
24
25        public Vector3 Center...   // Center of Sphere Bound
38        public float Radius...     // Radius of the bound
39
40        Drawing Support
51
52        //  Returns if the give aPoint is inside the sphere
53        public bool PointInSphere(Vector3 aPoint)
54        {
55            Vector3 diff = this.Center - aPoint;
56            return diff.magnitude <= this.Radius;
57        }
```

***Figure 3-6.** The MySphereBound class for creating and visualizing a spherical bound*

Next, examine the initialization of the variables in the Start() function:

```
void Start() {
     Debug.Assert(APoint != null);      // Ensure initialization
     Debug.Assert(TheCar != null);
     SphereBound = new MySphereBound();  // Visualize the bound
}
```

Besides verifying that APoint and TheCar variables are properly set up in the editor, the SphereBound variable is also instantiated. Lastly, take a look at the Update() function:

```
void Update() {
     // Step 1: Assume no collision
     SphereBound.BoundColor = MySphereBound.NoCollisionColor;
     // Step 2: Update the sphere bound
     SphereBound.Center = TheCar.transform.localPosition;
     SphereBound.Radius = CarBoundRadius;     // Set the radius
     SphereBound.DrawBound = DrawCarBound;    // Show/Hide bound
     // Step 3: Compute distance between APoint and SphereBound
     Vector3 diff = TheCar.transform.localPosition
                  - APoint.transform.localPosition;
     DistanceBetween = diff.magnitude;
     // Step 4: Testing and showing collision status
     bool isInside = (DistanceBetween <= CarBoundRadius);
     // TheCar.SetActive(!isInside);  // what does this do?
     if (isInside) {
         Debug.Log("Inside!! Distance:" + DistanceBetween);
         SphereBound.BoundColor = MySphereBound.CollisionColor;
         // The test is supported by MySphereCollider
         Debug.Assert(SphereBound.PointInSphere(
                      APoint.transform.localPosition));
     }
}
```

The Update() function performs the following four steps:

- Step 1: Set car sphere bound color to white, signifying that no collision has occurred.
- Step 2: Update the SphereBound parameters with the current user-specified values from the MyScript component on the Controller.
- Step 3: Calculate the distance between APoint and the center of the SphereBound.
- Step 4: Perform the sphere inside-outside test by comparing the computed distance to the radius of the SphereBound and update the color of the bound accordingly. Notice that as listed in Figure 36, the PointInSphere() function defined by the MySphereBound class implements the functionality of steps 3 and 4.

Takeaway from This Example

It is important to emphasize that the functionality of a sphere collider is implemented entirely in the Update() function and is independent of the MySphereBound class, for example, by defining the center and radius as the following:

```
Vector3 BoundCenter;  // Center of Sphere bound
float BoundRadius;    // Radius of Sphere bound
```

The exact same functionality, except the visualization of the sphere bound, can be implemented in the MyScript class without MySphereBound. Once again, make sure you focus on and understand the mathematical concepts and their implementation and not on how visualization is implemented.

Relevant mathematical concepts covered include

- Distance computation
- Sphere inside-outside test

Unity tools

- MySphereBound: A custom-defined class to support the visualization of a bounding sphere

EXERCISE

Modifying Game Behavior

Select the Controller object and toggle off the DrawCarBound flag. Run the game now and observe that the car sphere bound is now hidden. Manipulate CheckerSphere such that it is touching the car. Now, open the Console Window (label F of Figure 1-3) and look at the log messages generated and notice that the inside condition is still computed and detected even though the sphere bound is not being drawn. Next, stop the game, uncomment the following line in the Update() function, and then restart the game:

```
TheCar.SetActive(!isInside);   // what does this do?
```

Now, with the DrawCarBound flag being switched off, notice how the Car game object appears and disappears depending on how far away the CheckerSphere is. Imagine the CheckerSphere represents a projectile, then it would look as if the Car was being "hit" and destroyed when the projectile is in close proximity. Having another object collide with or being detected inside of a bounding sphere is a common reason for hiding (or destroying) objects in a game.

Collision of Bounding Spheres

The sphere inside-outside test can be generalized to determine if two spheres are colliding. Figures 3-7(a) and (b) show that the condition for collision between two spheres can be determined by comparing the distance between their centers to the sum of their radii. When the centers are further away than their radii summed, as illustrated in Figure 3-7(a), there is no intersection. Otherwise, as shown in Figure 3-7(b), the two spheres are colliding. Once again, this simple and straightforward computation results in bounding spheres being one of the most commonly used bounding geometries in 3D interactive graphical applications, including video games.

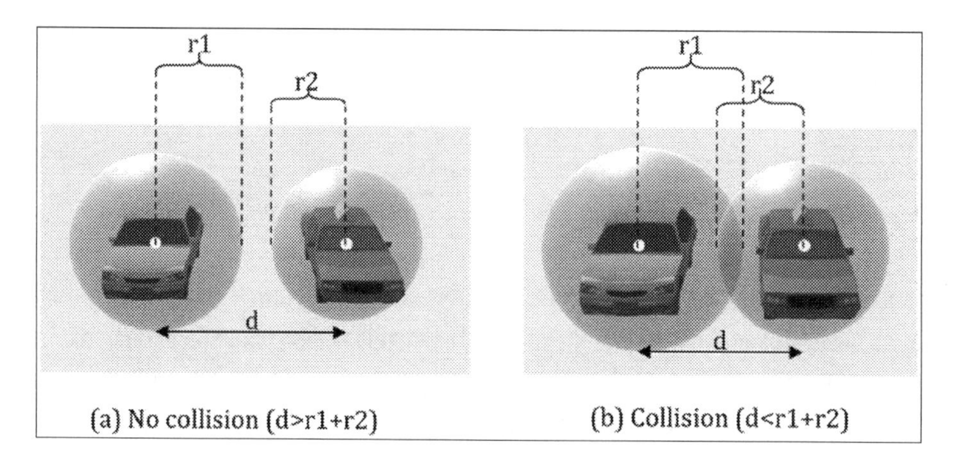

(a) No collision (d>r1+r2) (b) Collision (d<r1+r2)

Figure 3-7. *Calculating the collision between two spheres*

The Sphere Bound Intersections Example

This example demonstrates the generalization of the inside-outside test presented in the previous example to detect intersections or collisions of two spheres. Figure 3-8 shows a screenshot of running the `EX_3_3_SphereBoundIntersections` example.

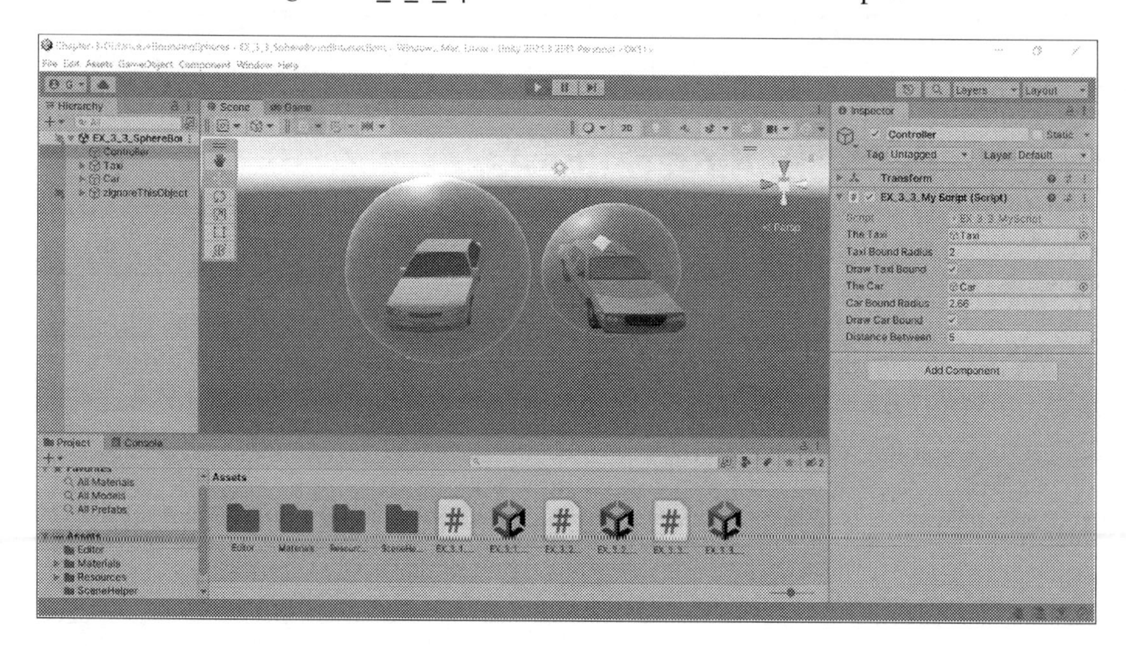

Figure 3-8. *Running the Sphere Bound Intersections example*

The goals of this project are for you to

- Understand how to intersect bounding spheres
- Examine and understand the implementation of bounding sphere intersection

Examine the Scene

Upon examining the scene, you will see that besides `Controller`, there are only two objects in the scene to pay attention to: `Car` and `Taxi`. This example builds a bounding sphere around each of these two vehicles and allows you to examine the details of bounding sphere intersection implementation.

Analyze Controller MyScript Component

The `MyScript` component on the `Controller` presents seven variables, two sets of three variables for each vehicle and then one for the both of them:

- Taxi
    - `TheTaxi`: The reference to the `Taxi` game object
    - `TaxiBoundRadius`: The radius of the sphere bounding the `Taxi` object
    - `DrawTaxiBound`: A toggle determining if the taxi bound should be drawn
- Car
    - `TheCar`: The reference to the `Car` game object
    - `CarBoundRadius`: The radius of the sphere bounding the `Car` object
    - `DrawCarBound`: A toggle determining if the car bound should be drawn
- `DistanceBetween`: The computed distance between the center of `TheCar` and the center of `TheTaxi`

Interact with the Example

Run the game and observe that each vehicle is almost completely bounded by its own transparent sphere. These are the bounding spheres for the corresponding vehicles. Manipulate the position of the vehicle, for example, move the Taxi in the positive x-direction, and observe the bounding sphere color change when the vehicles are sufficiently close to each other that the bounding spheres intersect or collide.

You can observe the effects of void space when the spheres trigger a collision event (when the spheres change color) without the two vehicles coming into contact. Change the bound radius of both the Taxi and the Car through their corresponding BoundRadius variable and observe the trade-off between the size of your void space and the likelihood of missing collisions.

Details of MyScript

Open MyScript in the IDE and observe the similarities of the code to those from the MyScript of the EX_3_2_SphereBounds example. The only significant difference is in the Update() function's sphere intersection computation in Step 5.

```
void Update() {
    // Step 1: Assume no intersection
        ...
    // Step 2: Update the Taxi sphere bound
        ...
    // Step 3: Update the Car sphere bound
        ...
    // Step 4: Compute distance as magnitude of a Vector3
    Vector3 diff = TaxiBound.Center - CarBound.Center;
    DistanceBetween = diff.magnitude;
    // Step 5: Testing and showing intersection status
    bool hasIntersection =
        DistanceBetween <= (TaxiBound.Radius + CarBound.Radius);
    if (hasIntersection) {
        Debug.Log("Intersect!! Distance:" + DistanceBetween);
        TaxiBound.BoundColor = MySphereBound.CollisionColor;
        CarBound.BoundColor = MySphereBound.CollisionColor;
```

```
            // functionality is also supported by MySphereCollider
            Debug.Assert(TaxiBound.SpheresIntersects(CarBound));
        }
}
```

In this example, as illustrated in Figure 3-7, Step 5 is accomplished by comparing the distance between two points to the sum of the two bounding sphere's radii to determine if a collision has occurred, instead of being compared to just one sphere's radius as it was in the Sphere Bounds example.

Takeaway from This Example

This example has been a straightforward generalization of the previous example in detecting whether a given position is inside or outside a sphere. In the previous example, you were able to detect if a position with a radius of zero entered a bounding sphere; in this example, you generalized that position to now have a radius of any value.

Relevant mathematical concepts covered include

- Testing for collision or intersection between two spheres

EXERCISE

Hierarchical Bounding Spheres

One way to remedy the potentially large void space shortcomings of bounding spheres is by defining a hierarchy of bounds. For example, define two more `SphereBounds` inside the given bound. These two `SphereBounds` should be located at the center of the front and back wheels, each with a radius about one-third of the outer bound. Now, when a position is inside the outer bound, you can perform the inside-outside test with the two inner bounds to decide if a collision has occurred. Try implementing this functionality. In general, a game object can be bounded by a hierarchy of bounding geometries, where the inner bounds will only be explored if the outer bound test returns a favorable result. Such a hierarchy can significantly increase the accuracy of collision approximation at a cost of increased computation and algorithmic complexities.

The Unity BoundingSphere Class

Unity API documents the BoundingSphere class as

> *Describes a single bounding sphere for use by a CullingGroup.*

You can think of a CullingGroup as a hierarchy of bounds. As it does not pertain to the math in this book, exactly how to implement a CullingGroup or use Unity's BoundingSphere class will not be discussed. Instead, they are mentioned here merely to verify that the bounding sphere is a widely used method for bounding objects. Unity BoundingSphere defines the following properties (https://docs.unity3d.com/ScriptReference/BoundingSphere.html):

- position: The center position of the BoundingSphere
- radius: The radius of the BoundingSphere

Notice how Unity's BoundingSphere class doesn't have any public methods. The MySphereBound class that you used throughout this chapter has additional functionality defined in the PointInSphere() and SpheresIntersects() functions. Due to the simplicity of these functions, it appears that Unity assumes the users of the BoundingSphere class will implement these tests themselves.

Summary

This chapter begins with reviewing how to apply the Pythagorean Theorem to compute distances between positions in a 3D Cartesian Coordinate System and then generalizes this knowledge to defining bounding spheres. Through working with the examples in this chapter, you have learned how to apply distance computation and use spheres as bounds in approximating collisions between geometrically complicated game objects. Your understanding of these concepts was gained based on your interaction with actual bounding spheres and improved upon by analyzing their implementation source code.

While straightforward to implement and widely used as a bounding geometry or collider, the major drawback of bounding spheres is the potentially significant void space within the bound. As you have observed in the case of cars, this issue of large void space can be especially profound for rectangular or elongated objects, like books, cars, or animals. Unfortunately, as discussed in the previous chapter, all bounding volumes have similar challenges in different degrees under different circumstances. The best ways to overcome the void space problem are to match your object to the best fitted bound or to use a hierarchy of bounds when one bound involves too much void space.

You have also learned more about the Unity `Vector3` class. The next chapter will cover vectors, the concept that the `Vector3` class is designed to support, in much more detail. The next chapter will build off what you have already seen and give you a greater understanding and appreciation of the usefulness and power of vectors in video games and computer graphics.

CHAPTER 4

Vectors

After completing this chapter, you will be able to

- Understand that a vector relates two positions to each other
- Recognize that all points in space are position vectors
- Comprehend that a vector encapsulates both a distance and a direction
- Perform basic vector algebra to scale, normalize, add, and subtract vectors
- Apply vectors to control the motions of game objects
- Implement simple game object behaviors like aiming and following
- Design and simulate simple external factors like wind conditions to affect object motion

Introduction

So far, you have reviewed some of the most elementary and ground laying mathematical concepts used in video game creation. These simple concepts that you have observed and interacted with can be developed further into a powerful and widely used tool set. This approach of introducing a simple concept and expanding it to solve real problems when designing a video game will be continued in this chapter with vectors and the fundamental algebra that accompanies them.

Vectors are entities that encapsulate point-to-point distance and direction. Vector algebra is the mechanism, or rules, for manipulating these two entities. It allows the user to, for example, increase the distance, change the direction, and combine, or detract,

K. Sung and G. Smith, *Basic Math for Game Development with Unity 3D*,
https://doi.org/10.1007/978-1-4842-9885-5_4

both the distance and direction at the same time. Vectors and their associated math concepts allow precise control and accurate prediction of basic game object movements as well as the support for many simple behaviors.

In many video games, object behaviors are often governed by their physical proximity to other objects, such as non-player characters changing from their predefined wandering pattern, for example, patrol path, and moving toward the approaching player. To support this simple scenario, you must be able to program the behavior of following a predefined route as well as the ability to detect and move toward the approaching player or character. Vectors, with their encapsulation of both distances and directions, are perfect for representing the motion of objects. Vector algebra complements this encapsulation with the ability to determine the relationships between the in-motion objects. Therefore, with just vectors and their accompanying mathematical operations, you as a game developer, at any moment in your game, can determine exactly what game behavior to invoke. Vectors and their associated algebra are one of the most fundamental tools in developing video games.

This chapter introduces vectors as a tool for controlling motion and computing spatial relationships between objects. In general, vectors are important for many, just as significant, applications that are unrelated to object motions. This is especially true for applications of vectors to fields outside of interactive graphical applications or video games, for example, applying vectors in machine learning for data cluster analysis. Even within the field of video games, vectors are important for other applications. Some of these other applications include predicting the exact intersection position between a motion path and a wall and computing the reflection direction after a collision, both of which will be discussed in future chapters.

This chapter begins by reviewing what you have learned from Chapter 3, but now with a focus on how vectors were used to perform the distance calculations you have experimented with and observed. The chapter then analyzes the details of the vector definition and the algebraic rules that govern the operations on vectors. Through these discussions, you will learn that the vector definition is independent of positions and that vectors can be scaled, normalized, and applied to represent velocities that define the motions of objects. The formal definition of vector algebra, the addition and subtraction operations, is presented toward the end of the chapter to conclude and verify the knowledge gained throughout the chapter.

Vectors: Relating Two Points

Vectors have been hinted at thus far in the book and even worked with in the previous chapter when you needed to compute the distance between positions, but now you will finally learn what they are and some of their applications. Please refer to Figure 4-1, which is identical to Figure 3-2 and copied here for convenience.

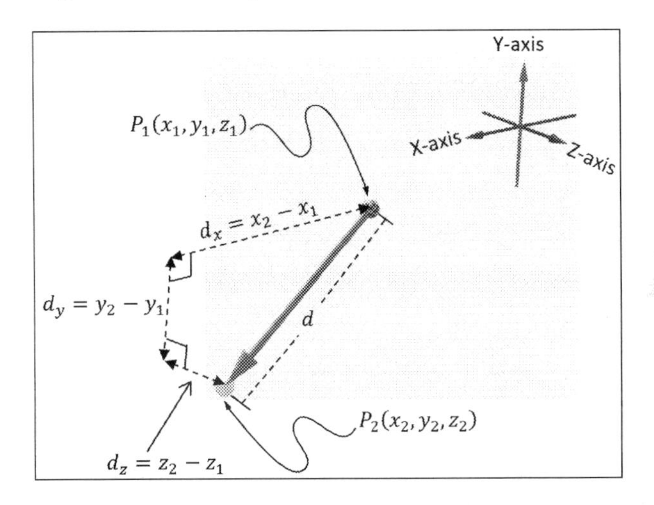

***Figure 4-1.** Calculating the distance between any two positions: P_1 and P_2 (same as Figure 3-2)*

Recall that in order to compute the distance between two positions, P_1 and P_2, the distances measured along the major axes must be computed.

- Distance along X-Axis: $d_x = x_2 - x_1$
- Distance along Y-axis: $d_y = y_2 - y_1$
- Distance along Z-axis: $d_z = z_2 - z_1$

You learned that the distance, d, between these positions can be derived by applying the Pythagorean Theorem twice to the two connecting right-angle triangles (see Figure 3-1 if you need a refresher). The derived formula is simply the square root of the summed squared distances measured along the major axes, which is listed as follows:

$$d = \sqrt{(x_2 - x_1)^2 + (y_2 - y_1)^2 + (z_2 - z_1)^2}$$

$$d = \sqrt{d_x^2 + d_y^2 + d_z^2}$$

This formula can be interpreted as the distance that is necessary to move an object from position P_1 to P_2. This displacement is defined by the shortest traveling distant, *d*, along the direction encoded by (d_x, d_y, d_z). This interpretation is reflected closely in the implementation of the `Update()` function in `EX_3_1_MyScript`, as copied and re-listed as follows for reference:

```
void Update() {
    // Update the sphere positions
    Checker.transform.localPosition = CheckerPosition;
    Stripe.transform.localPosition = StripePosition;

    // Apply Pythagorean Theorem to compute distance
    float dx = StripePosition.x - CheckerPosition.x;
    float dy = StripePosition.y - CheckerPosition.y;
    float dz = StripePosition.z - CheckerPosition.z;
    DistanceBetween = Mathf.Sqrt(dx*dx + dy*dy + dz*dz);

    // Compute the magnitude of a Vector3
    Vector3 diff = StripePosition - CheckerPosition;
    MagnitudeOfVector = diff.magnitude;

    #region Display the dx, dy, and dz
}
```

Pay attention to the last two lines of code once more, specifically, the `diff` variable which is the result of subtracting `CheckerPosition` (P_1) from `StripePosition` (P_2). As you learned from this example in the last chapter, the magnitude operator returns the distance, *d*, between the two positions. The same `diff` variable also defines the direction from P_1 to P_2. This entity, `diff`, that encodes those two pieces of information, distance and direction, is a **vector**. The line of code that computes `diff` can be expressed mathematically as follows:

$$\begin{aligned} \vec{V}_d &= P_2 - P_1 \\ &= (x_2 - x_1, y_2 - y_1, z_2 - z_1) \\ &= (d_x, d_y, d_z) \end{aligned}$$

Or simply, vector $\vec{V}_d = (d_x, d_y, d_z)$. There are a few interesting observations that can be made thus far:

- Symbol: The symbol for a vector, V, is shown as $\vec{V}$, with an arrow above the character V representing that it's a vector.
- Definition: A vector, $\vec{V} = P_2 - P_1$, describes the distance and direction to travel from P_1 to P_2.
- Notation: In 3D space, a vector is represented by a tuple of three floating-point values, signifying the displacements along each of the corresponding major axes. This notation is identical to that of a position in the Cartesian Coordinate System. In fact, given a tuple with three values, (x, y, z), without any context, it is impossible to differentiate between a position and a vector. This issue will be examined in the next section of this chapter.
- Representation: As illustrated in Figure 4-2, graphically, a vector $\vec{V} = (d_x, d_y, d_z)$ is drawn as a line that begins from a position, the **tail**, with an arrow pointing at the end position, the **head**, with the displacements of d_x, d_y, and d_z along the major axes. Note that in this case, d_y is a negative number because the y-displacement is in the negative direction of the Y-axis.
- Operations: You have already experienced working with the vector subtraction operator. This operator and others will be explored later in this chapter.

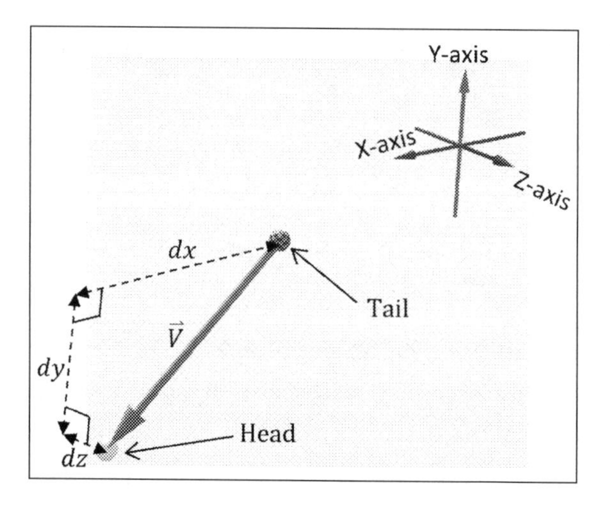

Figure 4-2. *A vector with its head and tail*

Position Vectors

For new learners of vectors, a common point of confusion is the position that defines a vector. For example, since the vector

$$\vec{V}_d = P_2 - P_1$$

defines the distance and direction from position P_1 to P_2, one may arrive at the wrong assumption that the vector $\vec{V}_d$, is "defined at position P_1." You will begin the exploration of vectors by analyzing this potentially confusing issue head-on and learn that vectors are defined independent of any specific position and, in fact, can be applied to any position.

Notice that the positions that define the vector $\vec{V}_d$, P_1 and P_2, are variables, indicating that this formula is true for any point located at any position. In the special case where $P_1^{'}$ is located at the origin of the Cartesian Coordinate System,(0, 0, 0), then,

$$\begin{aligned}\vec{V}_d^{'} &= P_2 - P_1^{'} \\ &= \left(x_2 - x_1^{'}, y_2 - y_1^{'}, z_2 - z_1^{'}\right) = \left(x_2 - 0, y_2 - 0, z_2 - 0\right) \\ &= \left(d_x, d_y, d_z\right) = \left(x_2, y_2, z_2\right)\end{aligned}$$

which shows that P_2 can be interpreted as a vector (x_2, y_2, z_2) from the origin. In fact, any position in the Cartesian Coordinate System at (x, y, z) can be interpreted as x-, y-, and z-displacements measured along the three major axes from the origin position and thus all positions in the Cartesian Coordinate System can be interpreted as vectors from the origin. In this way, the position of a point is also referred to as a **position vector**. In general, in the absence of a specific context, it is convenient to consider given tuples of three floats, for example, (x, y, z), as a position vector.

Note The origin position (0, 0, 0) is a special position vector and is referred to as the **zero vector**.

Following a Vector

Refer to Figure 4-1 again, recall that the detailed definition of vector $\vec{V}_d$ is as follows;

$$\begin{aligned}\vec{V}_d &= P_2 - P_1 \\ &= (x_2 - x_1, y_2 - y_1, z_2 - z_1) \\ &= (d_x, d_y, d_z)\end{aligned}$$

Remember that $\vec{V}_d$ defines the distance and direction from position P_1 to P_2. A subtle, but logical interpretation of this definition is that position P_2 can be arrived at if an object begins at position P_1 and travels along the X-axis by d_x, the Y-axis by d_y, and the Z-axis by d_z. This interpretation can be described as "following a vector" from P_1 to P_2 and can be verified mathematically as follows:

- P_2 x-position $= x_1 + d_x = x_1 + (x_2 - x_1) = x_2$
- P_2 y-position $= y_1 + d_y = y_1 + (y_2 - y_1) = y_2$
- P_2 z-position $= z_1 + d_z = z_1 + (z_2 - z_1) = z_2$

Not surprisingly, "following a vector" is expressed as

$$\begin{aligned} P_2 &= P_1 + \vec{V}_d \\ &= (x_1 + d_x,\ y_1 + d_y,\ z_1 + d_z) \\ &= (x_1 + x_2 - x_1,\ y_1 + y_2 - y_1,\ z_1 + z_2 - z_1) \\ &= (x_2, y_2, z_2) \end{aligned}$$

Graphically, you can imagine placing the tail of $\vec{V}_d$ at location P_1 and "follow the vector" to the head of the vector, to position P_2. This is how you can get from one position to another when you don't know the location of your next position, but you do have the distant and direction ($\vec{V}_d$) to get there.

Note You have seen the vector subtraction operator where the corresponding coordinate values are subtracted. Here you see vector addition operator, where the corresponding coordinate values are added. The details of vector subtraction and addition will be visited again later in this chapter.

Following a Vector from Different Positions

Following a vector, $\vec{V}_d$, from a given position, P_1, is also referred to as "applying the vector $\vec{V}_d$ at P_1." Since both $\vec{V}_d$ and P_1 are variables, the equation

$$P_2 = P_1 + \vec{V}_d$$

is true and applicable for any vector and any position. This concept is analyzed in detail in this section.

Figure 4-3 illustrates the alternative interpretations of the Cartesian Coordinate position, P_d, and the associated tuple of three floating-point values, (x_d, y_d, z_d).

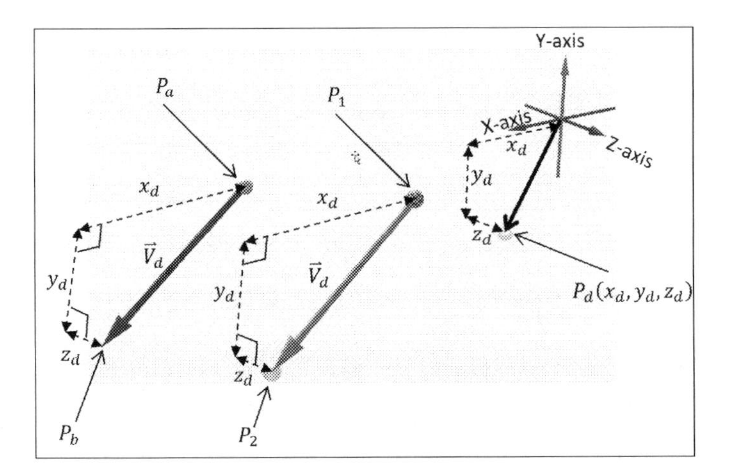

Figure 4-3. *Positions, position vectors, and applying vectors at different positions*

The top-right corner of Figure 4-3 illustrates that P_d is a position in 3D space located at distances x_d, y_d, and z_d from the origin. In this way, (x_d, y_d, z_d) is the position vector that identifies the location of the point P_d. The set of two spheres and the associated arrows on the left side of Figure 4-3 illustrate interpreting the three-float tuple, (x_d, y_d, z_d), as the vector $\vec{V}_d$. If you apply $\vec{V}_d$ to position P_1, you will arrive at position P_2. If you apply $\vec{V}_d$ to position P_a, then you will arrive at P_b. In this case, you know that the Cartesian Coordinate positions for P_1 and P_a are as follows:

$$P_1 = (x_1, y_1, z_1)$$

$$P_a = (x_a, y_a, z_a)$$

Then, the Cartesian Coordinate positions for P_2 and P_b must be as follows:

$$P_2 = P_1 + \vec{V}_d = (x_1 + x_d,\ y_1 + y_d,\ z_1 + z_d) = (x_2, y_2, z_2)$$

$$P_b = P_a + \vec{V}_d = (x_a + x_d,\ y_a + y_d,\ z_a + z_d) = (x_b, y_b, z_b)$$

These equations are true for any x-, y-, or z-values. This is to say that P_1 (and P_a) can be located at any position in the 3D Cartesian Coordinate System. In this way, a vector can indeed be applied to any position. In all cases, "following a vector" is simply placing the tail of the vector at the starting position, with the head of the vector always being located at the destination position.

Recall that when P_1 is located at the origin, or when

$$P_1' = (x_1', y_1', z_1') = (0,0,0)$$

then

$$P_2 = P_1' + \vec{V}_d = (0 + x_d, 0 + y_d, 0 + z_d) = (x_d, y_d, z_d) = P_d$$

Observe that when P_1' is located at the origin, then P_d is a coordinate position. This means that the associated tuple of three floating-point numbers, (x_d, y_d, z_d), can be interpreted as the vector $\vec{V}_d$ being applied to the origin, $(0, 0, 0)$. This is true for any coordinate position. For example, the tuple of three floats, (x_1, y_1, z_1), that defines the position P_1 also describes the vector $\vec{V}_1$ being applied to the origin. The reverse is also true that a given vector, $\vec{V}$, can be interpreted as the Cartesian Coordinate position, P, or a position vector. Without sufficient contextual information, such as the tail position, vectors are always depicted and visualized as a line segment with their tail located at the origin.

If you are given a three valued tuple, (x, y, z), without context, you can assume it is a position vector. If you are given a vector, $\vec{V}$, without context, you can assume it is a coordinate position (that it starts from the origin). The next example will cover the details of position vectors and help you understand working with a coordinate position and interpreting that position as a position vector.

The Position Vectors Example

The focus of this example is to allow you to visualize a position vector and then to apply that vector at different locations. This example allows you to adjust, examine, and verify that vectors are defined independent of any given position. Figure 4-4 shows a screenshot of running the `EX_4_1_PositionVectors` scene from the `Chapter-4-Vectors` project.

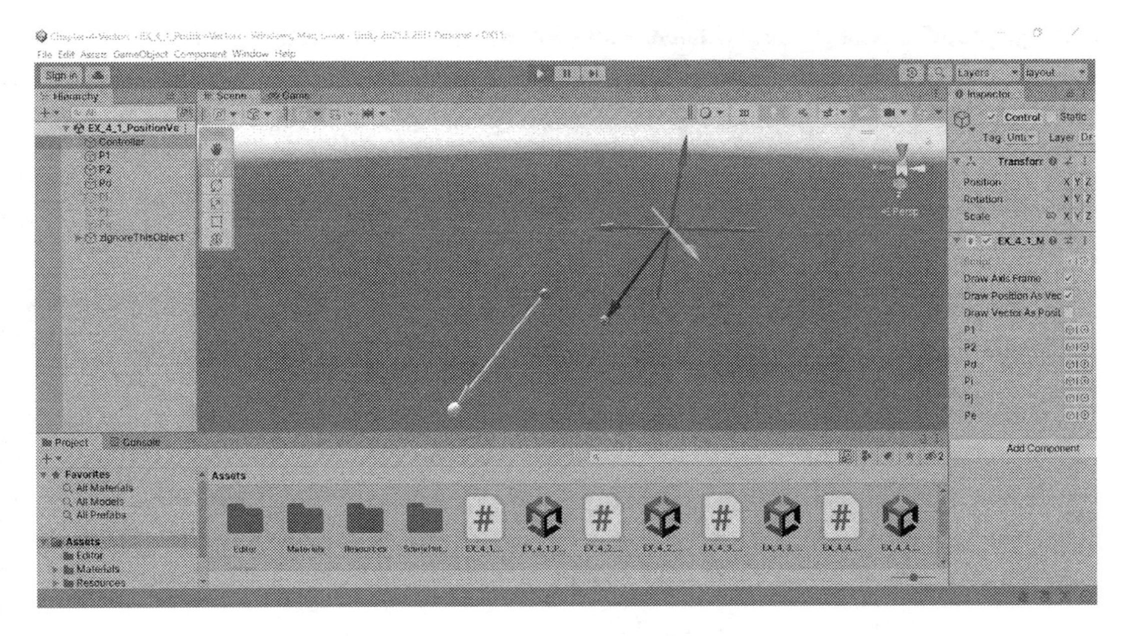

Figure 4-4. *Running the Position Vectors example*

The goals of this example are for you to

- Understand the relationship between positions, position vectors, and applying vectors at positions
- Manipulate a position and observe the position vector being applied at a different location
- Manipulate two positions to define a vector and observe the vector as a position vector
- Examine the implementation and application of vectors
- Increase familiarity with the `Vector3` class

Examine the Scene

Take a look at the `EX_4_1_PositionVectors` scene and observe the predefined game objects in the Hierarchy Window. There you will find the `Controller` and six other game objects that will assist in interpreting vectors from two alternative perspectives. These game objects are `P1`, `P2`, `Pd`, `Pi`, `Pj`, and `Pe`. This example will allow you to manipulate the head position of a position vector and to observe how the defined vector can be applied to any position. This example will also allow you to manipulate the positions of two points, observe how those two positions can define a vector, and how the defined vector can be shown as a position vector at the origin.

Analyze Controller MyScript Component

The `MyScript` component on the `Controller` presents nine variables that you can interact with. Three of these variables are toggle switches to control what you want to show and hide in the scene and the other six variables can be categorized into two sets of three variables each.

- Position vector:
    - `P1`: The reference to the `P1` game object
    - `P2`: The reference to the `P2` game object
    - `Pd`: The reference to the `Pd` game object
- Vector defined by two points:
    - `Pi`: The reference to the `Pi` game object
    - `Pj`: The reference to the `Pj` game object
    - `Pe`: The reference to the `Pe` game object
- Toggles:
    - `DrawAxisFrame`: A toggle determining if the axis frame should be drawn
    - `DrawPositionAsVector`: A toggle determining if a position should be drawn as a vector
    - `DrawVectorAsPosition`: A toggle determining if a vector should be drawn as a position

---

Note For convenience, whenever appropriate, the rest of the examples in this book will assign identical names to the game objects in the scene and the corresponding reference variables in `MyScript`.

Interact with the Example

Click the Play Button to run the example. Notice that by default, the DrawVectorAsPosition toggle is set to off and the corresponding game objects, Pi, Pj, and Pe, are not displayed. This is so you can focus on the position vector defined by Pd and apply it at position P1. Select Controller and ensure that the DrawAxisFrame is on to observe the axis frame in the scene. You only need to show this axis frame when you want to verify the location of the origin and the directions of the major axes. Feel free to hide the axis frame and to show it again whenever you need a reference.

Position Vector

First, verify that Pi, Pj, and Pe are not displayed by selecting these objects in the Hierarchy Window and confirming that they are inactive (the check box next to their name in the Inspector Window should be unchecked). Then, select P2 and try to manipulate its position. You will notice that whenever you change a value in P2's transform component in the Inspector Window, it reverts back to its old value. This is because P2's position is under the control of MyScript. Now select and manipulate the position of Pd and verify the following:

- Notice the thin red, green, and blue lines connecting from the origin to position Pd. Switch the DrawAxisFrame on and off to verify that these three lines are parallel to the corresponding X-, Y-, and Z-axes. The lengths of these three lines are x_d, y_d, and z_d, which are the corresponding values of the coordinate position of Pd.

- The position vector is the black vector with its tail at the origin and its head at the current Pd location. This vector represents interpreting the coordinate values of Pd, (x_d, y_d, z_d), as the x-, y-, and z-components of vector $\vec{V}_d$.

- Move Pd to a position close to the origin, for example, $(0.1, 0.1, 0.1)$, and notice that the black vector is now very small and difficult to observe. When Pd is moved to exactly the origin, the black vector becomes the zero vector and vanishes. The zero vector is a special case that describes a zero displacement. As you will learn, the definition of many vector operations specifically excludes the zero vector. These will be pointed out as you learn about them in future sections and chapters.

You have observed displaying a position as a position vector (a vector from the origin to the position) which demonstrates that all positions in the Cartesian Coordinate System can be interpreted as position vectors. Now, select and manipulate the position of `P1` and notice the following:

- Independent of the location of `P1`, the white vector is always identical to the black position vector where they are parallel and have the same length. The only difference between these vectors is that the white vector has its tail at `P1` and not the origin. You can verify this by observing that the thin red, green, and blues lines that connect `P1` to `P2` are the same length as the thin red, green, and blues lines that connect the origin to `Pd`.
- Position `P2` is always at the head of the white vector. In this case, `P2` is computed as follows:

$$P_2 = P_1 + \vec{V}_d$$

Through the application of a position vector at an arbitrary position (`P1`), you have observed that the position vector and the applied vector are indeed identical and that the only difference between them is that they are located, or applied, at different positions. This illustrates that vectors are independent of positions, meaning that once a vector is defined it can be applied to any position. It also demonstrates that a vector absent of any position information should be, and are, interpreted as position vectors—vectors originating from the origin. This part of the example has shown that a position in 3D space is simply a vector from the origin to that position.

Vector Defined by Two Points

Now, select the `Controller`, toggle off `DrawPositionAsVector`, and switch on `DrawVectorAsPosition`. Verify that `P1`, `P2`, and `Pd` are hidden by selecting them in the Hierarchy Window. Next, select and try to change the position of `Pe`. Note that just like with `P2`, `Pe`'s position is being set by `MyScript` and thus cannot be changed from the Inspection Window. Now, select and change the positions of `Pi` and `Pj` and notice the following:

- The pink vector, $\vec{V}_e = (x_e, y_e, z_e)$, is defined by the positions `Pi` (x_i, y_i, z_i) and `Pj` (x_j, y_j, z_j), where $\vec{V}_e = P_j - P_i$, or

    - $x_e = x_j - x_i$, which is the displacement along the X-axis (the length of the thin red line).
    - $y_e = y_j - y_i$, which is the displacement along the Y-axis (the length of the thin green line).
    - $z_e = z_j - z_i$, which is the displacement along the Z-axis (the length of the thin blue line).

- Independent of the locations of `Pi` and `Pj`, the pink and purple vectors are identical, having the same length, and are parallel to each other (they have same direction). The only difference between them is the location of their tail positions. The pink vector has a tail located at position `Pi` and the purple vector's tail is located at the origin.
- The purple vector's head position is always at Pe (x_e, y_e, z_e). Note how the coordinate component values are the same values as that of $\vec{V}_e$, indicating that Pe position is the position vector $\vec{V}_e$.

You have observed that any vector, $\vec{V}_e = (x_e, y_e, z_e)$, is equivalent to the coordinate position $P_e\ (x_e, y_e, z_e)$ and can be displayed as a position vector with tail at the origin.

Details of MyScript

Open `MyScript` and examine the source code in the IDE. The instance variables are as follows:

```
// For visualizing the two vectors
public bool DrawAxisFrame = true; // Draw or Hide the AxisFrame
public bool DrawPositionAsVector = true;
public bool DrawVectorAsPosition = true;

private MyVector ShowVd;      // From Origin to Pd
private MyVector ShowVdAtP1;  // Show Vd at P1
private MyVector ShowVe;      // From Origin to Pe
private MyVector ShowVeAtPi;  // Ve from Pi to Pj

// Support position Pd as a vector from P1 to P2
public GameObject P1;    // Position P1
public GameObject P2;    // Position P2
public GameObject Pd;    // Position vector: Pd
```

```
// Support vector defined by Pi to Pj, and show as Pe
public GameObject Pi;   // Position Pi
public GameObject Pj;   // Position Pj
public GameObject Pe;   // Position vector: Pe
```

All of the public variables for MyScript have been discussed when analyzing the Controller's MyScript component. The four private variables of MyVector data type are defined to support the visualization of the vectors as you have observed previously:

- ShowVd: Used for visualizing the position vector of Pd (the black vector)
- ShowVdAtP1: Used for visualizing the vector at position P1 (the white vector)
- ShowVe: Used for visualizing the position vector of Pe (the purple vector)
- ShowVeAtPi: Used for visualizing the vector at position Pi (the pink vector)

As in the case of the previous custom classes such as MyBoxBound and MySphereBound, MyVector is defined specifically for visualizing a vector and is irrelevant for understanding the math being discussed in this book. For example, you can always run the examples with all code concerning the MyVector data type removed, but the visualization of these vectors (black, white, pink, etc.) will no longer exist. You can see a screenshot of the MyVector class in Figure 4-5, which shows that MyVector is indeed defined for the drawing of a vector.

```
public class MyVector
{
    private functionality for drawing support

    public MyVector()...         // Constructor

    public float Magnitude...    // Size of the vector
    public Vector3 Direction...  // Direction of the vector
    public Vector3 VectorAt...   // The location to draw the vector

    // Drawing Support
    public bool DrawVector...      // Draw or Hide the interval
    public bool DrawVectorComponents ...
    public Color VectorColor...    // Color to draw

    // A vector from src to dst
    public void VectorFromTo(Vector3 src, Vector3 dst)...

    // A vector at src, with direction: dir and magnitude: len
    public void VectorAtDirLength(Vector3 pos, Vector3 dir, float len)...
}
```

Figure 4-5. *The MyVector class*

The `Start()` function for `MyScript` is listed as follows:

```
Void Start() {
    Debug.Assert(P1 != null);    // Ensure proper init
    Debug.Assert(P2 != null);
    Debug.Assert(Pd != null);
    Debug.Assert(Pi != null);
    Debug.Assert(Pj != null);
    Debug.Assert(Pe != null);

    // To support show position and vector at P1
    ShowVd = new MyVector {
        VectorColor = Color.black,
        VectorAt = Vector3.zero      // Vd from origin
    };
    ShowVdAtP1 = new MyVector {
        VectorColor = new Color(0.9f, 0.9f, 0.9f)
    };
```

```
    // To support show vector from Pi to Pj as position vector
    ShowVe = new MyVector {
        VectorColor = new Color(0.2f, 0.0f, 0.2f),
        VectorAt = Vector3.zero    // Ve from origin
    };
    ShowVeAtPi = new MyVector() {
        VectorColor = new Color(0.9f, 0.2f, 0.9f)
    };
}
```

The Start() function verifies proper public variable setup in the Hierarchy Window and instantiates and initializes the private MyVector variables to their respective colors. Note that ShowVd and ShowVe are defined to display position vectors and are therefore initialized to show the vectors starting from the origin (Vector3.zero). The Update() function is listed as follows:

```
Void Update()
{
    Visualization on/off: show or hide to avoid cluttering

    Position Vector: Show Pd as a vector at P1

    Vector from two points: Show Ve as the position Pe
}
```

The Update() function is divided into three separate #region areas according to the logic they perform and for readability. The details of these regions are explained in the next three sections.

Region: Visualization on/off

The code in this region, listed as follows, simply sets the active flag on the relevant game objects for displaying or hiding whichever game objects the user toggles via the MyScript component on the Controller:

```
#region  Visualization on/off: show or hide to avoid cluttering
AxisFrame.ShowAxisFrame = DrawAxisFrame; // Draw/Hide Axis Frame
P1.SetActive(DrawPositionAsVector);      // Position as vector
P2.SetActive(DrawPositionAsVector);
```

```
Pd.SetActive(DrawPositionAsVector);
Pi.SetActive(DrawVectorAsPosition);      // Vector as position
Pj.SetActive(DrawVectorAsPosition);
Pe.SetActive(DrawVectorAsPosition);
ShowVdAtP1.DrawVector = DrawPositionAsVector; // Draw or hide
ShowVd.DrawVector = DrawPositionAsVector;
ShowVeAtPi.DrawVector = DrawVectorAsPosition;
ShowVe.DrawVector = DrawVectorAsPosition;
#endregion
```

Region: Position Vector

The code in this region, listed as follows, is only active when the `DrawPositionAsVector` toggle is set to true:

```
#region Position Vector: Show Pd as a vector at P1
if (DrawPositionAsVector) {
    // Use position of Pd as position vector
    Vector3 vectorVd = Pd.transform.localPosition;

    // Step 1: take care of visualization for Vd
    ShowVd.Direction = vectorVd;
    ShowVd.Magnitude = vectorVd.magnitude;

    //         apply Vd at P1
    ShowVdAtP1.VectorAt = P1.transform.localPosition;
    ShowVdAtP1.Magnitude = vectorVd.magnitude;
    ShowVdAtP1.Direction = vectorVd;

    // Step 2: demonstrate P2 is indeed Vd away from P1
    P2.transform.localPosition =
                P1.transform.localPosition + vectorVd;
}
#endregion
```

In this case, as illustrated by the bolded font in the code listing, the position of Pd, `Pd.transform.localPosition`, is interpreted as a vector, `vectorVd`, or $\vec{V}_d$. In Step 1, vectorVd is drawn via the ShowVd variable. Recall that ShowVd is initialized to be drawn at the origin. For this reason, ShowVd is simply drawing vectorVd, or the coordinate values of Vd, as a position vector. In order to show the same vector at position P1, the magnitude (length) and direction of ShowVdAtP1 are assigned the corresponding values from vectorVd and are then displayed at the location of P1, P1.transform.localPosition, instead of the origin like that of vectorVd. In Step 2, once again shown in bolded font, P2's position is set as $P_2 = P_1 + \vec{V}_d$ which will always place P2 at the head of $\vec{V}_d$. This repeated updating of P2's position is the reason why when you interacted with this example, you were not able to move the P2 game object.

In the Cartesian Coordinate System, positions are defined by three-float tuples. So far, this example shows that the same three-float tuple can be interpreted as a vector. This alternative interpretation allows vectors to be used as a tool for describing physical behaviors, like object movements. This topic will be covered in detail in a later section of this chapter.

Region: Vector from Two Points

The code in this region, listed as follows, is only active when the DrawVectorAsPosition toggle is set to true:

```
#region Vector from two points: Show Ve as the position Pe
if (DrawVectorAsPosition) {
    // Use from Pi to Pj as vector for Ve
    Vector3 vectorVe = Pj.transform.localPosition -
                       Pi.transform.localPosition;

    // Step 1: Take care of visualization
    //         for Ve: from Pi to Pj
    ShowVeAtPi.VectorFromTo(Pi.transform.localPosition,
                            Pj.transform.localPosition);
    //         Show as Ve at the origin
    ShowVe.Direction = vectorVe;
    ShowVe.Magnitude = vectorVe.magnitude;

    // Step 2: demonstrate Pe is indeed Ve away from the origin
    Pe.transform.localPosition = vectorVe;
}
#endregion
```

As illustrated by the bolded font in the code listing, the vector `vectorVe`, or $\vec{V}_e$, is computed based on the positions of `Pi` and `Pj` according to the formula

$$\vec{V}_e = P_j - P_i$$

In Step 1, `ShowVeAtPi` is set to be drawn as a vector between `Pi` and `Pj`'s positions. `ShowVe`'s direction and magnitude are assigned by the corresponding values of `vectorVe`. Recall that the draw position of `ShowVe` was initialized to the origin, and thus `ShowVe` is showing `vectorVe` as a position vector. In Step 2, again shown in bolded font, the position of `Pe` is set to the corresponding x-, y-, and z-component values of `vectorVe`, literary showing `vectorVe` as a coordinate position. Similar to the case of `P2`'s position, in this case, `Pe` is continuously updated by the script and thus the user has no control over the position of `Pe` while the scene is running.

In general, the ability to interpret a given vector as a position allows all vectors to be plotted as position vectors from the origin, supporting straightforward visualization and comparisons across multiple vectors. You have completed the cycle of interpreting positions as vectors and now vectors as positions. This entire discussion is designed to demonstrate that once defined, a vector is an entity that can be analyzed and applied at any position because its definition is independent of any specific position.

Note The vector from `Pi` to `Pj` is computed by subtracting `Pi` from `Pj`:

$$\vec{V}_e = P_j - P_i$$

The order of subtraction is important. Reversing the subtraction order, $P_i - P_j$, computes a vector from `Pj` to `Pi`. Vector subtraction will be discussed in detail later in this chapter.

Takeaway from This Example

This example presents you with two ways to define, manipulate, and interpret a vector. The first method is based on initializing a starting point (e.g., the origin) and then selecting the ending position. The second method is based on defining a vector between two explicitly controlled positions. In all interactions, all four vectors describe how to move from one position to another: from origin to `Pd` (black), from `P1` to `P2` (white), from `Pi` to `Pj` (pink), and from origin to `Pe` (purple).

You have seen that it does not matter where a vector is applied (or drawn), if the encoded distances and direction information are the same, the underlying vectors are the same. You have also witnessed that a vector can be treated as a position, and a position can be treated as a vector.

Relevant mathematical concepts covered include

- A vector describes the movement from one position to another.
- The vector between two given positions is defined by the differences between the corresponding coordinate values in the x-, y-, and z-components.
- The Cartesian Coordinate values for any position $P(x, y, z)$ describes the displacements from the origin to the position P. For this reason, the (x, y, z) values of any position can be interpreted as a vector between the origin and the position. This interpretation of the coordinate position is referred to as position vector.
- All positions in the Cartesian Coordinate system can be interpreted as position vectors.
- The zero vector is the position vector of the origin. This vector describes a displacement with zero distance, or a position moving back onto itself. This is a special vector where many vector operations cannot operate or do not work on the zero vector.
- Vectors are independent of positions; thus, once defined, a vector can be applied to any position.
- In the absence of position information, vectors are often drawn as a position vector, a line segment from the origin to the coordinate position defined by the x-, y-, and z-component values of that vector.

Unity tools

- `MyVector`: A custom-defined class to support the visualization of vectors
- `AxisFrame.ShowAxisFrame`: A Boolean flag to control the showing of the Cartesian Coordinate origin and axes' directions

Note The Unity `Vector3` data type closely encapsulates the concept of a vector. From the code listing in the `Update()` function, you can observe the power and convenience of working with proper data abstraction. With the Unity `Vector3` abstraction, you can avoid the nuisance of retyping similar code for individual values of each major axis when computing distances between positions, or when following a vector. For the rest of this book, with very few exceptions, such as when analyzing the detailed definitions of vector operations, you will work with the `Vector3` class and will not work with the values of the individual coordinate axes.

EXERCISES

Contrast the Creation of $\vec{V}_d$ and $\vec{V}_e$

Note that $\vec{V}_d$ is created via a single position being interpreted as a position vector, while $\vec{V}_e$ is created by subtracting two positions explicitly. Nevertheless, both methods can accomplish the creation of the same vector. For example, move the position of `Pi` to overlap `P1`. This can be accomplished by running the game, selecting `P1` in the Hierarchy Window, taking note of the position values of the `Transform` component of `P1`, and copying these values to be the position values of `Pi`'s `Transform` component. You can now adjust `Pj`, or Pd, to try to align $\vec{V}_e$ with $\vec{V}_d$.

Switch Vector Creation Methods

You can take advantage of the observation that both position vector and the difference between two points can create the same vector. Edit `MyScript` and remove Pe, `Pi`, and `Pj` variables. Instead, include a new Boolean flag `CreateWithPositionVector` which will allow `P1`, `P2`, and Pd to behave as Pe, `Pi`, and `Pj` did.

- When `CreateWithPositionVector` is true, let the user manipulate Pd to create the vector and show the vector at `P1`. In this case, `P2` is computed based on the vector defined and the user will not be able to adjust `P2`.
- When `CreateWithPositionVector` is false, let the user manipulate both `P1` and `P2` and use the difference between these two points to compute the position vector to `Pd`. In this case, Pd is computed based on the vector defined and the user will not be able to adjust `Pd`.

Note the "two ways to define a vector" logic is similar to that of the "two ways to define a bounding box." You can refer to the `Update()` function of the `EX_2_2_BoxBounds_IntervalsIn3D` scene of `Chapter-2-Examples` project for a template of the control logic required for this exercise.

Verify Vector Size, or Length, or Magnitude

A vector describes the movement from one position to another; it encapsulates both the distance and the direction to travel. You have seen the distance being referred to as "magnitude"; it is also commonly referred to as the "size" or "length" of the vector. Edit `MyScript` to print the size of each of the vectors, either via public float variables or via `Debug.Log()` function calls. Verify that both `ShowVd` and `ShowVdAtP1` and `ShowVe` and `ShowVeAtPi` are indeed two sets of vectors with identical lengths.

Manipulate Vector Lengths

Manipulate the two vectors in this example such that $\vec{V}_d = (2,0,0)$ and $\vec{V}_e = (0,2,0)$. Notice that in this case, $\vec{V}_d$ and $\vec{V}_e$ have the same lengths of 2.0. However, the two vectors are pointing toward drastically different directions: toward positive X-axis and Y-axis. Notice that it is possible to define two vectors with identical length but with very different directions.

Verify Vector Directions

You can verify two vectors are the same by printing out the values of the x-, y-, and z-components. Edit `MyScript` to print the coordinate values of `ShowVe` and `ShowVeAtPi` to verify that these two vectors are indeed exactly the same. With previous exercises on vector size, the obvious question is, "is it possible to manipulate the two vectors such that they are pointing in the same direction but with different lengths?" The short answer is yes. For example, consider vectors, (1, 0, 0) and (2, 0, 0). Both are pointing toward the positive x-direction, but the lengths are 1 and 2. The general consideration for this question is slightly more involved and is the topic for the next section.

Vector Algebra: Scaling

A vector encodes both a distance and a direction, describing how an object can move from position P_1 (x_1, y_1, z_1), in a straight line, and arrive at P_2 (x_2, y_2, z_2). You know that a vector, $\vec{V}_a$, that describes this movement can be defined as follows:

$$\begin{aligned}\vec{V}_a &= P_2 - P_1 \\ &= (x_2 - x_1, y_2 - y_1, z_2 - z_1) \\ &= (x_a, y_a, z_a)\end{aligned}$$

The distance, *d*, between the two points is referred to as the size (or magnitude, or length) of the vector and is labeled with the symbol $\vec{V}_a$. The size of a vector is defined as follows:

$$d = \|\vec{V}_a\| = \sqrt{x_a^2 + y_a^2 + z_a^2}$$

The size of a vector can be scaled. For example, if there is a vector $\vec{V}_b = (x_b, y_b, z_b) = (5x_a, 5y_a, 5z_a)$, then

$$\begin{aligned}\|\vec{V}_b\| &= \sqrt{x_b^2 + y_b^2 + z_b^2} \\ &= \sqrt{(5x_a)^2 + (5y_a)^2 + (5z_a)^2} \\ &= \sqrt{25(x_a^2 + y_a^2 + z_a^2)} \\ &= 5\sqrt{x_a^2 + y_a^2 + z_a^2} \\ &= 5\|\vec{V}_a\|\end{aligned}$$

Note that in general, the observed relationship is true for any floating-point number, *s*. That is, if

$$\vec{V}_a = (x_a, y_a, z_a)$$

and

$$\vec{V}_b = (sx_a, sy_a, sz_a)$$

then

$$\|\vec{V}_b\| = s\|\vec{V}_a\|$$

The length or magnitude of $\vec{V}_b$ is s times that of $\vec{V}_a$. In this case, $\vec{V}_b$ is described as "scaling $\vec{V}_a$ by a factor s," or simply, "scaling $\vec{V}_a$ by s," and is expressed as

$$\vec{V}_b = s\vec{V}_a$$

Note While it is always true that if $\vec{V}_b = s\vec{V}_a$, then $\|\vec{V}_b\| = s\|\vec{V}_a\|$. The reverse is not always true. For example, if $\vec{V}_a = (1,0,0)$ and $\vec{V}_b = (0,s,0)$, then in this case, it is true that $\|\vec{V}_b\| = s\|\vec{V}_a\|$, but $\vec{V}_b = s\vec{V}_a$ is certainly not true.

Figure 4-6 illustrates an example where $\vec{V}_a = (x_a,0,0)$, $\vec{V}_b = 1.5\vec{V}_a$, and $\vec{V}_c = \frac{1}{x_a}\vec{V}_a$.

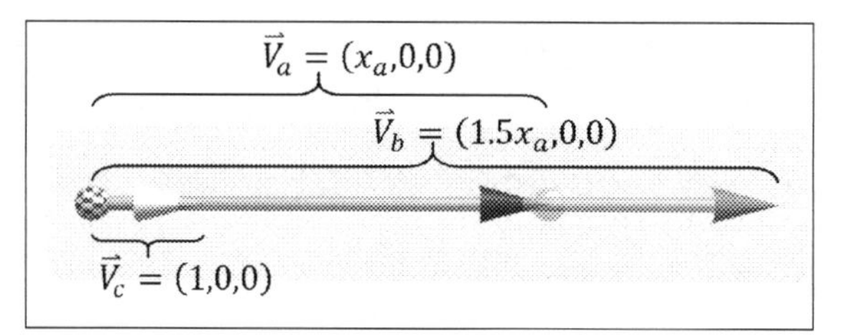

Figure 4-6. *Scaling of a vector that is in the x-direction*

Referring to Figure 4-6, you now know that

- $\vec{V}_b = 1.5\vec{V}_a = (1.5x_a,0,0)$
- $\vec{V}_c = \frac{1}{x_a}\vec{V}_a = \left(\frac{1}{x_a}x_a,0,0\right) = (1,0,0)$

Additionally, you know when x_a is a positive number, the lengths of the three vectors in Figure 4-6 are as follows:

$$\|\vec{V}_a\| = \sqrt{{x_a}^2 + 0^2 + 0^2} = x_a$$

$$\|\vec{V}_b\| = 1.5\|\vec{V}_a\| = 1.5x_a$$

$$\|\vec{V}\|_c = \frac{1}{x_a}\|\vec{V}_a\| = 1$$

Lastly, and very importantly, based on your knowledge of the Cartesian Coordinate System and so far in this chapter, you know that although the vectors in Figure 4-6 have different lengths, the three vectors overlap perfectly and are all pointing in the positive X-axis direction. This overlap shows that scaling a vector only changes the distance that it encodes and does not affect the direction. It turns out, as illustrated in Figure 4-7, this statement is true for any direction.

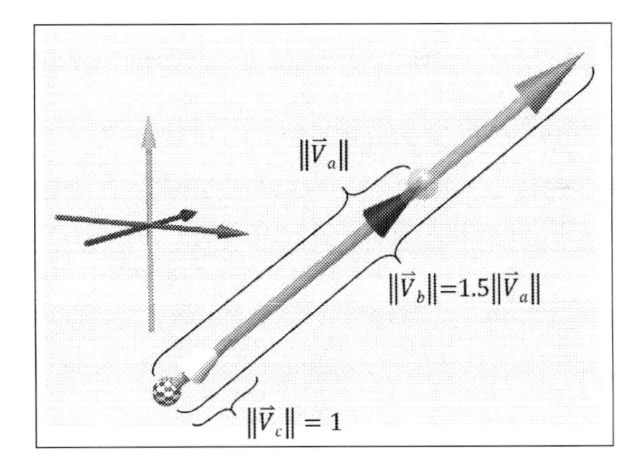

Figure 4-7. *Scaling of an arbitrary vector*

Figure 4-7 shows three vectors with the same lengths as of those in Figure 4-6:

- Vector $\vec{V}_a$ with magnitude $\|\vec{V}_a\|$
- Vector $\vec{V}_b = 1.5\vec{V}_a$ with magnitude $1.5\|\vec{V}_a\|$
- Vector $\vec{V}_c = \frac{1}{\|\vec{V}_a\|}\vec{V}_a$ with magnitude of 1.0

Notice that in exactly the same manner as the vectors in the X-axis direction (Figure 4-6), these three vectors all point in the same direction as each other. In all cases, scaling a vector only affects its size and not the direction. In general, scaling a vector by any positive number will result in a vector that is in the same direction, while scaling by a negative number will flip the direction of that vector. This means when a positive x-direction vector is scaled by a negative value, the resulting vector will point in the negative x-direction. Scaling by a negative number is left as an exercise for you to complete in the next example.

Similar to how multiplying scaling factors to the number zero will produce a result of zero, scaling a zero vector has no effect and will result in the same zero vector.

Normalization of Vectors

Vector $\vec{V}_c$ in Figure 4-7 is the result of scaling an existing vector by the inverse of the length of that vector. This is interesting because with such a specific scaling factor, the magnitude of $\vec{V}_c$ is guaranteed to be 1. As you will see frequently in the rest of this book, and is true in general, vectors with a magnitude of 1 are important as they enable convenient computations in many situations.

A vector with a magnitude of 1 is so important that it has its own symbol, $\hat{V}$, which is the same as the original symbol for a vector, but replaces the arrow above the "V" with a cap. This vector has a special name, **normalized vector** or **unit vector**. The process of computing a normalized vector is referred to as **vector normalization**. In general, it is always the case that for any nonzero vector, $\vec{V} = (x, y, z)$:

- Magnitude of vector $\vec{V}$

$$\|\vec{V}\| = \sqrt{x^2 + y^2 + z^2}$$

- Normalization of vector $\vec{V}$

$$\hat{V} = \frac{1}{\|\vec{V}\|}\vec{V}$$

$$= \frac{1}{\sqrt{x^2 + y^2 + z^2}}\vec{V}$$

$$= \left(\frac{x}{\sqrt{x^2 + y^2 + z^2}}, \frac{y}{\sqrt{x^2 + y^2 + z^2}}, \frac{z}{\sqrt{x^2 + y^2 + z^2}}\right)$$

Notice that normalization is a division by length. Recall that a zero vector has a length of zero, and from basic algebra, that division by zero is an undefined operation. This means that the zero vector cannot be normalized. This is the first case you encounter, but certainly not the last, that a vector operation is not applicable to the zero vector.

Note The vector normalization process involves a division by a square root. Though with modern hardware this computation cost is becoming less of a concern, it is still a good practice to pay attention to the need for normalization in general. For example, the Unity `Vector3` class defines the `sqrMagnitude` property to return the squared of a vector length, $\|\vec{V}^2\|$, which can be used when information on vector length is needed, but not normalization. For example, when performing size comparisons, for example, determining which vector is longer.

Direction of Vectors

The magnitude of a vector can be simply and effectively conveyed by a number. In contrast, the direction of a vector must be expressed in relation to a "frame of reference." For example, "in the x-direction" uses the X-axis as the frame of reference. In the 3D Cartesian Coordinate System, a direction can be described by using the X-, Y-, and Z-axes as references. Such a description involves a reference direction and a rotation. For example, a direction that is defined by a rotation of the Y-axis about the Z-axis in the X-axis direction by 15 degrees. If you find that description difficult to follow, you are not alone. Fortunately, there are alternatives to describing the direction of a vector.

Recall that as illustrated in Figure 4-7, the direction of a vector does not change when the vector is scaled. This means that a unit vector uniquely identifies the direction of all vectors with different lengths in that direction. For simplicity, both representationally and computationally, this book chooses to identify the direction of a vector by referring to its unit vector. For example, for a given vector, $\vec{V}$, this book refers to its magnitude as $\|\vec{V}\|$ and its direction as $\hat{V}$. In the rest of this book, you will encounter phrases like "the direction of $\vec{V}$" or "the direction of $\hat{V}$"; both refer to the direction of the vector $\hat{V}$.

Since the normalized zero vector is undefined, a zero vector has no direction.

The Vector Scaling and Normalization Example

This example demonstrates the results of scaling a vector and defining a vector with separate input for magnitude and direction. It allows you to adjust and examine the effects of changing the vector scaling factor, as well as control the creation of a vector via specifying its magnitude and direction. Figure 4-8 shows a screenshot of running the `EX_4_2_VectorScaling` scene from the `Chapter-4-Vectors` project.

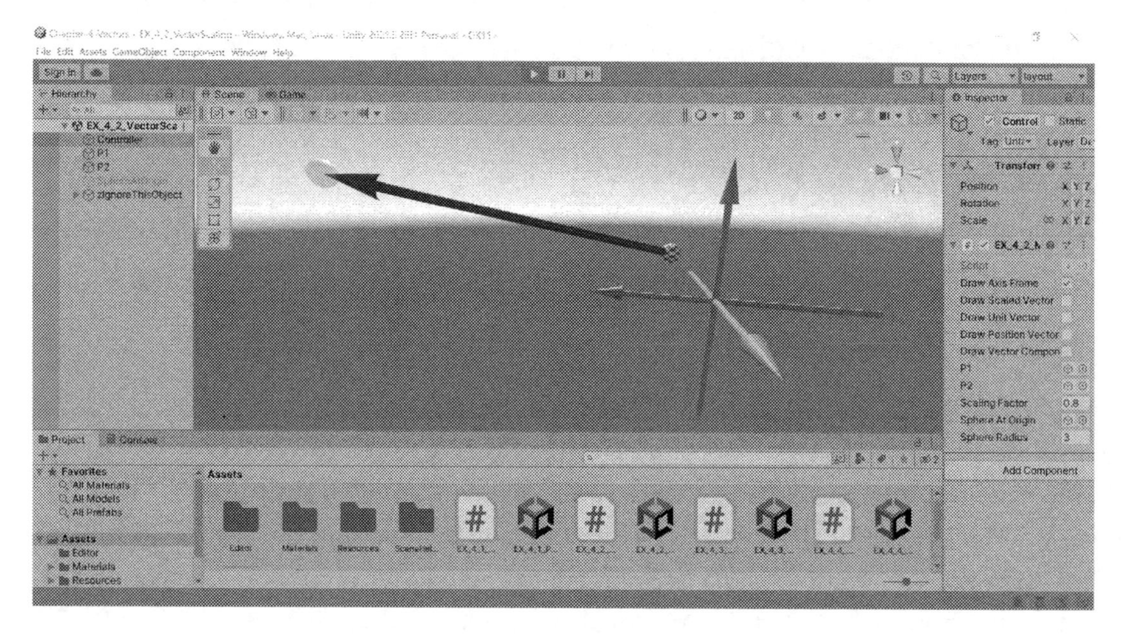

Figure 4-8. *Running the Vector Scaling example*

The goals of this example are for you to

- Interact with and examine the effects of scaling vectors
- Experience defining vectors based on specifying their magnitude and direction
- Understand the effects of separately changing the magnitude and direction of a vector
- Examine the implementation of working with vectors

Examine the Scene

Take a look at the `Example_4_2_VectorScaling` scene and observe, besides `Controller`, the three predefined game objects in the Hierarchy Window: `P1`, `P2`, and `SphereAtOrigin`. As in the previous example, `P1` and `P2` together will allow you to define a vector, $\vec{V}_a$. The `SphereAtOrigin` is a transparent sphere located at the origin, where you will create a position vector in the same direction as $\hat{V}_a$, with a magnitude that just touches the surface of this transparent sphere.

Analyze Controller MyScript Component

The `MyScript` component on the `Controller` shows ten variables that can be categorized into three groups:

- Drawing control: Allows you to show or hide different information relevant to a vector
    - `DrawAxisFrame`: Shows or hides the Cartesian Coordinate origin and reference axis frame.
    - `DrawScaledVector`: Shows or hides the scaled version of $\vec{V}_a$.
    - `DrawUnitVector`: Shows or hides the unit vector $\hat{V}_a$.
    - `DrawPositionVector`: Shows or hides the position vector that touches the `SphereAtOrigin` surface.
    - `DrawVectorComponents`: Shows or hides the x-, y-, and z-displacements of each vector. Notice that for clarity, when displayed, the position vector always draws its vector components.
- Definition of $\vec{V}_a$: Defines and allows manipulation of the vector $\vec{V}_a$
    - `P1`: The reference to the `P1` game object
    - `P2`: The reference to the `P2` game object
    - `ScalingFactor`: The factor to scale the vector $\vec{V}_a$ by

- Definition of a position vector: Defines and allows manipulation of the position vector
  - SphereAtOrigin: The reference to the SphereAtOrigin game object
  - SphereRadius: The radius of the SphereAtOrigin sphere and the length of the position vector that will be parallel to $\hat{V}_a$

Interact with the Example

Click the Play Button to run the example. Notice that by default, except DrawAxisFrame, all vector drawing toggles are off so you should only be observing the axis frame and vector $\vec{V}_a$, the vector being drawn between positions P1 and P2. Now select the Controller and get ready to toggle drawing options and observe the following.

Scaled Vector

Toggle on the drawing option for DrawScaledVector to observe a slightly shorter pink vector in the same direction as $\vec{V}_a$. Now adjust the ScalingFactor variable and watch as the pink vector changes size. This pink vector is displaying the vector $\vec{V}_s$

$$\vec{V}_s = ScalingFactor \times \vec{V}_a$$

Notice three interesting intervals:

- $0 < ScalingFactor < 1$: $\vec{V}_s$ has a length shorter than $\vec{V}_a$ and is thus displayed as a vector embedded in $\vec{V}_a$.
- $ScalingFactor > 1$: $\vec{V}_s$ has a magnitude larger than $\vec{V}_a$ and is thus a vector that extends beyond $\vec{V}_a$.
- $ScalingFactor < 0$: $\vec{V}_s$ points in the reversed direction of $\vec{V}_a$. Note that the two vectors are drawn at the same position, P1, and that the two vectors do indeed extend in the exact opposite directions.

Normalized or Unit Vector

Toggle on the drawing option for `DrawUnitVector` to observe a short white vector embedded in $\vec{V}_a$. This is $\vec{V}_a$ normalized, or $\hat{V}_a$. Recall that $\hat{V}_a$ is computed by scaling $\vec{V}_a$ by the inverse of its magnitude, $\frac{1}{\|\vec{V}_a\|}$. Initially, $\vec{V}_a$ has a magnitude of 5, so if you adjust `ScalingFactor` to the value of $\frac{1}{5}=0.2$, you will observe that the pink ($\vec{V}_s$) and white vectors overlap exactly. This overlap will stop once you adjust the `ScalingFactor`. Remember, $\vec{V}_s$ has a length that is `ScalingFactor` times the current $\|\vec{V}_a\|$, yet the size of $\hat{V}_a$ is always 1.

Manipulate and set the positions of `P1` and `P2` to be identical, for example, by copying values of `P1`'s `Transform` component to that of `P2`. Now, notice error messages in the Console Window about NaN and that the normalized white vector now points in an arbitrary direction. When positions of `P1` and `P2` are identical, $\vec{V}_a$ becomes the zero vector and $\hat{V}_a$ is undefined. Later, when you examine the implementation, you will notice that the zero vector condition is not checked. Here, you are observing the results of a common coding error: performing a vector operation without verifying if the operation is defined for the given vector. A responsible developer should always invoke precondition checking before performing the corresponding vector operations.

Position Vector from Direction and Magnitude

Toggle on the drawing option for `DrawPositionVector` to observe a navy-blue position vector, $\vec{V}_p$, that is parallel to $\hat{V}_a$ and has a magnitude that is defined by the `SphereRadius` variable:

$$\vec{V}_p = SphereRadius \times \hat{V}_a$$

You can verify this by adjusting `SphereRadius` and noting that the `SphereAtOrigin` game object (the transparent sphere) changes size, and $\vec{V}_p$, while maintaining the direction of $\hat{V}_a$, adjusts its magnitude such that its tip touches the sphere surface. You can toggle off and hide the axis frame via `DrawAxisFrame` to observe the thin red, green, and blue vector components of $\vec{V}_p$, verifying that this vector does indeed just touch the sphere surface, indicating that the length of the vector is indeed the radius of the sphere.

This interaction shows that you can create a direction and a magnitude separately and combine them to create a desired vector. Note that since $\hat{V}_a$ is a unit vector, the size of $\vec{V}_p$, or $\|\vec{V}_p\|$, is simply SphereRadius. An important observation is that if a vector is defined by a size and a unit vector, then this size is the magnitude property of that vector. In the next section, you will see how this simple observation can be applied to implement the behavior of an object following a target.

Summary of Interaction

Four vectors are created and examined in this example:

- $\vec{V}_a$: Vector between two user control positions, P1 and P2.
- $\vec{V}_s = ScalingFactor \times \vec{V}_a$: A vector in the same or opposite direction as $\vec{V}_a$.
- $\hat{V}_a = \frac{1}{\|\vec{V}_a\|} \times \vec{V}_a$: The normalized vector of $\vec{V}_a$; since this vector is always scaled by the inverse of its magnitude, it has a constant size of 1.
- $\vec{V}_p = SphereRadius \times \hat{V}_a$: A constructed vector based on a size and a direction.

Details of MyScript

Open MyScript and examine the source code in the IDE. The instance variables are as follows:

```
// Toggle of what to draw
public bool DrawAxisFrame = false;
public bool DrawScaledVector = false;
public bool DrawUnitVector = false;
public bool DrawPositionVector = false;
public bool DrawVectorComponents = false;

// For defining Va and Vs (ScaledVector)
public GameObject P1 = null;   // Position P1
public GameObject P2 = null;   // Position P2
public float ScalingFactor = 0.8f;
```

```
// For defining Vp (PositionVector)
public GameObject SphereAtOrigin = null;    // sphere at origin
public float SphereRadius = 3.0f;

// For visualizing all vectors
private MyVector ShowVa;                  // Vector Va
private MyVector ShowVaScaled;            // Scaled Va
private MyVector ShowNorm;                // Normalized Va
private MyVector ShowPositionVector;      // Position vector
```

All the public variables for MyScript have been discussed when analyzing the Controller's MyScript component. The four private variables of the MyVector data type are for visualizing the four vectors: $\vec{V}_a$, $\vec{V}_s$, $\hat{V}_a$, and $\vec{V}_p$, respectively. The Start() function for MyScript is listed as follows:

```
void Start(){
   Debug.Assert(P1 != null);    // Check for proper setup in the editor
   Debug.Assert(P2 != null);
   Debug.Assert(SphereAtOrigin != null);

   // To support visualizing the vectors
   ShowVa = new MyVector {
       VectorColor = Color.black };
   ShowNorm = new MyVector {
       VectorColor = new Color(0.9f, 0.9f, 0.9f)};
   ShowVaScaled = new MyVector {
       VectorColor = new Color(0.9f, 0.4f, 0.9f) };
   ShowPositionVector = new MyVector {
       VectorColor = new Color(0.4f, 0.9f, 0.9f),
       VectorAt = Vector3.zero      // Position Vector at origin
    };
}
```

The Debug.Assert() calls ensure proper setup regarding referencing the appropriate game objects via the Inspector Window, while the MyVector variables are instantiated and initialized with the proper colors. The Update() function is listed as follows:

```
void Update()
{
    Visualization on/off: show or hide to avoid cluttering

    Vector Va: Compute Va and setup the drawing for Va

    if (DrawScaledVector) ...

    if (DrawUnitVector) ...

    if (DrawPositionVector) ...
}
```

The Update() function is logically structured into five steps: handling the drawing toggles and then computing and showing $\vec{V}_a$, $\vec{V}_s$, $\hat{V}_a$, and $\vec{V}_p$, respectively. The details in each step are presented next in separate subsections. While reading the code, note the exact one-to-one match between the derived formula to compute each vector and the corresponding listed code. This is an important and elegant characteristic of vector-based game object behavior; the implementation often closely resembles the underlying mathematical derivation.

Visualization on/off

The code in this region sets the game object's active state for displaying or hiding according to user's toggle settings. This code is listed as follows:

```
#region  Visualization on/off: show or hide to avoid cluttering
AxisFrame.ShowAxisFrame = DrawAxisFrame;     // Draw or Hide Axis Frame
ShowVaScaled.DrawVector = DrawScaledVector; // Display or hide the vectors
ShowNorm.DrawVector = DrawUnitVector;
ShowVa.DrawVectorComponents = DrawVectorComponents;
ShowVaScaled.DrawVectorComponents = DrawVectorComponents;
ShowNorm.DrawVectorComponents = DrawVectorComponents;
ShowPositionVector.DrawVector = DrawPositionVector;
SphereAtOrigin.SetActive(DrawPositionVector);
#endregion
```

Vector Va

The code in this region computes $\vec{V}_a$ based on the current P1 and P2 positions and sets up the ShowVa variable for visualizing the vector. This code is listed as follows:

```
#region Vector Va: Compute Va and setup the drawing for Va
Vector3 vectorVa = P2.transform.localPosition -
                   P1.transform.localPosition;
// Show the Va vector at P1
ShowVa.Direction = vectorVa;
ShowVa.Magnitude = vectorVa.magnitude;
ShowVa.VectorAt = P1.transform.localPosition;
#endregion
```

The variable vectorVa is $\vec{V}_a = P_2 - P_1$. The ShowVa variable receives the corresponding direction and size values from vectorVa and is set to display the vector at position P1.

DrawScaledVector

When this toggle is set to true, $\vec{V}_s$ is computed and shown. The code to accomplish this is listed as follows:

```
if (DrawScaledVector) {
    Vector3 vectorVs = ScalingFactor * vectorVa;
    ShowVaScaled.Direction = vectorVs;
    ShowVaScaled.Magnitude = vectorVs.magnitude;
    ShowVaScaled.VectorAt = P1.transform.localPosition;
}
```

The variable vectorVs is $\vec{V}_s = ScalingFactor \times \vec{V}_a$. The ShowVaScaled is properly set up to display vectorVs at P1.

DrawUnitVector

When this toggle is set to true, $\hat{V}_a$ is computed and shown. The code to accomplish this is listed as follows:

```
if (DrawUnitVector) {
    // scale Va by its inversed size
    Vector3 unitVa = (1.0f / vectorVa.magnitude) * vectorVa;
    // Vector3 dirVa = vectorVa.normalized;
                             // Alternate way to normalized Va
    ShowNorm.Direction = unitVa;
    ShowNorm.Magnitude = unitVa.magnitude;
    ShowNorm.VectorAt = P1.transform.localPosition;
}
```

The variable unitVa is $\hat{V}_a = \frac{1}{\left\|\vec{V}_a\right\|} \times \vec{V}_a$. Notice the alternative way commented out below this line of code, Vector3.normalized, to compute a unit vector.

Here you can observe a coding error, where vectorVa.magnitude is used as the denominator in the normalization computation without first being verified that its value is not zero. Once again, a zero vector will have a length of zero and therefore cannot be normalized. In this case, the logic should check if vectorVa is equal to the zero vector, and if so, simply skip the drawing of ShowNorm.

Note In general, it is not advisable to compare computation results to floating-point constants. For example, it is unwise to attempt to detect the zero vector condition by performing

```
if (vectorVa.magnitude == 0.0f)
```

The chance of the results of a floating-point computation being exactly zero is almost nonexistent. In this case, you should check for the condition of smaller than a "very small" number. The C# programming language defines the float.Epsilon for this purpose. In this case, the condition to check for zero vector should be

```
if (vectorVa.magnitude < float.Epsilon)
    // vectorVa is, for all practical purposes, a zero vector
```

DrawPositionVector

When this toggle is set to true, $\vec{V}_p$ is computed and shown. The code to accomplish this is listed as follows:

```
if (DrawPositionVector) {
    Vector3 vectorVp = SphereRadius * vectorVa.normalized;
    ShowPositionVector.Direction = vectorVp;
    ShowPositionVector.Magnitude = vectorVp.magnitude;
    ShowPositionVector.VectorAt =
                    SphereAtOrigin.transform.localPosition;

    // Set the radius of the sphere at the origin
    SphereAtOrigin.transform.localScale =
                    new Vector3(2.0f * SphereRadius,
                                2.0f * SphereRadius,
                                2.0f * SphereRadius);
}
```

The variable vectorVp is $\vec{V}_p = SphereRadius \times \hat{V}_a$. Note that in this case, $\hat{V}_a$ is computed based on the Unity Vector3.normalized utility. The last line of code scales the sphere by setting the Unity Transform.localScale. Notice that the scaling factor for the sphere is its diameter, or 2 times the radius. This is because localScale adjusts the scale of a sphere based on its diameter, not its radius.

Takeaway from This Example

Note that the entire implementation for this example, the code in the Update() function that performs useful computation, is actually just four lines: one line for each of the vectors, $\vec{V}_a$, $\vec{V}_s$, $\hat{V}_a$, and $\vec{V}_p$, respectively. The rest of the code is there to support user interaction and to set up the four toggle variables for visualizing the vectors. This example shows that when working with vector-based logic, the code can be rather compact with the implementation closely resembling the actual math involved to compute such results.

Relevant mathematical concepts covered include

- All scaled vectors are along exactly the same direction as their reference vector.
- The unit vector, or normalized vector, is a special case of the scaled vector; it is a vector scaled by the inverse of the size of its reference vector.
- The normalized vector, or unit vector, always has a length of one and does indeed uniquely and consistently represent the direction of vectors with different scaling factors.
- The zero vector cannot be normalized. Proper coding should include specific conditional checks before invoking the normalization computation.
- A vector can be defined based on a magnitude and a direction. An interesting implication of this fact is that any vector can be decomposed into a unit vector with a scale.

Unity tools

- `Transform.localScale`: To change the size of game objects
- `Sphere` primitive: The scale value is the diameter of the sphere

EXERCISES

Verify the Directions of vectorVa and vectorVp

Make sure that $\vec{V}_a$, $\vec{V}_s$, and $\hat{V}_a$ are in the exact same direction by setting `ScalingFactor` to a positive value. Next, verify the $\vec{V}_p$ vector is also in the same direction by moving `P1` to the origin. Interestingly, you can also move the position of the `SphereAtOrigin` to `P1` by changing the value of `SphereAtOrigin.Transform.localPosition`.

Properly Handle the Zero Vector

Implement the detection and handling of the zero vector condition to avoid the normalization process when necessary.

Work with Unit Vector and MyVector

A unit vector always has a size of 1 and can be a convenient reference for defining vectors of different lengths. For example, edit `MyScript` to display 5 different vectors with lengths of 1, 2, 3, 4, and 5 in the $\hat{V}_a$ direction. Display these vectors at the X-axis locations that correspond to their length, length 1 at (1,0,0), length 2 at (2,0,0), etc. The easiest solution to this problem would be to compute $\hat{V}_a$ and loop from 1 to 5, scaling each vector accordingly and working with `MyVector` to display the vectors at their proper positions.

Application of Vector: Velocity

When riding in a traveling car, you move at the speed and direction of that car. On a per-unit time basis, you will cover the "speed" amount of distance in the direction of the car. For example, during rush hours, a taxi traveling at 1.4 miles per hour toward the northeast will cover 1.4 miles in the northeast direction each hour. In this way, a velocity is speed in a specific direction, or simply, a vector. Figure 4-9 illustrates the example of that taxi ride.

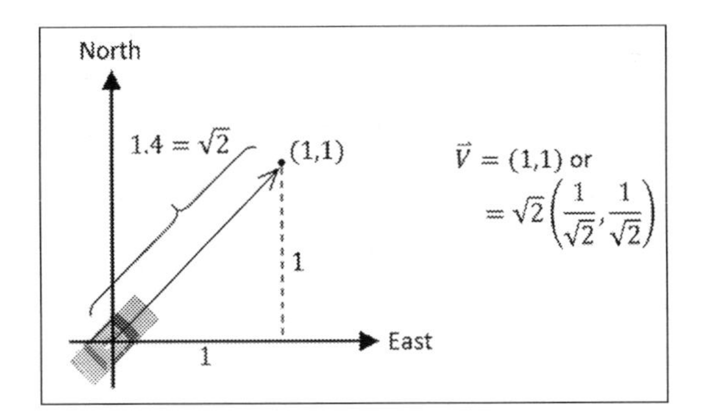

***Figure 4-9.** Driving at 1.4 miles per hour toward the northeast*

As illustrated in Figure 4-9, the 1.4 miles per hour speed of the taxi describes the total distance covered per hour and is actually the magnitude of the vector. In this case, a velocity of

$$\vec{V}_t = (1,\ 1)\,miles\,/\,hour$$

will, in an hour, cover a distance of

$$\|\vec{V}\|_t = \sqrt{1^2 + 1^2} = \sqrt{2} \approx 1.4 \text{ miles}$$

and the traveling direction is indeed toward the northeast (assuming north is the positive y-direction and east is the positive x-direction). Notice in this description the distance covered is separated from the movement direction of the taxi ride. When discussing velocities, it is important to identify the speed and the direction of travel. In terms of implementation, this means that it is convenient to express a velocity, $\vec{V}_t$, as

$$\vec{V}_t = Speed \times \hat{V}_t$$

In the case of Figure 4-9,

- Speed = 1.4
- $\hat{V}_t = \left(\frac{1}{\sqrt{2}}, \frac{1}{\sqrt{2}}\right)$

Recall that you have worked with vectors in this format in the `DrawPositionVector` portion of the previous example, `EX_4_2_VectorScaling`. Representing vectors in this way supports independent adjustments to the magnitude and the direction. In the context of velocity, this representation supports the independent adjustments to the speed (Figure 4-10) and the traveling direction (Figure 4-11).

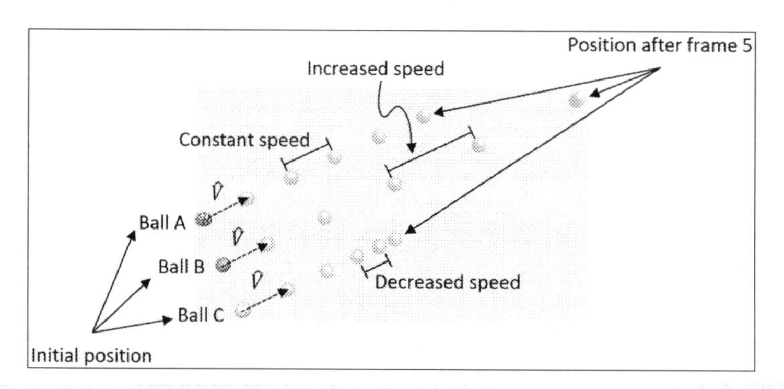

Figure 4-10. *Adjusting the speed while maintaining the direction of travel*

Figure 4-10 shows three balls, A, B, and C, traveling in the same direction, $\hat{V}$, at constant, increasing, and decreasing speeds, respectively. Notice how the balls continue to travel parallel to each other but end up at very different locations along their parallel paths after a few updates.

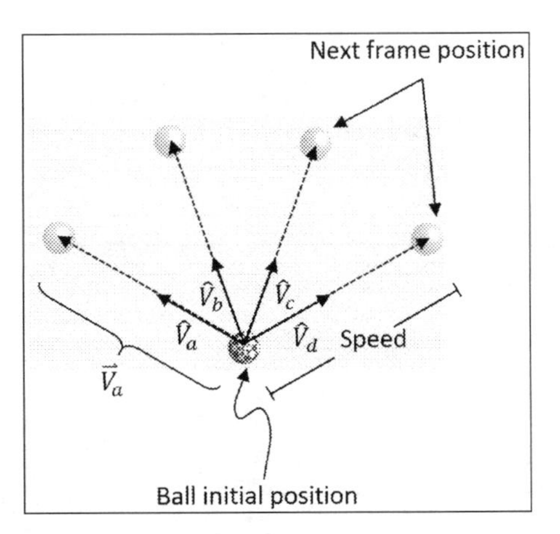

Figure 4-11. *Adjusting the direction of travel while maintaining a constant speed*

In contrast to Figure 4-10, Figure 4-11 shows how the traveling direction of an object can be adjusted without altering its speed. In this case, after subsequent updates, the objects would travel a constant distance from the original position but will end up at very different locations. In all cases, mathematically, the position of an object will change or "travel" by "following the velocity vector," $\vec{V}_t$. If

P_{init}: Initial Position

then at the end of the time unit, the object would travel "following the vector $\vec{V}_t$" and arrive at

$$P_{final} = P_{init} + \left(\vec{V}_t \times elapsedTime\right)$$

This further illustrates the fact that velocity can be perfectly represented as a vector where the vector's magnitude is speed and direction is the direction of travel. This representation of velocity as a vector is convenient for game development and will be showcased in the next example.

The Velocity and Aiming Example

This example demonstrates the manipulation of object velocity and simple aiming functionality based on the vector concepts you have learned in the previous sections. The example allows you to separately adjust the speed, direction, and the traveling distance of an object. This example also allows you to examine the implementation of these factors. Figure 4-12 shows a screenshot of running the `EX_4_3_VelocityAndAiming` scene from the `Chapter-4-Vectors` project.

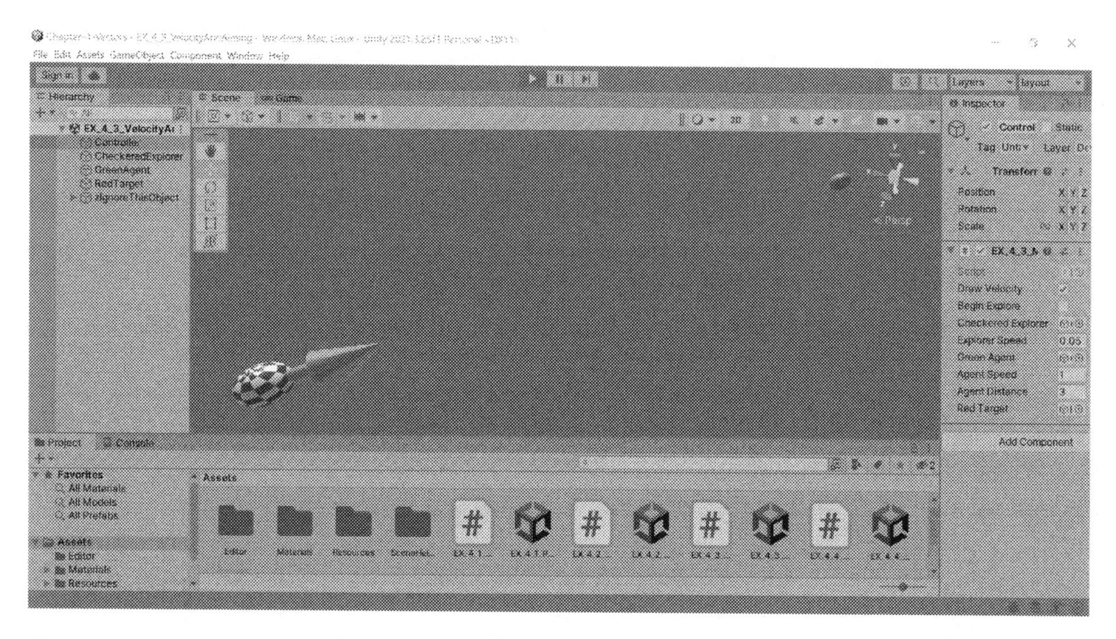

***Figure 4-12.** Running the Velocity and Aiming example*

The goals of this example are for you to

- Understand the distinction between speed and direction of a velocity
- Experience controlling a velocity by manipulating its speed and direction separately
- Examine a simple aiming behavior
- Examine the implementation of vector-based motion control

Examine the Scene

Take a look at the `Example_4_3_VelocityAndAiming` scene and observe the predefined game objects in the Hierarchy Window. In addition to the `Controller`, there are three objects in this scene: `CheckeredExplorer`, `GreenAgent`, and `RedTarget`. Select these objects in the Hierarchy Window to note that the `CheckeredExplorer` is the checkered sphere, the `GreenAgent` is the small green sphere, and the `RedTarget` is the red sphere.

As in all previous examples, these game objects represent positions where only their transform.localPosition are referenced. When the game begins to run and the BeginExplore toggle is true, the CheckeredExplorer position will move slowly toward the position of the RedTarget while continuously sending out the GreenAgent toward the RedTarget as well, but at a faster speed.

Analyze Controller MyScript Component

The MyScript component on the Controller shows four sets of variables:

- Control toggles: Toggles drawing on or off, or allows object movement
    - DrawVelocity: Shows or hides the velocity of the CheckeredExplorer
    - BeginExplore: Enables the movement of the CheckeredExplorer and the GreenAgent
- Support for the CheckeredExplorer:
    - CheckeredExplorer: The reference to the CheckeredExplorer game object
    - ExplorerSpeed: The traveling speed of the CheckeredExplorer
- Support for the GreenAgent:
    - GreenAgent: The reference to the GreenAgent game object
    - AgentSpeed: The traveling speed of the GreenAgent
    - AgentDistance: The distance that the GreenAgent should travel before returning to base and restarting the exploration
- Support for the RedTarget:
    - RedTarget: The reference to the RedTarget game object

The velocity direction for both the CheckeredExplorer and the GreenAgent is implicitly defined by their relative position to the RedTarget because that is the target position that both the CheckeredExplorer and GreenAgent are moving toward.

Interact with the Example

Click the Play Button to run the example. Initially the BeginExplore toggle is set to false and there will thus be no movement in the scene. The green vector you observe extending from the CheckeredExplorer represents the velocity of the CheckeredExplorer object if it were allowed to move. Since you know the vector from the CheckeredExplorer to the RedTarget is, $\vec{V}_{ET}$, then assuming the CheckeredExplorer object is located at $P_{Explorer}$ and the RedTarget object is located at P_{Target}, then

$$\vec{V}_{ET} = P_{Target} - P_{Explorer}$$

Both the CheckeredExplorer and the GreenAgent will be traveling, with their respective speeds of ExplorerSpeed and AgentSpeed, toward the RedTarget. The velocities of these two objects, $\vec{V}_{Explorer}$ and $\vec{V}_{Agent}$, are defined as

$$\vec{V}_{Explorer} = ExplorerSpeed \times \hat{V}_{ET}$$

$$\vec{V}_{Agent} = AgentSpeed \times \hat{V}_{ET}$$

Note that the two velocities are in the same direction, unit vector $\hat{V}_{ET}$, but with different magnitudes, or speeds. Additionally, in both cases, the speeds are under user control and yet the velocity direction is implicitly defined by the RedTarget position.

The green vector you observed represents $\vec{V}_{Explorer}$. Now, adjust ExplorerSpeed in the MyScript component of the Controller object and notice the green vector's length changes accordingly. Since this vector's length is determined by ExplorerSpeed, you can expect the CheckeredExplorer object to move quicker when the green vector is long and slower when it is short. Now, enable the BeginExplore toggle and observe the following:

- The CheckeredExplorer follows slowly behind the repeating and faster traveling GreenAgent. You can adjust the speed of the CheckeredExplorer via the ExplorerSpeed variable and observe, as mentioned previously, that the speed is proportional to the length of the green vector.

- The GreenAgent continuously repeats the quick motion of traveling from the CheckeredExplorer toward the RedTarget. Try adjusting the AgentSpeed variable and observe how the GreenAgent's speed changes accordingly.

- The AgentDistance variable dictates how far the GreenAgent can travel from the CheckeredExplorer before its position is reset and it starts over. If $\vec{V}_{EA}$ is the vector from GreenAgent to the CheckeredExplorer, then

$$\vec{V}_{EA} = P_{Agent} - P_{Explorer}$$

  The current distance between the two is simply the magnitude of this vector, $\|\vec{V}_{EA}\|$. Now, try altering the value of AgentDistance to observe the green sphere traveling that corresponding distance from the checkered sphere before restarting.

- The RedTarget is stationary, but you can manipulate its position via its transform components, and since

$$\vec{V}_{ET} = P_{Target} - P_{Explorer}$$

  when the RedTarget position, P_{Target}, is changed, the vector $\vec{V}_{ET}$ is updated accordingly. The velocity direction, $\hat{V}_{ET}$, of both the CheckeredExplorer and GreenAgent is also updated. In this way, both of these objects are always aiming at and moving toward the RedTarget.

Notice that when the CheckeredExplorer arrives at a location that is very close to the RedTarget, the green vector that represents its velocity will rapidly flip back and forth. As you will find out when analyzing the implementation, there is no logic involved for checking the stop condition of the CheckeredExplorer. Therefore, you are observing the CheckeredExplorer continuously moving pass the RedTarget, flipping its velocity, and then moving pass the RedTarget again. The logic to stop the CheckeredExplorer's motion is left as an exercise at the end of this example.

Details of MyScript

Open MyScript and examine the source code in the IDE. The instance variables are as follows:

```
// Drawing control
public bool DrawVelocity = true;
public bool BeginExplore = false;
```

```
public GameObject CheckeredExplorer = null;// CheckeredExplorer
public float ExplorerSpeed = 0.05f;        // units per second

public GameObject GreenAgent = null;       // GreenAgent
public float AgentSpeed = 1.0f;            // units per second
public float AgentDistance = 3.0f;         // explore distance

public GameObject RedTarget = null;        // RedTarget

private MyVector ShowVelocity = null;  // Show Explorer velocity

private const float kSpeedScaleForDrawing = 15f;
```

All public variables for MyScript have been discussed when analyzing the Controller's MyScript component. The private variable ShowVelocity is to support the visualization of the CheckeredExplorer velocity where the kSpeedScaleForDrawing is a constant value meant to scale this vector such that it is visible. The Start() function for MyScript is listed as follows:

```
void Start() {
    Debug.Assert(CheckeredExplorer != null);
    Debug.Assert(RedTarget != null);
    Debug.Assert(GreenAgent != null);

    ShowVelocity =  new MyVector() {
        VectorColor = Color.green;
    }

    // initially Agent is resting inside the Explorer
    GreenAgent.transform.localPosition =
                CheckeredExplorer.transform.localPosition;
}
```

As in all previous examples, the Debug.Assert() calls ensure proper setup regarding referencing the appropriate game objects via the Inspector Window, while the ShowVelocity variable is properly instantiated. Lastly, the initial position of GreenAgent is set to that of the CheckeredExplorer. The Update() function is listed as follows:

```
void Update() {
    Vector3 vET = RedTarget.transform.localPosition -
                  CheckeredExplorer.transform.localPosition;

    ShowVelocity.VectorAt =
                   CheckeredExplorer.transform.localPosition;
    ShowVelocity.Magnitude =
                   ExplorerSpeed * kSpeedScaleForDrawing;
    ShowVelocity.Direction = vET;
    ShowVelocity.DrawVector = DrawVelocity;

    if (BeginExplore) {
        float dToTarget = vET.magnitude;  // Distance to target
        if (dToTarget < float.Epsilon)
            return; // Avoid normalizing a zero vector
        Vector3 vETn = vET.normalized;

        Process the Explorer (checkered sphere)

        Process the Agent (small green sphere)
    }
}
```

The first line of the Update() function computes $\vec{V}_{ET} = P_{Target} - P_{Explorer}$, and the next four lines set up the ShowVelocity variable for visualizing the CheckeredExplorer's velocity as a vector with its tail located at the position of CheckeredExplorer. Note that because of CheckeredExplorer's slow speed (ExplorerSpeed's value), the ShowVelocity.Magnitude is scaled by kSpeedScaleForDrawing in order to properly display the vector for visual inspection.

When BeginExplore is enabled, the magnitude of $\vec{V}_{ET}$, or $\|\vec{V}_{ET}\|$, is checked to avoid the normalization of a zero vector. Next, $\hat{V}_{ET}$ is computed and stored in the variable vETn. The two regions that process the CheckeredExplorer and the GreenAgent are explained in the following subsections.

Process the Explorer

The code in this region, listed as follows, computes the velocity of the explorer,

$$\vec{V}_{Explorer} = ExplorerSpeed \times \hat{V}_{ET}$$

and updates CheckeredExplorer.transform.localPosition accordingly.

```
#region Process the Explorer (checkered sphere)
Vector3 explorerVelocity = ExplorerSpeed * vETn;
CheckeredExplorer.transform.localPosition +=
        explorerVelocity * Time.deltaTime; // update position
#endregion
```

Remember that displacement, or distance, is velocity traveled over time, or Velocity × elapsedTime. In Unity, the per-update elapsed time is recorded in the Time.deltaTime property. The very last line in this region computes the total displacement over time and updates CheckeredExplorer's position with the computed displacement, ensuring smooth movement.

Process the Agent

As illustrated in the following code, similar to processing the movement of CheckeredExplorer, the first two lines of code in this region compute the velocity of the agent,

$$\vec{V}_{Agent} = AgentSpeed \times \hat{V}_{ET}$$

and update GreenAgent.transform.localPosition accordingly. Note that, as mentioned previously, because $\vec{V}_{Explorer}$ and $\vec{V}_{Agent}$ are both computed based on scaling the same unit vector, the CheckeredExplorer and GreenAgent are traveling in the exact same direction, $\hat{V}_{ET}$, with different speeds, ExplorerSpeed and AgentSpeed.

```
#region Process the Agent (small green sphere)
Vector3 agentVelocity = AgentSpeed * vETn; // define velocity
GreenAgent.transform.localPosition +=
        agentVelocity * Time.deltaTime;   // update position
```

```
Vector3 vEA = GreenAgent.transform.localPosition -
              CheckeredExplorer.transform.localPosition;
if (vEA.magnitude > AgentDistance)
    GreenAgent.transform.localPosition =
              CheckeredExplorer.transform.localPosition;
#endregion
```

The last three lines of code compute the vector between the explorer and the agent,

$$\vec{V}_{EA} = P_{Agent} - P_{Explorer}$$

compare the magnitude of this vector, $\|\vec{V}_{EA}\|$, to the user-specified `AgentDistance`, and then reset the agent's position when it is too far away from the explorer, or when $\|\vec{V}_{EA}\| > AgentDistance$.

Takeaway from This Example

This example demonstrates the application of vector concepts learned in modeling the simple object behaviors of aiming at and moving toward a target position. You have observed that the velocity of objects can be described by scaling a unit vector with speed and that velocities computed based on the same unit vector will move objects in exactly the same direction. Lastly, you have experienced once again that the distance between two objects can be easily computed as the magnitude of the vector defined between these two objects.

Relevant mathematical concepts covered include

- The velocity of an object can be represented by a vector.
- A velocity can be composed by scaling a direction, or unit vector, with speed.
- The distance between two objects is the magnitude of the vector that is defined by the positions of those two objects.

EXERCISES

Stop the CheckeredExplorer When It Reaches the RedTarget

Recall that the motion of CheckeredExplorer never terminates and that it tends to overshoot the RedTarget followed by turning around and overshooting it again. This cycle continues, causing the CheckeredExplorer to swing back and forth around the RedTarget. Modify MyScript to define a bounding box around the RedTarget and stop the CheckeredExplorer when it is inside the bounding box. Notice that in this case, it is actually easier and more accurate to treat the RedTarget as a bounding sphere and to stop the motion of the CheckeredExplorer when it is inside the bounds of the sphere.

Reset the GreenAgent When It Reaches the RedTarget

Run the game and increase the AgentDistance to some large value, for example, 15. Now set BeginExplore to true and observe how the GreenAgent passes through the RedTarget and continues to move forward until its position is more than 15 units from the CheckeredExplorer, in which case it finally resets. With the bound you defined in the previous exercise, modify MyScript to reset the GreenAgent's position as soon as it is inside the RedTarget's bounds.

Invert the GreenAgent's Velocity Direction

Modify MyScript such that when the GreenAgent is too far away from the CheckeredExplorer, instead of resetting the position, the GreenAgent would simply move toward the CheckeredExplorer as though it is now the target. In this way, the GreenAgent would move continuously between the CheckeredExplorer and the RedTarget. This example allows you to gain experience with reversing the direction of a given vector.

Vector Algebra: Addition and Subtraction

Although it has not yet been formally defined, based on observing the relative positions in the Cartesian Coordinate System, you have worked with vector addition and subtraction for quite a while now. For example, you have learned that the statement "position P_1 can be reached by following a vector $\vec{V}_1$ at position P_0" is expressed mathematically as

$$P_1 = P_0 + \vec{V}_1$$

In this case, by interpreting P_0 and P_1 as position vectors, the "+" operator has two vector operands and produces a position vector as the result of the operation. You have also learned that the statement "the vector $\vec{V}_1$ is a vector with its tail at position P_0 and head at position P_1" is expressed mathematically as

$$\vec{V}_1 = P_1 - P_0$$

Once again, with P_0 and P_1 interpreted as position vectors, the "–" operation also has two vector operands and produces a vector as the result of the operation.

Rules of Vector Addition and Subtraction

You have learned and experienced that in both vector addition and subtraction, the resulting vectors are simply the addition and subtraction of the corresponding x-, y-, and z-component values. These observations are summarized in Table 4-1.

Table 4-1. *Vector addition and subtraction*

Operation	Operand 1	Operand 2	Result
+:Addition	$\vec{V}_1 = (x_1, y_1, z_1)$	$\vec{V}_2 = (x_2, y_2, z_2)$	$\vec{V}_1 + \vec{V}_2 = (x_1 + x_2, y_1 + y_2, z_1 + z_2)$
–: Subtraction	$\vec{V}_1 = (x_1, y_1, z_1)$	$\vec{V}_2 = (x_2, y_2, z_2)$	$\vec{V}_1 - \vec{V}_2 = (x_1 - x_2, y_1 - y_2, z_1 - z_2)$

Note that the given definition in Table 4-1 states that the following is always true:

$$\vec{V} + \vec{V} = 2\vec{V}$$

$$\vec{V} - \vec{V} = Zero\,Vector$$

Because the operators add and subtract the corresponding coordinate component values, the familiar floating-point arithmetic addition and subtraction properties are obeyed. The properties of commutative, associative, and distributive with a floating-point scaling factor, *s*, are summarized in Table 4-2.

Table 4-2. *Properties of vector addition and subtraction*

Properties	Vector Addition	Vector Subtraction
Commutative	$\vec{V}_1 + \vec{V}_2 = \vec{V}_2 + \vec{V}_1$	$\vec{V}_1 - \vec{V}_2 \neq \vec{V}_2 - \vec{V}_1$ **[not a property]**
Associative	$(\vec{V}_1 + \vec{V}_2) + \vec{V}_3 = \vec{V}_1 + (\vec{V}_2 + \vec{V}_3)$	$(\vec{V}_1 - \vec{V}_2) - \vec{V}_3 = \vec{V}_1 - (\vec{V}_2 - \vec{V}_3)$
Distributive	$s(\vec{V}_1 + \vec{V}_2) = s\vec{V}_1 + s\vec{V}_2$	$s(\vec{V}_1 - \vec{V}_2) = s\vec{V}_1 - s\vec{V}_2$

As illustrated in the first-row, right column of Table 4-2, just as with floating-point subtraction, vector subtraction is not commutative. In fact, similar to floating-point subtraction, vector subtraction is anti-commutative, or

$$\vec{V}_1 - \vec{V}_2 = -1 \times (\vec{V}_2 - \vec{V}_1) = -\vec{V}_2 + \vec{V}_1$$

$$= \vec{V}_1 - \vec{V}_2$$

Addition and Subtraction with the Zero Vector

As in the case of floating-point arithmetic, vector addition and subtraction with the zero vector behave as expected.

$$\vec{V}_1 + ZeroVector = ZeroVector + \vec{V}_1 = \vec{V}_1$$

$$\vec{V}_1 - ZeroVector = \vec{V}_1$$

$$ZeroVector - \vec{V}_1 = -\vec{V}_1$$

Vectors in an Equation

Vectors behave just like floating-point values in an equation. For example, if

$$\vec{V}_3 = \vec{V}_1 + \vec{V}_2,$$

then adding a $-\vec{V}_2$ to both sides of the equation:

$$\vec{V}_3 + \left(-\vec{V}_2\right) = \vec{V}_1 + \vec{V}_2 + (-\vec{V}_2)$$

$$\vec{V}_3 - \vec{V}_2 = \vec{V}_1$$

$$\vec{V}_1 = \vec{V}_3 - \vec{V}_2.$$

This little example helps demonstrate that vector algebra obeys the basic algebraic equation rule that a term can be moved across the equality by flipping its sign.

Geometric Interpretation of Vector Addition and Subtraction

Fortunately, there are intuitive diagrammatic interpretations for the essential rules of vector addition and subtraction. Please refer to Figure 4-13, where vectors $\vec{V}_1$ and $\vec{V}_2$ are defined by the three given positions, P_0, P_1, and P_2. These two vectors are defined as

$$\vec{V}_1 = P_1 - P_0$$

$$\vec{V}_2 = P_2 - P_1$$

Figure 4-13 shows vector $\vec{V}_1$ with its tail at P_0 and vector $\vec{V}_2$ with its tail at P_1.

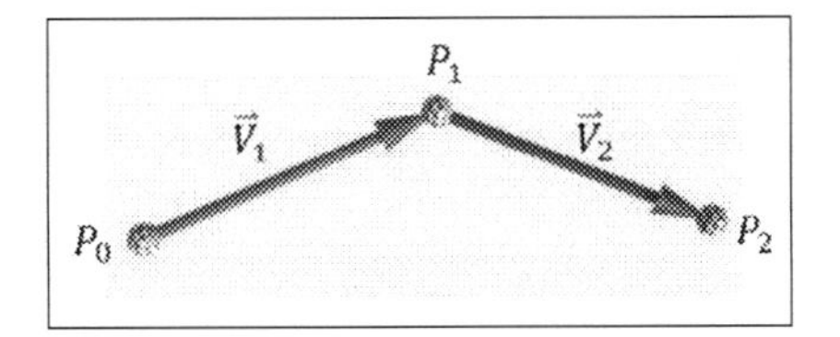

Figure 4-13. *Two vectors defined by three positions*

Vector Addition

Figure 4-14 shows the result of vector addition geometrically. Notice that the result of adding the two vectors

$$\vec{V}_{sum} = \vec{V}_1 + \vec{V}_2$$

is a vector with its tail located at the tail of $\vec{V}_1$, P_0, and its head located at the head of $\vec{V}_2$, P_2. This can be interpreted geometrically as $\vec{V}_{sum}$ is the combined results of "following $\vec{V}_1$ then $\vec{V}_2$." Except that in case this, instead of following the two vectors sequentially, the summed vector, $\vec{V}_{sum}$, will take you directly from the beginning to the end along the shortest path. This observation is true in general; the result of summing vectors is always a vector that combines the results of following all of the operand vectors sequentially and is then the shortest path from the beginning location to the final destination location.

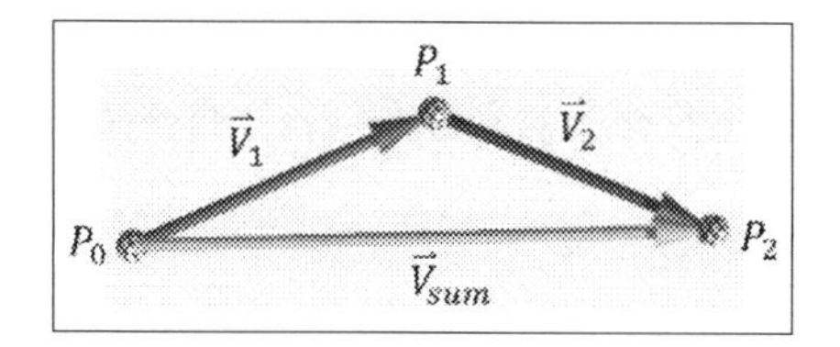

Figure 4-14. *Vector addition*

Commutative Property of Vector Addition

Figure 4-15 illustrates the commutative property of vector addition:

$$\vec{V}_{sum} = \vec{V}_1 + \vec{V}_2 = \vec{V}_2 + \vec{V}_1$$

Note the difference in the order of operations; the top half of Figure 4-14 applies $\vec{V}_1$ at P_0 followed by applying $\vec{V}_2$ at the head of $\vec{V}_1$, while the latter applies $\vec{V}_2$ at P_0 followed by applying $\vec{V}_1$ at the head of $\vec{V}_2$. Observe that in both cases, the result is identical; $\vec{V}_{sum}$ has its tail located at P_0 and its head at P_2.

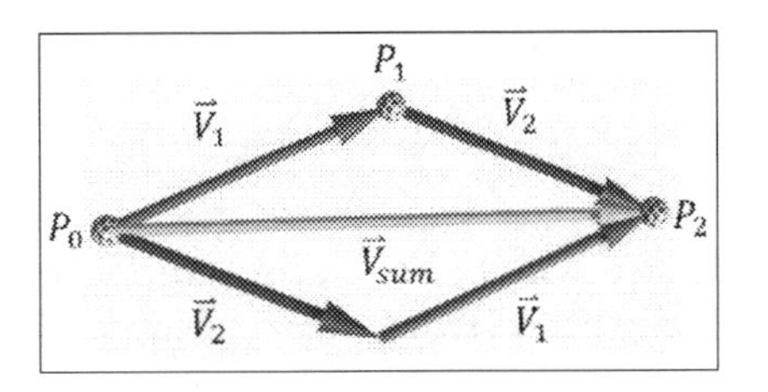

Figure 4-15. *The commutative property of vector addition*

Figure 4-14 shows that, geometrically, vector addition depicts a triangle where the first two edges are the operands and the third is the resulting sum. In Figure 4-15, the two $\vec{V}_1$ are of the same length and are parallel and so are the two $\vec{V}_2$ vectors. For this reason, the depiction in Figure 4-15 is a parallelogram. These observations are true in general—that vector addition and the commutative property always depict a triangle and parallelogram, respectively. Though these observations do not result in direct applications in video games, they provide insights into relationships between different fields of mathematics, in this case, linear algebra and geometry.

Vector Subtraction

Figure 4-16 shows the result of vector subtraction geometrically. The two vectors with tails at position P_1 are $\vec{V}_2$ and a scaling of $\vec{V}_2$ by a factor of -1 resulting in $-\vec{V}_2$, or $\vec{V}_{n2}$, a vector with same length in the opposite direction to $\vec{V}_2$. This figure shows that subtracting a vector is essentially the same as using the opposite direction of that vector in a vector addition. In this case, $\vec{V}_1 - \vec{V}_2$ can be understood as travel along $\vec{V}_1$, followed by traveling along the opposite direction of $\vec{V}_2$. This interpretation can be verified mathematically as follows. Notice that just as floating-point algebra, the subtraction of the two vectors

$$\vec{V}_{sub} = \vec{V}_1 - \vec{V}_2$$

can be written as an addition

$$\vec{V}_{sub} = \vec{V}_1 + \vec{V}_{n2}$$

where

$$\vec{V}_{n2} = -\vec{V}_2$$

or simply

$$\vec{V}_{sub} = \vec{V}_1 - \vec{V}_2 = \vec{V}_1 + \left(-\vec{V}_2\right)$$

Notice the perfect correspondence between the expression, $\vec{V}_1 + \left(-\vec{V}_2\right)$, and the description, "travel along $\vec{V}_1$, followed by traveling along the opposite direction of $\vec{V}_2$."

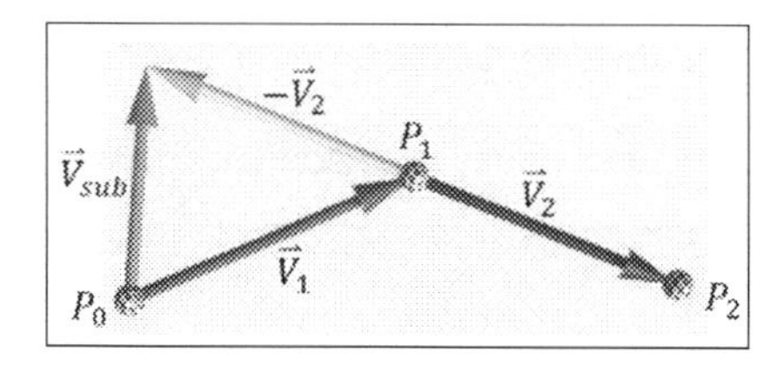

Figure 4-16. *Vector subtraction*

The Vector Add and Sub Example

This example demonstrates the results of and allows you to interact with the vector addition and subtraction operations. This example also serves as a review and reaffirmation that vectors can be located at any position as their definition does not link them to a specific position. Figure 4-17 shows a screenshot of running the `EX_4_4_VectorAddandSub` example from the `Chapter-4-Vectors` project.

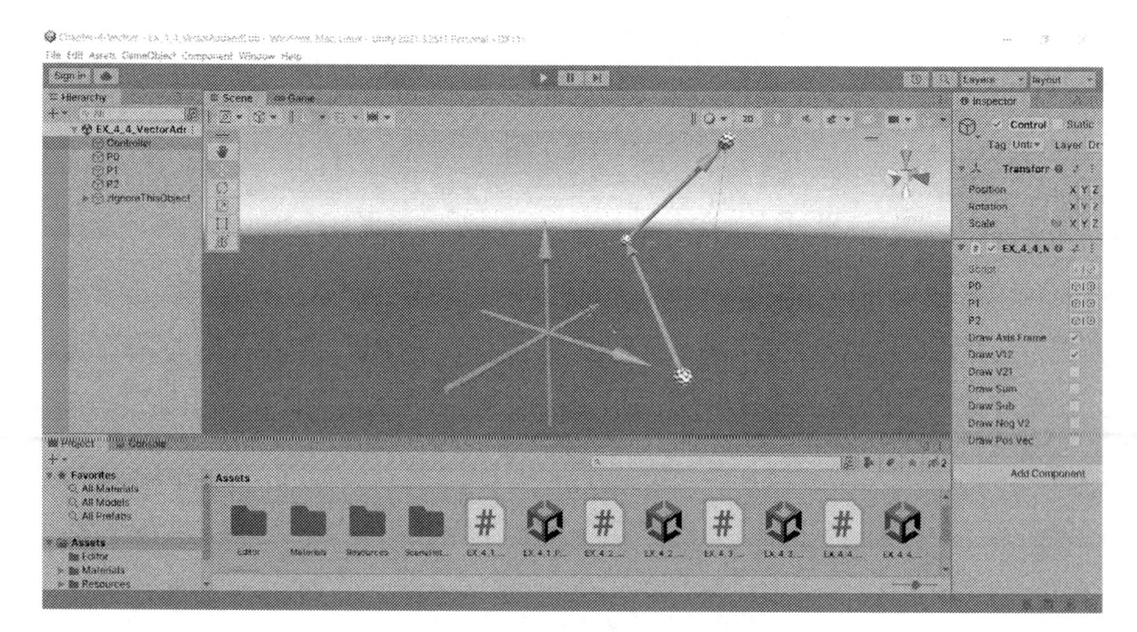

Figure 4-17. *Running the Vector Add and Sub example*

The goals of this example are for you to

- Examine and gain understanding of vector addition and subtraction
- Understand that vector subtraction is simply vector addition with a negative vector as the second operand
- Review that all vectors are defined independent of any position

Examine the Scene

Look at the `Example_4_4_VectorAddandSub` scene and observe the predefined game objects in the Hierarchy Window. In addition to the `Controller`, there are three objects in this scene: `P0`, `P1`, and `P2`. Each of these objects references one of the spheres in the scene which in turn represent a position in the Cartesian Coordinate System. In this example you can manipulate these three positions to define two vectors, where the results of adding and subtracting these two vectors are shown at those positions and at the origin as position vectors.

Analyze Controller MyScript Component

The `MyScript` component on the `Controller` shows two sets of variables:

- The three positions:
  - `P0`: The reference to the `P0` game object.
  - `P1`: The reference to the `P1` game object.
  - `P2`: The reference to the `P2` game object.

  The `transform.localPosition` of these objects will provide the positions defining the two vectors:

$$\vec{V}_1 = P_1 - P_0$$

$$\vec{V}_2 = P_2 - P_1$$

- Draw control: There are seven toggles for showing or hiding the following.
  - `DrawAxisFrame`: Shows or hides the axis frame; the axis frame serves as a reference for showing position vectors.
  - `DrawV12`: Shows or hides vector $\vec{V}_1$ at position P_0 and $\vec{V}_2$ at the head of $\vec{V}_1$. This is convenient for examining $\vec{V}_1 + \vec{V}_2$.
  - `DrawV21`: Shows or hides vector $\vec{V}_2$ at position P_0 and $\vec{V}_1$ at the head of $\vec{V}_2$. This is convenient for examining $\vec{V}_2 + \vec{V}_1$.
  - `DrawSum`: Shows or hides the vectors $\vec{V}_{sum} = \vec{V}_1 + \vec{V}_2$ and $\vec{V}_{sum} = \vec{V}_2 + \vec{V}_1$.
  - `DrawSub`: Shows or hides the vector $\vec{V}_{sub} = \vec{V}_1 - \vec{V}_2$.
  - `DrawNegV2`: Shows or hides the vector $-\vec{V}_2$.
  - `DrawPosVec`: Shows or hides currently visible vector(s) as position vector(s).

The purpose of this example is for you to manipulate the `P0`, `P1`, and `P2` positions and toggle each of the preceding drawing options to closely examine each of the corresponding vectors.

Interact with the Example

Click the Play Button to run the example. Initially, both `DrawAxisFrame` and `DrawV12` are enabled so you should observe the axis frame and the two vectors $\vec{V}_1$ (in red) and $\vec{V}_2$ (in blue) connecting the checkered spheres `P0`, `P1`, and `P2`. Now, enable `DrawPosVec` to observe vectors $\vec{V}_1$ and $\vec{V}_2$ drawn at the origin as position vectors. At any point in the following interaction, feel free to toggle on `DrawAxisFrame` for referencing. For now, please toggle it off to avoid cluttering the scene.

Vector Addition and the Commutative Property

With `DrawPosVec` on, switch on both `DrawV12` and `DrawV21` toggles to show these two sets of vectors. Select and manipulate position `P1` to observe how the two sets of vectors change. Now toggle `DrawSum` on and continue with the manipulation of position `P1`. Observe that since $\vec{V}_{sum} = \vec{V}_1 + \vec{V}_2 = \vec{V}_2 + \vec{V}_1$ is a vector from `P0` to `P2`, changing `P1` has absolutely no effect on $\vec{V}_{sum}$. Next, select and manipulate P0 to observe how the red $\vec{V}_1$ and green $\vec{V}_{sum}$ vectors change together while the blue $\vec{V}_2$ remains constant. Repeat the manipulation for P2 and observe $\vec{V}_2$ and $\vec{V}_{sum}$ altering while $\vec{V}_1$ remains constant.

Through these interactions, you have verified that vector addition is indeed accumulating the results of individual operands and that the operation does indeed obey the commutative property. You were also reminded, through turning on the DrawPosVec toggle, that vectors are independent of positions as all three vectors were identical to their corresponding color partner except for their tail location.

Vector Subtraction

Reset all toggles to off and switch on DrawPosVec, DrawV12, and DrawNegV2. You should observe three sets of vectors: $\vec{V}_1$ (in red), $\vec{V}_2$ (in blue), and $-\vec{V}_2$ (in yellow). Manipulate the Scene View camera to observe that the yellow vectors are indeed the same length and in opposite directions as the blue vectors. Select and manipulate P1 to observe the two sets of three vectors changing in sync. If you manipulate P2, it will only affect $\vec{V}_2$ (in blue) and $-\vec{V}_2$ (in yellow) vectors. Now switch on the DrawSub toggle to observe the gray $\vec{V}_{sub}$ vector as the sum of the red and yellow vector, $\vec{V}_{sub} = \vec{V}_1 + \left(-\vec{V}_2\right)$.

Through these interactions, you have verified that vector subtraction is indeed the same as vector addition with the second operand being negated. In fact, every operand after the first operand, if originally being subtracted, can instead be added after it's been negated, just like with floating-point arithmetic.

Position Vector

With DrawPosVec toggle on, every computed vector is displayed at the origin as a position vector. For example, while $\vec{V}_{sum}$ was computed by $\vec{V}_1 + \vec{V}_2$ and the geometric depiction suggests that $\vec{V}_{sum}$ must always have its tail at P0, this is not the case. Once again, a vector is a length and a direction; this definition holds true independent of any specific position, even when a position is used initially to define that vector.

Details of MyScript

Open MyScript and examine the source code in the IDE. The instance variables are as follows:

```
public GameObject P0, P1, P2;              // V1=P1-P0 and V2=P2-p1

// For visualizing the vectors
private MyVector
    ShowV1atP0, ShowV2atV1, // Show V1 at P0 and V2 at head of V1
    ShowV2atP0, ShowV1atV2, // Show V2 at P0 and V1 at head of V2
    ShowSumV12, ShowSumV21, // V1+V2, and V2+V1
    ShowSubV12,             // V1-V2
    ShowNegV2;              // -V2
// Show as position vectors
private MyVector PosV1, PosV2, PosSum, PosSub, PosNegV2;

// Toggles for drawing/hiding corresponding vectors
public bool DrawAxisFrame = true;
public bool DrawV12 = false, DrawV21 = false;
public bool DrawSum = false;
public bool DrawSub = false, DrawNegV2 = false;
public bool DrawPosVec = false;
```

All public variables for MyScript have been discussed when analyzing the Controller's MyScript component. The large number of private MyVector variables is for visualizing the corresponding vectors. The Start() function for MyScript is listed as follows:

```
void Start() {
    Debug.Assert(P0 != null);
    Debug.Assert(P1 != null);
    Debug.Assert(P2 != null);

    ShowV1atP0 = new MyVector() { // Show V1 vectors
        VectorColor = Color.red      };
    ShowV1atV2 = new MyVector() {
        VectorColor = Color.red      };
```

```
    PosV1 = new MyVector()       { // Show V1 as position vector
        VectorAt = Vector3.zero,  // always show at the origin
        VectorColor = Color.red     };

    ShowV2atP0 = new MyVector() { // Show V2 vectors
       VectorColor = Color.blue     };
    ShowV2atV1 = new MyVector() {
        VectorColor = Color.blue    };
    PosV2 = new MyVector()       { // Show V2 as position vector
        VectorAt = Vector3.zero,
        VectorColor = Color.blue    };

    ShowSumV12 = new MyVector() { // Show V1 + V2
        VectorColor = Color.green    };
    ShowSumV21 = new MyVector() { // Show V2 + V1
        VectorColor = Color.green    };
    PosSum = new MyVector()       { // Show sum as position vector
        VectorAt = Vector3.zero,
        VectorColor = Color.green    };

    ShowSubV12 = new MyVector() { // Show V1 - V2
        VectorColor = Color.gray     };
    PosSub = new MyVector()       { // Show as position vector
        VectorAt = Vector3.zero,
        VectorColor = Color.gray     };

    ShowNegV2 = new MyVector()    { // Show -V2
        VectorColor = new Color(0.9f, 0.9f, 0.2f, 1.0f) };
    PosNegV2 = new MyVector()      {
        VectorAt = Vector3.zero,
        VectorColor = new Color(0.9f, 0.9f, 0.2f, 1.0f) };
}
```

As in all previous examples, the `Debug.Assert()` calls ensure proper setup regarding referencing the appropriate game objects via the Inspector Window. The rest of the `Start()` function instantiates the many `MyVector` variables for visualization, setting their colors and display positions. The `Update()` function is listed as follows:

```
void Update() {
    Vector3 V1 = P1.transform.localPosition -
                 P0.transform.localPosition;
    Vector3 V2 = P2.transform.localPosition -
                 P1.transform.localPosition;
    Vector3 sumV12 = V1 + V2;
    Vector3 sumV21 = V2 + V1;
    Vector3 negV2 = -V2;
    Vector3 subV12 = V1 + negV2;

    Draw control: switch on/off what to show

    V1: show V1 at P0 and head of V2

    V2: show V2 at P0 and head of V1

    Sum: show V1+V2 and V2+V1

    Sub: show V1-V2

    Negative vector: show -V2
}
```

The `Update()` function first computes all the relevant vectors:

- $\vec{V}_1 = P_1 - P_0$
- $\vec{V}_2 = P_2 - P_1$
- $\vec{V}_{sum12} = \vec{V}_1 + \vec{V}_2$
- $\vec{V}_{sum21} = \vec{V}_2 + \vec{V}_1$
- $\vec{V}_{n2} = -\vec{V}_2$
- $\vec{V}_{sub12} = \vec{V}_1 - \vec{V}_2$

Then it sets up the corresponding `MyVector` variables for display based upon their values and if their toggle switches are true. The details of this visualization code are independent of the vector operations being studied and are therefore not discussed here. You can explore the code in these regions at your own leisure.

Takeaway from This Example

This example demonstrates the details of vector addition and subtraction where the commutative property of vector addition is verified and vector subtraction is presented as vector addition with a negated vector. Equally important is the review of a vector's independence of positions.

Relevant mathematical concepts covered include

- Vector addition results in a vector that accumulates the operand vectors.
- Vector addition is indeed commutative.
- Vector subtraction is simply an addition with the second operand being negated.
- Reviewed that vectors are independent of any particular position.

EXERCISES

Verify Vector Addition Accumulates in General

Modify the scene and `MyScript` to include a fourth position, `P3`, and a vector, $\vec{V}_3$.

$$\vec{V}_3 = P_3 - P_2$$

Now, define $\vec{V}_{sum}$

$$\vec{V}_{sum} = \vec{V}_1 + \vec{V}_2 + \vec{V}_3$$

Verify that it is always true that if the tail of $\vec{V}_{sum}$ is located at `P0`, then its head will be located at `P3`.

Verify the Associative Property of Addition and Subtraction

With the fourth position, `P3`, and vector $\vec{V}_3$, verify

$$\left(\vec{V}_1 + \vec{V}_2\right) + \vec{V}_3 = \vec{V}_1 + \left(\vec{V}_2 + \vec{V}_3\right)$$

and

$$\left(\vec{V}_1 - \vec{V}_2\right) - \vec{V}_3 = \vec{V}_1 - \left(\vec{V}_2 - \vec{V}_3\right)$$

by computing and displaying each as a different `MyVector` object.

Application of Vector Algebra

Although seldom applied directly, the indirect applications of vector algebra in video games are ubiquitous and vital. For example, you have already experienced working with vector subtraction in defining a vector between two positions for distance computation and vector addition in computing movements when applying a velocity to an object.

A straightforward application of vector addition is in simulating velocity under a constant external factor, for example, an airplane flying or a ship sailing under a constant wind condition. Please refer to Figure 4-18 where a traveling ball is progressing toward a target with a velocity of $\vec{V}_T$. Under the wind condition, $\vec{V}_{wind}$, the effective velocity experienced by the ball then becomes $\vec{V}_A$:

$$\vec{V}_A = \vec{V}_T + \vec{V}_{wind}$$

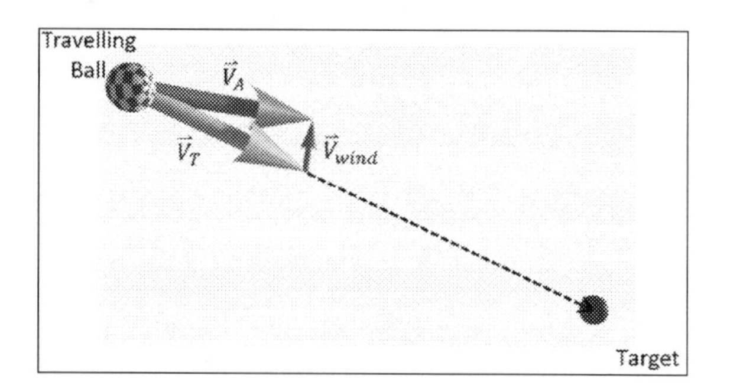

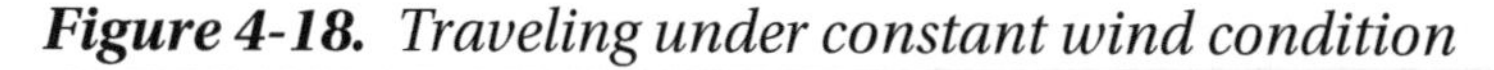

Figure 4-18. *Traveling under constant wind condition*

With your knowledge of vectors and vector addition, this wind condition is straightforward to simulate and is examined in the next example.

The Windy Condition Example

This example uses vector addition to simulate an object traveling under a constant wind condition. The example allows you to adjust all the parameters of this simulation, including the speed of the traveling object and the wind, the direction of the wind, and if the wind condition should affect the traveling object. Figure 4-19 shows a screenshot of running the `EX_4_5_WindyCondition` example from the `Chapter-4-Vectors` project.

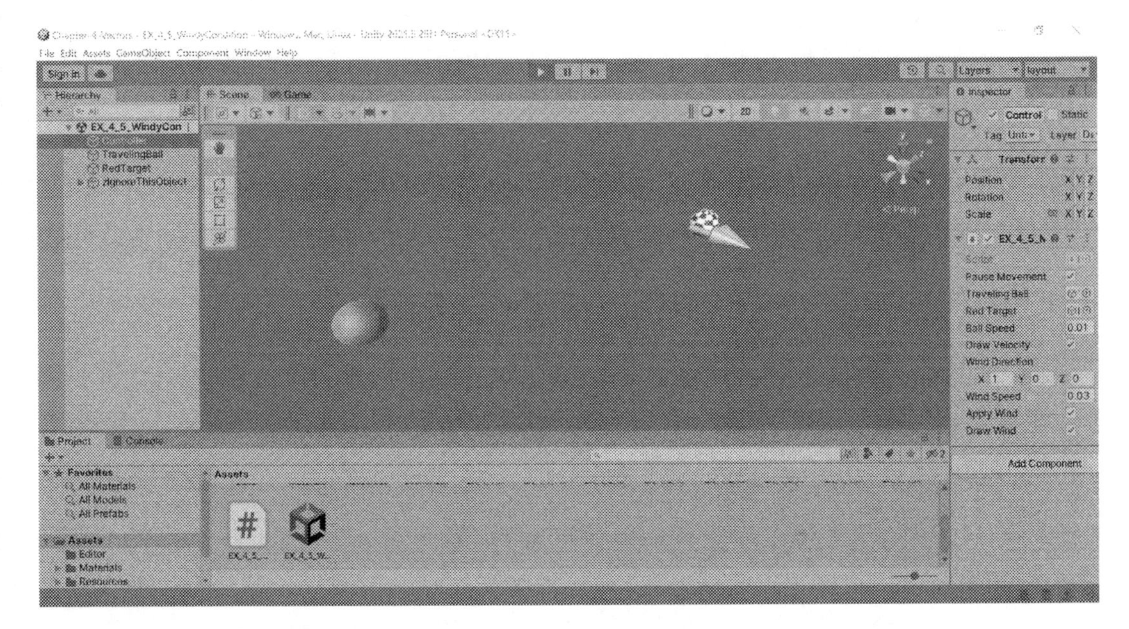

Figure 4-19. Running the Windy Condition example

The goals of this example are for you to

- Experience a straightforward example of applying vector addition to affect object behavior
- Examine and understand the simple implementation of how velocity can be affected under a constant wind condition

Examine the Scene

Take a look at the `Example_4_5_WindyCondition` scene and observe the predefined game objects in the Hierarchy Window. In addition to the `Controller`, there are two objects in this scene: `TravelingBall` and `RedTarget`. This example simulates the `TravelingBall` progressing toward the `RedTarget` under a constant wind condition that affects its velocity.

Analyze Controller MyScript Component

The `MyScript` component on the `Controller` shows four sets of variables:

- Simulation control: Variables that control the simulation
  - `PauseMovement`: The toggle that stops the simulation and the movements of the objects in the scene, allowing for careful examination of the scene.
- The objects: The objects in the scene that you can interact with
  - `TravelingBall`: The reference to the `TravelingBall` game object
  - `RedTarget`: The reference to the `RedTarget` game object
- Traveling ball speed: Variables that affect the speed of the traveling ball
  - `BallSpeed`: The speed at which the ball is traveling without any wind. Note that the direction of ball's velocity is along the vector defined by the ball and the target positions. Assuming P_B and P_T are the positions of the ball and the target, respectively, then

    $$\vec{V}_T = BallSpeed \times (P_T - P_B).Normalized$$

  - `DrawVelocity`: A toggle to hide or show the ball's velocity vector, $\vec{V}_T$.
- Wind condition: The variables that control the wind condition in the simulation
  - `WindDirection`: Determines the direction of the wind velocity, $\vec{V}_{wind}$
  - `WindSpeed`: Determines the speed of the wind velocity, $\vec{V}_{wind}$
  - `ApplyWind`: Toggles the effect of the wind on or off
  - `DrawWind`: A toggle to hide or show the wind's velocity vector

Interact with the Example

Click the Play Button to run the example. Note that initially PauseMovement is enabled and the traveling ball does not move. The three vectors you observe are explained as follows. The green vector pointing from the TravelingBall toward the RedTarget is the ball's current velocity, $\vec{V}_T$. The red vector is the wind's velocity, $\vec{V}_{wind}$. Lastly, the blue vector is the path that the ball will take, the resulting vector, $\vec{V}_A$, where

$$\vec{V}_A = \vec{V}_T + \vec{V}_{wind}$$

Increase the BallSpeed and WindSpeed to observe the corresponding green and red vectors increase in length. Select and move the RedTarget to verify that the direction of the green vector, $\vec{V}_T$, always points toward the RedTarget. Next, select and change the components of the WindDirection variable to verify that the direction of the red vector changes accordingly.

Now, switch off PauseMovement toggle to allow the simulation to proceed. Try increasing WindSpeed, for example, to 0.05, and observe $\vec{V}_T$ being affected while the TravelingBall proceeds and drifts toward the RedTarget. Note that when WindSpeed and WindDirection are unfavorable, for example, a speed of 0.15 in the direction of (1,0,0), the TravelingBall will drift away from and never reach the RedTarget.

Details of MyScript

Open MyScript and examine the source code in the IDE. The instance variables are as follows:

```
public bool PauseMovement = true;

public GameObject TravelingBall = null;
public GameObject RedTarget = null;

public float BallSpeed = 0.01f;          // units per second
public bool DrawVelocity = false;
private float VelocityDrawFactor = 20f; // To see the vector

public Vector3 WindDirection = Vector3.zero;
public float WindSpeed = 0.01f;
public bool ApplyWind = false;
public bool DrawWind = false;
```

```
private MyVector ShowVelocity = null;
private MyVector ShowWindVector = null;
private MyVector ShowActualVelocity = null;
```

All public variables for MyScript have been discussed when analyzing Controller's MyScript component. The private variable VelocityDrawFactor is for scaling the small magnitude velocity vectors such that they can be visible. The MyVector data type private variables are to visualize the three vectors, $\vec{V}_T$, $\vec{V}_{wind}$, and $\vec{V}_A$. The Start() function for MyScript is listed as follows:

```
void Start() {
    Debug.Assert(TravelingBall != null);
    Debug.Assert(RedTarget != null);

    ShowVelocity = new MyVector() {
        VectorColor = Color.green,
        DrawVectorComponents = false };
    ShowWindVector = new MyVector() {
        VectorColor = new Color(0.8f, 0.3f, 0.3f, 1.0f),
        DrawVectorComponents = false  };
    ShowActualVelocity = new MyVector() {
        VectorColor = new Color(0.3f, 0.3f, 0.8f, 1.0f),
        DrawVectorComponents = false    };
}
```

As in all previous examples, the Debug.Assert() calls ensure proper setup regarding referencing the appropriate game objects via the Inspector Window, while the rest of the function instantiates the MyVector variables for proper visualization of the vectors. The Update() function is listed as follows:

```
void Update() {
    Vector3 vDir = RedTarget.transform.localPosition -
                   TravelingBall.transform.localPosition;
    float distance = vDir.magnitude;

    if (distance > float.Epsilon) { // if not at the target
        vDir.Normalize();
        WindDirection.Normalize();
```

```
        Vector3 vT = BallSpeed * vDir;
        Vector3 vWind = WindSpeed * WindDirection;
        Vector3 vA = vT + vWind;

        // Display the vectors

        if (PauseMovement)
            return;

        if (ApplyWind)
            TravelingBall.transform.localPosition +=
                                        vA * Time.deltaTime;
        else
            TravelingBall.transform.localPosition +=
                                        vT * Time.deltaTime;
    } // if (distance < float.Epsilon)
}
```

The Update() function first computes the vector from TravelingBall toward the RedTarget, $\vec{V}_{dir}$. Next, the magnitude of $\vec{V}_{dir}$, distance, is computed and checked to ensure that this is not a very small number. This checking accomplishes two important objectives. First, a small distance value means that the TravelingBall object is closed to or has reached the RedTarget object and further simulation is no longer required. Second, when distance is approximately zero, $\vec{V}_{dir}$ is approximately a zero vector and thus cannot be normalized. When distance is larger than approximately zero, the following velocity vectors are computed:

$$\vec{V}_T = BallSpeed \times \hat{V}_{dir}$$

$$\vec{V}_{wind} = WindSpeed \times \widehat{WindDirection}$$

$$\vec{V}_A = \vec{V}_T + \vec{V}_{wind}$$

When the simulation condition is true, depending on if the user wants to observe the effects of the wind, the TravelingBall position is updated by either $\vec{V}_T \times elapsedTime$ or $\vec{V}_A \times elapsedTime$.

Takeaway from This Example

This example demonstrates the straightforward application of vector addition by simulating traveling under a constant, external effect, like a wind condition. You have observed that such a condition can be simulated as a velocity vector being added to the traveling velocity.

Relevant mathematical concepts covered include

- Model constant wind breeze as a velocity
- Changing an object's velocity by the addition of an object's own velocity with that of external velocities

EXERCISES

Compensate for the Wind Conditions

Note that if the wind velocity, $\vec{V}_{wind}$, is available during the computation of an object's velocity, $\vec{V}_T$, then it is possible to compensate for the wind condition. Instead of moving toward the target, $\hat{V}_{dir}$, the traveling velocity should point toward the target only after $\hat{V}_{dir}$ is affected by the wind condition, or

$$BallSpeed \times \hat{V}_{dir} = \vec{V}_T + \vec{V}_{wind}$$

So

$$\vec{V}_T = BallSpeed \times \hat{V}_{dir} - \vec{V}_{wind}$$

Implement this compensation and observe a smoother `TravelingBall` movement. You have observed that it is possible to compensate and largely remove the external wind factor by not traveling directly toward the final destination.

Travel Under Multiple External Factors

Support a strong wind gust which occurs probabilistically (or pseudo-randomly). In addition to speed and direction, allow your user to adjust the occurrence frequency and duration of this wind gust. Now, as the `TravelingBall` moves toward its target, it may get blown off course some of the times. You now know how to add simple environmental factors into a game.

Summary

This chapter introduces vectors by relating to your understanding of measurement and distance computations in the Cartesian Coordinate System. You have learned the following:

- A vector is a size and a direction that can relate two positions.
- The vector definition is independent of any particular position.
- All positions in the Cartesian Coordinate System can be considered as position vectors.
- Scaling a vector by a floating-point number changes its size but not its direction.
- A normalized or unit vector has a size of 1 and is convenient for representing the direction of a vector.
- Vectors are ideal for representing the velocities of objects.
- It is convenient to represent a velocity by separately storing its speed and direction of movement.
- Vector addition and subtraction rules follow closely to those of floating-point algebra.

The examples presented in this chapter allowed you to interact with and examine the details of vectors and their operations. Based on vector concepts, you have examined the simple object behaviors of following, or aiming, at a target and the environmental affects you can create by disturbing an object's motion with an external velocity.

Through this chapter, you have gained the basic knowledge of what a vector is, its basic rules, and how it can be used to model simple object behaviors and environmental effects. You are now ready to examine the more advanced operations of vectors, like the dot product, which determines the relationship of two given vectors.

Before you continue, it is important to remember that the applications of vector related concepts go far beyond interactive graphical applications like video games. In fact, in many cases it is impossible to depict or visualize the vectors being used in different applications. For example, a vector in n-dimensional space where n is significantly large than 100! It is important to remember that you are learning one flavor of vector usage: applications in interactive graphics. In general, vectors can be applied to solve problems in a wide variety of disciplines.

CHAPTER 5

Vector Dot Products

After completing this chapter, you will be able to

- Understand the vector dot product definition, its properties, and its geometric interpretation
- Recognize how the vector dot product relates two vectors by their subtended angle and relative projection sizes
- Comprehend how a vector represents a line segment
- Apply the dot product to allow the interpretation of a line segment as an interval
- Perform the simple inside-outside test for a point and an arbitrary interval
- Apply the vector dot product to determine the shortest distance between a point and a line
- Apply the vector dot product to compute the closest distance between two lines

Introduction

In Chapter 4 you learned that a vector is defined by the relationships between two positions in the Cartesian Coordinate System: the direction from one position to another and the distance between them. Though simple, the vector, or the concept and the associated rules of relating two positions, is demonstrated to be a powerful tool that is capable of representing object velocity and simple environmental effects for video game development. This chapter continues with this theme and introduces the vector dot product to relate two vectors.

K. Sung and G. Smith, *Basic Math for Game Development with Unity 3D*,
https://doi.org/10.1007/978-1-4842-9885-5_5

Vectors are defined by their direction and magnitude, and thus when relating two vectors, it is essential to include descriptions of how these two quantities are measured with respect to each other. The vector dot product relates vector directions by calculating the cosine of the subtended angle, or the angle between two vectors where their tails are connected, and the vectors' magnitudes by computing the respective projected sizes, or one vector's magnitude when measured along the direction of the other vector. These ways of relating vectors are some of the most fundamental tools in analyzing the proximity and connections between positions and directions in 3D space. The results of applying the vector dot product provide the basis for predicting and controlling object behaviors in almost all video games.

In video games it is often necessary to analyze the spatial relationships, such as distances and intersections, of traveling objects and then predicting what events will occur. For example, detecting and hinting to the player the situation where the pathway of their explorer will pass within a hidden treasure's proximity. To model this situation mathematically, as you have learned from the previous chapters, the pathway of the explorer is a function of their traveling velocity and can be represented as a vector. Then, the hidden treasure can be wrapped by a bounding volume, that is, bounding sphere. In this way, the problem to solve is to compute the closest distance between the vector and bounding sphere center and determine if that distance is closer than the bounding sphere radius. As you will learn from this chapter, the vector dot product can provide a solution for this situation that is elegant and straightforward to implement. In fact, the vector dot product is the best tool for determining distances between positions and line segments.

This chapter begins by introducing the vector dot product, what it is, how it is computed, and the rules for working with the operation. The chapter then moves on to explain how to geometrically interpret the dot product results as the angle between vectors and as projected lengths along these vectors. The inside-outside test of a 1D interval along a major axis discussion from Chapter 2 is then cast and generalized as an inside-outside problem based on vector line segments and projections. The two application areas of the vector dot product that are examined specifically are the line to point and the line to line distances. These types of applications play many roles in video game development as well as other interactive graphical applications. Finally, this chapter concludes by reviewing what you have learned about the vector dot product and its many applications.

Vector Dot Product: Relating Two Vectors

Recall that the vector definition is independent of any position. In other words, a vector can have its tail located at any position. This knowledge is important because when you analyze the relationship between two vectors, it is convenient to depict the tails of the vectors at the same location. Figure 5-1 shows a drawing of two arbitrary vectors, $\vec{V}_1$ and $\vec{V}_2$, with the same tail position, P_0. As you can see, the shared tail position allows the two vectors to be in close proximity and facilitates convenient visual comparison. By placing two vectors at the same location, it becomes easier to analyze, understand, and quantify the relationship between them.

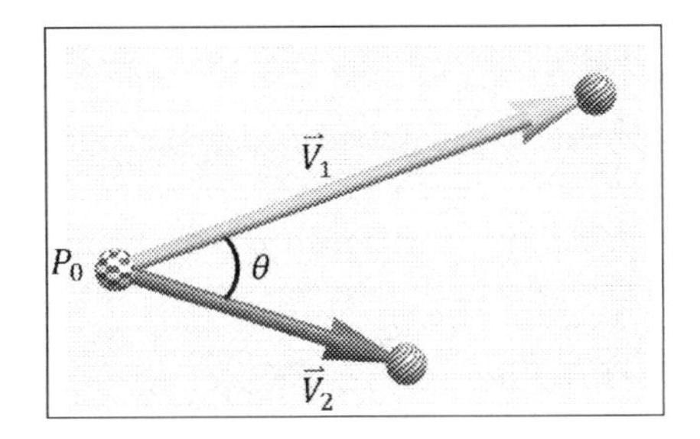

***Figure 5-1.** Relationship between two given vectors*

Notice in Figure 5-1 that although the two vectors could be in any direction with any magnitude in 3D Cartesian Coordinate Space, the two vectors together can always be properly depicted on a 2D plane. In fact, the 2D plane that these vectors are depicted on may or may not be parallel to any major axes. In general, it is true that given any two arbitrary vectors in 3D space, as long as the two vectors are not parallel, there is always a 2D plane where both of the vectors in 3D space can be drawn. This observation is what allows two vectors in 3D space to be drawn and analyzed on a 2D plane, as depicted in Figure 5-1.

The second observation from Figure 5-1 is that, recalling that a vector is comprised of a direction and a magnitude, that the relationship between two vectors can be characterized by the angle between the vectors, θ, and by the relative sizes of the two vectors. The vector dot product is the operation that can provide definitive answers to both of these characteristics.

Definition of Vector Dot Product

Given two vectors in 3D space

$$\vec{V}_1 = (x_1, y_1, z_1)$$

$$\vec{V}_2 = (x_2, y_2, z_2)$$

the **dot product**, or vector dot product, between the two vectors is defined as the sum of the product of the corresponding coordinate components, or

$$\vec{V}_1 \cdot \vec{V}_2 = x_1 x_2 + y_1 y_2 + z_1 z_2$$

Notice that

- Symbol: The symbol for the dot product operation, "·", is literally a "dot".
- Operands: The operation expects two vector operands.
- Result: The result of the operation is a floating-point number.

It is especially important to pay attention to the last point. Similar to vector addition and subtraction, the dot product operates on two vector operands. However, unlike the other two operations, the result of the dot product is not a vector but a simple floating-point number. It is this floating-point number that encodes the angle between the two operand vectors and the relative sizes of the two operand vectors. How these values are encoded in this single floating-point number and what you can do with it are the topics that will be explored in the following subsections. However, before you begin that journey, you will first need to explore and understand the rules and properties of the dot product.

Note The dot product is also referred to as the **inner product** or the **scalar product** in different disciplines of mathematics. This book will refer to the operation as dot product exclusively.

Do not confuse the dot product symbol, "·", for a multiplication sign. When multiplications are involved in vector expressions, there will be no symbol between the operands, such as $s\vec{V}_2$. Since one cannot multiply two vectors, you will never see $\vec{V}_1\vec{V}_2$, and therefore you can safely assume that if you see a "·" between two vectors, the dot product is the operation to perform and not multiplication.

Properties of Vector Dot Product

The vector dot product properties of commutative, associative, and distributive over a floating-point scaling factor *s* and other vector operations are summarized in Table 5-1.

Table 5-1. *Properties of vector dot product*

Properties	**Vector Dot Product**
Commutative	$\vec{V}_1 \cdot \vec{V}_2 = \vec{V}_2 \cdot \vec{V}_1$
Associative	$\left(\vec{V}_1 \cdot \vec{V}_2\right) \cdot \vec{V}_3$ **[Undefined!]**
Distributive over vector operation	$\left(\vec{V}_1 \cdot \vec{V}_2\right)\left(\vec{V}_A + \vec{V}_B\right) = \left(\vec{V}_1 \cdot \vec{V}_2\right)\vec{V}_A + \left(\vec{V}_1 \cdot \vec{V}_2\right)\vec{V}_B$
Distributive over scale factor, *s*	$s\left(\vec{V}_1 \cdot \vec{V}_2\right) = \left(s\vec{V}_1\right) \cdot \vec{V}_2 = \vec{V}_1 \cdot \left(s\vec{V}_2\right)$

Take note of the undefined associative property. In this situation, it can be helpful to remember that the result of the dot product operation is a floating-point number, so it is possible to let

$$\left(\vec{V}_1 \cdot \vec{V}_2\right) = f$$

then it becomes obvious that

$$\left(\vec{V}_1 \cdot \vec{V}_2\right) \cdot \vec{V}_3 = f \cdot \vec{V}_3$$

is an undefined operation since the first operand is not a vector but a floating-point number. In general, please pay attention to the subtle differences in the notation. While

$$\left(\vec{V}_1 \cdot \vec{V}_2\right)\vec{V}_A = f\vec{V}_A$$

is scaling vector $\vec{V}_A$ by the result of the dot product,

$$\left(\vec{V}_1 \cdot \vec{V}_2\right) \cdot \vec{V}_A = f \cdot \vec{V}_A$$

is attempting to perform a dot product between a floating-point number, *f*, and the vector, $\vec{V}_A$, and is therefore an undefined operation. The only difference is in the single "·" symbol! If you continue to use *f* to represent the result of $\vec{V}_1$ dot $\vec{V}_2$, then you can rewrite the distributive property over vector addition as

$$\left(\vec{V}_1 \cdot \vec{V}_2\right)\left(\vec{V}_A + \vec{V}_B\right) = f\left(\vec{V}_A + \vec{V}_B\right) = f\vec{V}_A + f\vec{V}_B$$

which is the distributive property of vector addition over a scaling factor, *f*. This means that the distributive property also applies over vector subtraction

$$\left(\vec{V}_1 \cdot \vec{V}_2\right)\left(\vec{V}_A - \vec{V}_B\right) = \left(\vec{V}_1 \cdot \vec{V}_2\right)\vec{V}_A - \left(\vec{V}_1 \cdot \vec{V}_2\right)\vec{V}_B$$

or

$$\left(\vec{V}_1 \cdot \vec{V}_2\right)\left(\vec{V}_A - \vec{V}_B\right) = f\left(\vec{V}_A - \vec{V}_B\right) = f\vec{V}_A - f\vec{V}_B$$

The vector dot product distributive property over a scale factor, *s*, is worth some special attention. Notice that the scale factor *s* is only applied to one of the operands and not both. At first glance, this may seem counterintuitive; however, it makes perfect sense if you consider distributive property over a scale factor, *s*, of a floating-point multiplication between *a* and *b*

$$s \times (a \times b) = (s \times a) \times b = a \times (s \times b)$$

Now, recall that the magnitude of a vector, $\vec{V} = (x,\ y,\ z)$, is

$$\left\|\vec{V}\right\| = \sqrt{x^2 + y^2 + z^2}$$

For this reason, a vector dotted with itself is its magnitude squared

$$\vec{V}_1 \cdot \vec{V}_1 = x_1 x_1 + y_1 y_1 + z_1 z_1 = x_1^{\ 2} + y_1^{\ 2} + z_1^{\ 2} = \left\|\vec{V}_1\right\|^2$$

Lastly, the dot product between any vector with the zero vector always results in a zero vector

$$\vec{V}_1 \cdot ZeroVector = ZeroVector \cdot \vec{V}_1 = ZeroVector$$

The Angle Between Two Vectors

This section derives a formula that computes the angle θ between the vectors $\vec{V}_1$ and $\vec{V}_2$ in Figure 5-1. As illustrated in Figure 5-2, this formula derivation begins by subtracting the two given vectors

$$\vec{V}_3 = \vec{V}_1 - \vec{V}_2$$

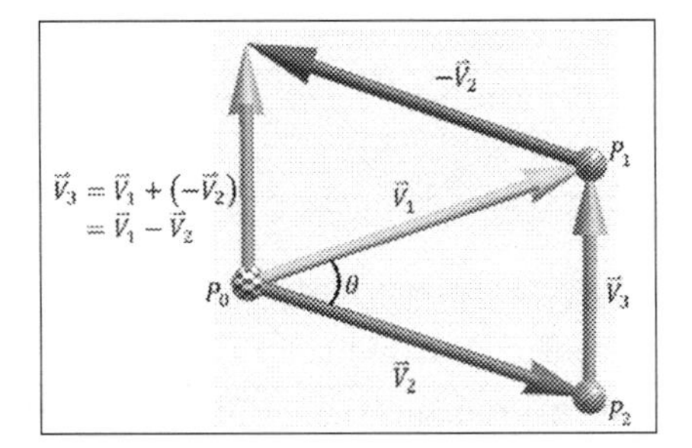

Figure 5-2. *Subtracting the given two vectors*

In Figure 5-2, similar to Figure 5-1, both $\vec{V}_1$ and $\vec{V}_2$ have their tails located in the lower-left corner at position P_0. Notice the $-\vec{V}_2$ vector with its tail at position P_1 and that the vector $\vec{V}_3$ with its tail at P_0 is the result of adding $\vec{V}_1$ with $-\vec{V}_2$, or

$$\vec{V}_3 = \vec{V}_1 + \left(-\vec{V}_2\right) = \vec{V}_1 - \vec{V}_2$$

Figure 5-2 also depicts vector $\vec{V}_3$ with its tail at P_2 to create triangle $P_0P_1P_2$. Recall that the Laws of Cosine from trigonometry states that

$$\left\|\vec{V}_3\right\|^2 = \left\|\vec{V}_1\right\|^2 + \left\|\vec{V}\right\|_2^2 - 2\left\|\vec{V}_1\right\|\left\|\vec{V}_2\right\|\cos\theta$$

In this case, you know that

$$\vec{V}_1 = \left(x_1, y_1, z_1\right)$$

$$\vec{V}_2 = \left(x_2, y_2, z_2\right)$$

$$\vec{V}_3 = \left(x_1 - x_2, y_1 - y_2, z_1 - z_2\right)$$

With algebraic simplification left as an exercise, you can show that

$$\cos\theta = \frac{x_1x_2 + y_1y_2 + z_1z_2}{\sqrt{x_1^2 + y_1^2 + z_1^2}\sqrt{x_2^2 + y_2^2 + z_2^2}}$$

this equation says that

$$x_1x_2 + y_1y_2 + z_1z_2 = \sqrt{x_1^2 + y_1^2 + z_1^2}\sqrt{x_2^2 + y_2^2 + z_2^2}\cos\theta = \left\|\vec{V}_1\right\|\left\|\vec{V}_2\right\|\cos\theta$$

or simply

$$\vec{V}_1 \cdot \vec{V}_2 = x_1x_2 + y_1y_2 + z_1z_2 = \left\|\vec{V}_1\right\|\left\|\vec{V}_2\right\|\cos\theta$$

You have just shown that the dot product definition, the sum of the products of the corresponding coordinate components, actually computes a floating-point number that is equal to the product of the magnitude of the two vectors and the cosine of the angle between these two vectors. By normalizing $\vec{V}_1$ and $\vec{V}_2$, $\left\|\vec{V}_1\right\|$ and $\left\|\vec{V}_2\right\|$ both become 1.0, so that

$$\hat{V}_1 \cdot \hat{V}_2 = \left\|\hat{V}_1\right\|\left\|\hat{V}_2\right\|\cos\theta = \cos\theta$$

This formula says that the dot product of two normalized vectors is the cosine of the angle between the vectors.

It is important to note that the angle between two vectors is the one subtended by the two vectors (the smaller angle). As illustrated in Figure 5-3, if θ in Figure 5-1 was 45°, then the angle between the two vectors is 45° and not 315°. The key to remember is that the angle subtended by two vectors, or two lines, is always between 0° and 180°.

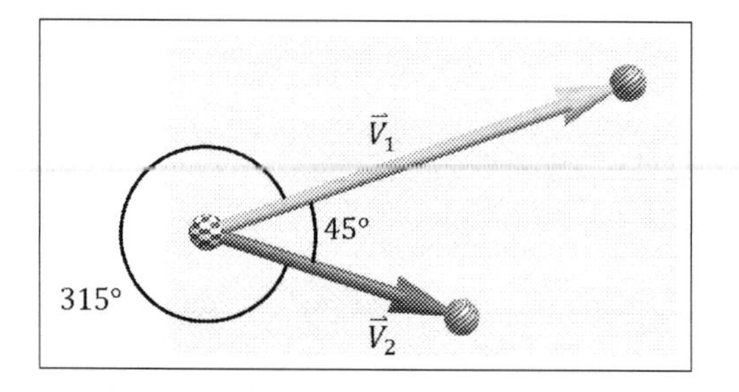

Figure 5-3. *The angle subtended by vectors $\vec{V}_1$ and $\vec{V}_2$*

Figure 5-4 depicts the angle measurements θ_2 to θ_6 between vector $\vec{V}_1$ and vectors $\vec{V}_2$ to $\vec{V}_6$, respectively. In this case, $\vec{V}_3$ is perpendicular to $\vec{V}_1$ and $\vec{V}_5$ is in the opposite direction to $\vec{V}_1$; thus $\theta_3 = 90°$, while $\theta_5 = 180°$. Notice the measurement of the angle θ_6, the angle between vectors $\vec{V}_1$ and $\vec{V}_6$, is the angle subtended by these two vectors and is not an accumulation from the angle θ_5. Once again and very importantly, the angle subtended by two vectors is always an angle between 0° and 180°.

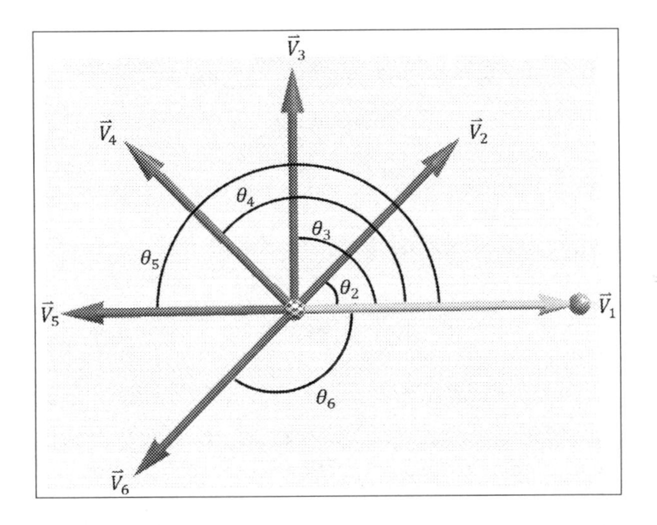

***Figure 5-4.** The angles between vectors*

Figure 5-5 is a simple plot and a reminder of the cosine function. Recall that the results of $\cos\theta$ are positive between 0° and 90° and become negative between 90° and 180°. With the dot product of two normalized vectors being the cosine of the subtended angle, you can now determine the relative directions of vectors with a simple dot product calculation. In particular, when the subtended angle is less than 90°, the cosine is positive, and thus you can conclude that the vectors are pointing along a similar direction. Conversely, when the subtended angle is more than 90°, the cosine is negative, and thus you can conclude that the vectors are pointing away from each other.

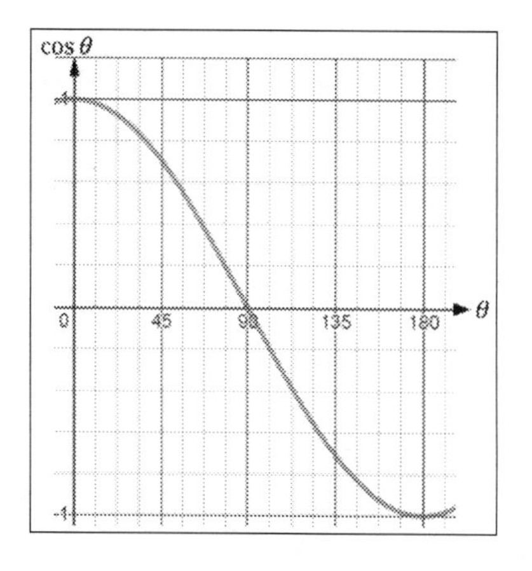

Figure 5-5. *Simple plot of the y = cos θ function*

In the cases of Figure 5-4, you know

$$\hat{V}_1 \cdot \hat{V}_2 = \cos\theta_2 = \text{a positive number} \quad \text{because } \theta_2 < 90°$$

$$\hat{V}_1 \cdot \hat{V}_3 = \cos\theta_3 = 0 \quad \text{because } \theta_3 = 90°$$

$$\hat{V}_1 \cdot \hat{V}_4 = \cos\theta_4 = \text{a negative number} \quad \text{because } \theta_4 > 90°$$

$$\hat{V}_1 \cdot \hat{V}_5 = \cos\theta_5 = -1 \quad \text{because } \theta_5 = 180°$$

$$\hat{V}_1 \cdot \hat{V}_6 = \cos\theta_6 = \text{a negative number} \quad \text{because } \theta_6 > 90°$$

These observations can be summarized in Table 5-2 for any given vectors, $\vec{V}_1$ and $\vec{V}_2$.

Table 5-2. *Dot product results and subtended angles*

Dot Product Results	Subtended Angle θ	Conclusions
$\hat{V}_1 \cdot \hat{V}_2 = \cos\theta = 1$	$\theta = 0°$	The vectors are in the exact same direction, $\hat{V}_1 = \hat{V}_2$
$\hat{V}_1 \cdot \hat{V}_2 = \cos\theta = 0$	$\theta = 90°$	The vector directions are perpendicular to each other
$\hat{V}_1 \cdot \hat{V}_2 = \cos\theta > 0$	$\theta < 90°$	The vectors are pointing along similar directions
$\hat{V}_1 \cdot \hat{V}_2 = \cos\theta < 0$	$\theta > 90°$	The vectors are pointing along similar, but opposite directions
$\hat{V}_1 \cdot \hat{V}_2 = \cos\theta = -1$	$\theta = 180°$	The vectors are in the exact opposite direction, $\hat{V}_1 = -\hat{V}_2$

The Angle Between Vectors Example

This example allows you to manipulate three positions that define two vectors. The example computes and displays the angle between the two vectors and enables you to verify the conclusions gathered from Table 5-2. Additionally, this example demonstrates that as long as the two given vectors are not parallel, a 2D plane can always be found for drawing the two vectors. Figure 5-6 shows a screenshot of running the `EX_5_1_AngleBetweenVectors` example from the `Chapter-5-DotProducts` project.

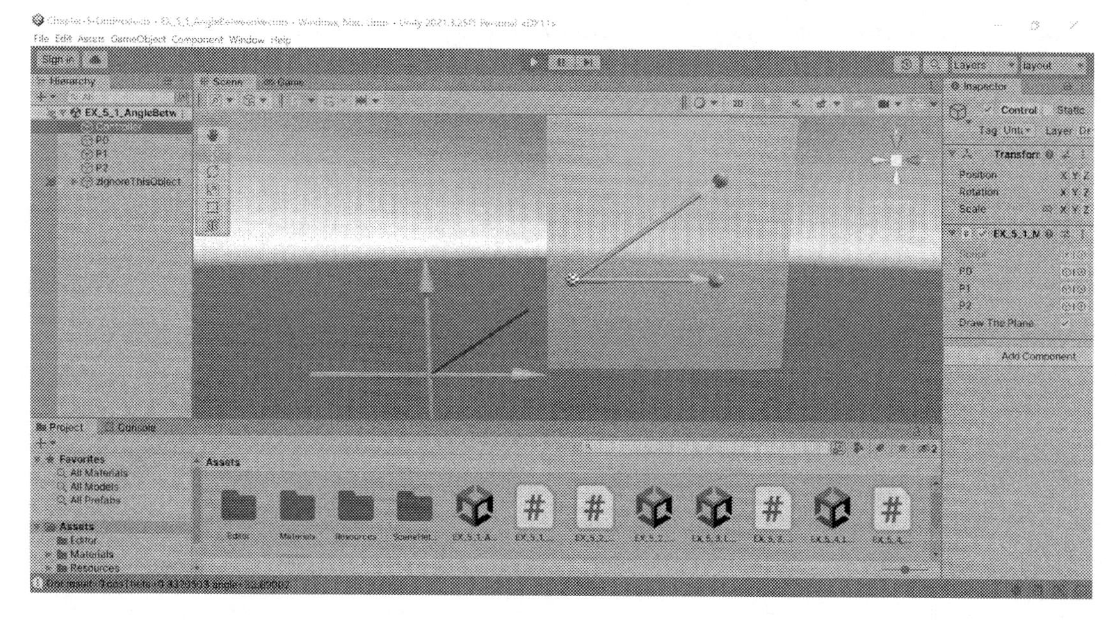

Figure 5-6. Running the Angle Between Vectors example

The goals of this example are for you to

- Experience manipulating the angle subtended by two vectors and observe the results of the dot product
- Verify that a 2D plane can always be found for drawing two non-parallel vectors
- Examine the implementation of and appreciate the subtleties of vector normalization when computing dot products

Examine the Scene

Take a look at the `Example_5_1_AngleBetweenVectors` scene and observe the predefined game objects in the Hierarchy Window. In addition to the `Controller`, there are three objects in this scene: the checkered sphere (`P0`) and the stripped spheres (`P1` and `P2`). These three game objects, with their corresponding `transform.localPosition`, will be referenced to define the two vectors for performing dot product calculations.

Analyze Controller MyScript Component

The `MyScript` component on the `Controller` shows four variables: `P0`, `P1`, `P2`, and `DrawThePlane` toggle. The toggle is for showing or hiding the 2D plane where the two vectors are drawn, while the other three variables are defined for accessing the game objects with their corresponding names. In this example, you will manipulate the positions of the three game objects and examine the dot product resulting from the vectors, $\vec{V_1}$ and $\vec{V_2}$, defined accordingly

$$\vec{V_1} = P_1 - P_0$$

$$\vec{V_2} = P_2 - P_0$$

Interact with the Example

Click the Play Button to run the example. In the Scene View window, you will observe two vectors with tail positions located at the checkered sphere, `P0`, and a greenish plane where the two vectors are drawn. The two vectors are the $\vec{V_1}$ and $\vec{V_2}$ and are defined by the positions of `P0`, `P1`, and `P2` game objects. Also visible in the Scene View window is the 2D axis frame with the red X-axis and green Y-axis vectors. On the axis frame, extending from the origin is a black line segment. The angle subtended by this black line segment and the red X-axis is the same angle subtended by vectors $\vec{V_1}$ and $\vec{V_2}$, and the length of this black line is proportional to the cosine of that angle, scaled by 1.5 times for easier visual analysis. Lastly, take a look at the Console Window to observe the text output reporting the computed angle between vectors $\vec{V_1}$ and $\vec{V_2}$.

Now that you have looked over the scene, you will manipulate and observe the cosine of the angle subtended by the two vectors and notice how the angle itself changes. Please switch off the `DrawThePlane` toggle as the 2D plane can be distracting. Next, select `P1` and change its x- and y-component values to vary the angle between the two vectors. In the Console Window, you can verify the values of the subtended angle and the cosine of this angle. Observe how the black line segment, with its length changing, rotates toward or away from the red X-axis direction, corresponding to the angle changes you are making.

Since the length of the black line segment is proportional to the cosine of the subtended angle, from the plot in Figure 5-5 and Table 5-2, you can verify that when the subtended angle increases, up to 90°, the cosine of the angle decreases and thus the length of the line also decreases. The opposite is also true, as the angle decreases (between 90° and 0°), the cosine of the angle, and thus the length of the black line, increases. In fact, you should notice that the length of the black line is maximized when the subtended angle approaches zero and that the length of the line approaches zero when the two vectors are approximately perpendicular. You can observe this behavior by decreasing the `P1` x-component value such that the subtended angle approaches 90°. When doing so, notice how the length of black line segment also approaches zero, corresponding to $\cos 90° = 0$.

When you increase the subtended angle beyond 90°, you will notice the color of the black line segment changes to red, indicating that the sign of the dot product result has turned into a negative number. Now, decrease the y-component value of `P1` to continue to increase the subtended angle and notice that the red line segment continues to grow in length once more as it rotates away from the positive X-axis direction. When $\vec{V}_1$ and $\vec{V}_2$ are approximately in the opposite direction, the red line segment will achieve maximum length and should be on top of the negative red X-axis line, indicating the angle between the two vectors is about 180° and that $\cos 180° = -1$. Now, notice that any attempt to increase the subtended angle beyond 180° will cause the red line segment to rotate back toward the positive X-axis direction. This is similar to cases of vectors that are between $\vec{V}_5$ and $\vec{V}_6$ in Figure 5-4. This exercise is to reaffirm that subtended angles are always between 0° and 180° and to visually demonstrate what Table 5-2 showcases.

Next, you will verify that a 2D plane can always be defined to draw two vectors that are not parallel. Please switch on the `DrawThePlane` toggle and rotate the camera to see that the two vectors are indeed drawn on the greenish plane by examining that the plane slices through the two arrows representing $\vec{V}_1$ and $\vec{V}_2$, respectively. You can manipulate any of the `P0`, `P1`, or `P2` positions to observe the vectors change accordingly and more importantly observe that the green plane also changes accordingly: it always cuts through both vectors. Now, adjust `P1` to the exact location of `P2`. One way you can do this is by copying the values from `P2`'s transform components in the Inspector Window and pasting them onto that of `P1`'s corresponding transform components. Once done, notice that the 2D plane disappears. In this case, since the two vectors are pointing in the exact same direction, there are an infinite number of 2D planes that can cut through the vectors and thus none are shown.

Details of MyScript

Open MyScript and examine the source code in the IDE. The instance variables are as follows:

```
// Three positions to define two vectors: P0->P1 and P0->P2
public GameObject P0 = null;   // Position P0
public GameObject P1 = null;   // Position P1
public GameObject P2 = null;   // Position P2

public bool DrawThePlane = true;

#region For visualizing the vectors
#endregion
```

All the public variables for MyScript have been discussed when analyzing the Controller's MyScript component. The code in the "For visualizing the vectors" region is specific to drawing the vectors and as usual does not pertain to the math being discussed in this section.

Note By now, you have observed and may even have worked with some of the visualization code. From here on, the visualization portion of MyScript will become increasingly complex and involved. To avoid unnecessary distractions, beginning from this example, the code for visualization will be separated into collapsed hidden regions. The details of these regions will not be explained or brought up as they can be tedious and in all cases are irrelevant to the concepts being discussed. You are very welcome to explore these at your leisure.

The Start() function for MyScript is listed as follows:

```
void Start() {
    Debug.Assert(P0 != null);   // Ensure proper init
    Debug.Assert(P1 != null);
    Debug.Assert(P2 != null);

    #region For visualizing the vectors
    #endregion
}
```

As in all previous examples, the Debug.Assert() calls ensure proper setup regarding referencing the appropriate game objects via the Inspector Window. The region "For visualizing the vectors," which contains the details of initializing the visualization variables for the vectors in the scene, is once again irrelevant to the math being discussed and can be distracting. Therefore, this region will not be discussed. The Update() function is listed as follows:

```
void Update() {
    float cosTheta = float.NaN;
    float theta = float.NaN;
    Vector3 v1 = P1.transform.localPosition -
                 P0.transform.localPosition;
    Vector3 v2 = P2.transform.localPosition -
                 P0.transform.localPosition;
    float dot = Vector3.Dot(v1, v2);
    if ((v1.magnitude > float.Epsilon) &&
        (v2.magnitude > float.Epsilon)) {
        cosTheta = dot / (v1.magnitude * v2.magnitude);
        // Alternatively,
        // costTheta = Vector3.Dot(v1.normalize, v2.normalize)
        theta = Mathf.Acos(cosTheta) * Mathf.Rad2Deg;
    }
    Debug.Log("Dot result=" + dot +
             " cosTheta=" + cosTheta + " angle=" + theta);
    #region  For visualizing the vectors
    #endregion
```

The first three lines of the Update() function compute

$$\vec{V}_1 = P_1 - P_0$$

$$\vec{V}_2 = P_2 - P_0$$

$$dot = \vec{V}_1 \cdot \vec{V}_2$$

The `if` condition ensures that neither of the vectors are the zero vector, which as you have learned does not have a length, cannot be normalized, and thus, cannot subtend angles. When both of the vectors are properly defined, the cosine of the angle between them can be computed by recognizing that

$$dot = \vec{V}_1 \cdot \vec{V}_2 = \|\vec{V}_1\|\|\vec{V}_2\|\cos\theta$$

which means that the cosine of the subtended angle is simply

$$\cos\theta = \frac{dot}{\|\vec{V}_1\|\|\vec{V}_2\|}$$

Finally, theta, the subtended angle value, can be derived by the arccosine function. Note that alternatively, $\cos\theta$ can be computed by performing the dot operation with the normalized version of the two vectors. The dot products between vectors that are normalized will be examined in more detail in the following sections.

Takeaway from This Example

This example verifies that when given two non-parallel vectors, a 2D plane can always be derived to draw the two vectors. This fact allows the examination of the two arbitrary vectors, which may not be aligned with any major axes, to be drawn, examined, and analyzed. You have interacted with and closely examined the angle subtended by two vectors and that this angle is always between 0° and 180°. Finally, you have observed that the cosine of a subtended angle can be computed by dividing the dot product of the two vectors with their magnitudes or, alternatively, from the dot product of the two vectors after they have been normalized.

$$\cos\theta = \frac{dot}{\|\vec{V}_1\|\|\vec{V}_2\|} = \hat{V}_1 \cdot \hat{V}_2$$

Relevant mathematical concepts covered include

- The dot product of normalized vectors is the cosine of their subtended angle.
- The value of the dot product provides insights into the relative directions of the operand vectors (see Table 5-2).
- A unique 2D plane can be derived from two non-parallel vectors such that both vectors can be drawn on the plane.

Unity tools

- The `Mathf` library can be used for mathematical functions.
- `Rad2Deg`: The scale factor for radian to degree conversion.
- `Acos` can be used to compute arccosine.
- The `Mathf.Acos` function returns the angle in units of radian and not degree.

EXERCISES

Derive the Dot Product Formula

Given that

$$\|\vec{V}_3\|^2 = \|\vec{V}_1\|^2 + \|\vec{V}_2\|^2 - 2\|\vec{V}_1\|\|\vec{V}_2\|\cos\theta$$

and that

$$\vec{V}_1 = (x_1, y_1, z_1)$$

$$\vec{V}_2 = (x_2, y_2, z_2)$$

$$\vec{V}_3 = (x_1 - x_2, y_1 - y_2, z_1 - z_2)$$

show that

$$\cos\theta = \frac{x_1x_2 + y_1y_2 + z_1z_2}{\sqrt{x_1^2 + y_1^2 + z_1^2}\sqrt{x_2^2 + y_2^2 + z_2^2}}$$

Verify the Need for Normalization

When computing theta in `MyScript`,

```
cosTheta = dot / (v1.magnitude * v2.magnitude);
theta = Mathf.Acos(cosTheta) * Mathf.Rad2Deg;
```

replace these two lines of code with the non-normalized vectors' version

```
theta = Mathf.Acos(Vector3.Dot(v1, v2)) * Mathf.Rad2Deg;
```

Try running your game and observe the error messages. Now, include proper normalization

```
theta = Mathf.Acos(Vector3.Dot(v1.normalize, v2.normalize)) *
        Mathf.Rad2Deg;
```

Try running this latest version and observe the same results as the original. This simple exercise shows that it is vital to normalize your vectors when computing the angle between them.

Verify the Dot Product Formula

When computing theta in `MyScript`,

```
cosTheta = dot / (v1.magnitude * v2.magnitude);
```

replace this line of code with the explicit dot product computation

```
cosTheta = (v1.x*v2.x + v1.y*v2.y + v1.z*v2.z) /
           (v1.magnitude * v2.magnitude);
```

Verify that runtime results are identical.

Vector Projections

You have learned that the dot product between two vectors, $\vec{V}_1$ and $\vec{V}_2$, computes the product of the vector magnitudes and the cosine of the angle subtended by the two vectors

$$\vec{V}_1 \cdot \vec{V}_2 = \left\|\vec{V}_1\right\| \left\|\vec{V}_2\right\| \cos\theta$$

In the previous example, you have verified that by normalizing both of the vectors beforehand, ensuring that

$$\left\|\hat{V}_1\right\| = \left\|\hat{V}_2\right\| = 1.0$$

the dot product now simply computes the cosine of the angle between the given vectors

$$\hat{V}_1 \cdot \hat{V}_2 = \cos\theta$$

Now, you can examine the two remaining ways of computing the dot product between two given vectors—with only one of the vectors being normalized or

$$\hat{V}_1 \cdot \vec{V}_2 = \left\|\vec{V}_2\right\| \cos\theta$$

$$\vec{V}_1 \cdot \hat{V}_2 = \left\|\vec{V}_1\right\| \cos\theta$$

Figure 5-7 depicts the geometric interpretation of these two dot product computations.

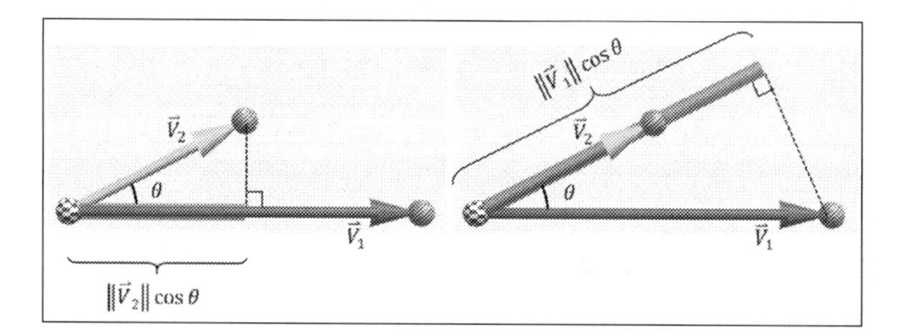

Figure 5-7. *Computing dot products between two vectors with one being normalized*

It is important to note that the left and right images of Figure 5-7 are both based on exactly the same two vectors, $\vec{V}_1$ and $\vec{V}_2$. The left image of Figure 5-7 shows that with vector $\hat{V}_1$ normalized, the dot product computed

$$\hat{V}_1 \cdot \vec{V}_2 = \left\|\vec{V}_2\right\| \cos\theta$$

is the length of $\vec{V}_2$ when measured along direction of the $\vec{V}_1$ vector. This is also referred to as the projected length of $\vec{V}_2$ on the $\vec{V}_1$ vector. Notice that with the tails of the two vectors located at the same position, the head of $\vec{V}_2$ is projected perpendicular to and onto the $\vec{V}_1$ vector, as evident by the dotted line with the right-angle indicator. This projected length can also be interpreted through trigonometry. You can treat $\vec{V}_2$ as the hypotenuse that subtends the angle, θ, with the base direction being $\hat{V}_1$ and the last side being the dotted line, thus forming a right-angle triangle. In this case, the length of the base of the right-angle triangle is $\left\|\vec{V}_2\right\|\cos\theta$.

The right image of Figure 5-7 shows the same two vectors, $\vec{V}_1$ and $\vec{V}_2$, where the dot product is computed with vector $\hat{V}_2$ being normalized instead of $\hat{V}_1$

$$\vec{V}_1 \cdot \hat{V}_2 = \left\|\vec{V}_1\right\|\cos\theta$$

In this case, the dot product computes the exact complement of the previous case—the length of $\vec{V}_1$ when measured along the direction of $\vec{V}_2$, or the projected length of $\vec{V}_1$ on the $\vec{V}_2$ vector, or the length of the base of the right-angle triangle that is in the $\vec{V}_2$ direction and subtends an angle, θ, with the hypotenuse $\vec{V}_1$, and the dotted line as its final side. The right image of Figure 5-7 also illustrates a case where the length of the base of the right-angle triangle extends beyond the head of the vector $\vec{V}_2$. This example shows that the projected size can be larger than the magnitude of the vector that it is being projected onto or

$$\left\|\vec{V}_1\right\|\cos\theta > \left\|\vec{V}_2\right\|$$

Finally, remember that $\cos\theta$ is negative for $\theta > 90°$, and therefore, $\left\|\vec{V}\right\|\cos\theta$, or the projected size of a vector can actually be a negative value. In such cases, you know that the vector being projected onto is more than 90° away from the vector being projected. This turns out to be important information with many applications, some of which will be elaborated in later subsections.

This discussion shows that with the appropriate vector normalized, the dot product can compute the projection of the length of one vector onto the direction of the other vector and can provide a way to relate the lengths of these two vectors. In other words, the dot product allows you the capability to project one vector onto another. Observe that the normalized operand is the vector being projected onto. These projections are examined in the next example. The actual applications of the vector dot product will be discussed after the next section.

The Vector Projections Example

This example allows you to interact with and examine the results of vector projections. You will manipulate the definition of two vectors and examine the results of projecting these two vectors onto each other. Figure 5-8 shows a screenshot of running the `EX_5_2_VectorProjections` scene from the `Chapter-5-DotProducts` project.

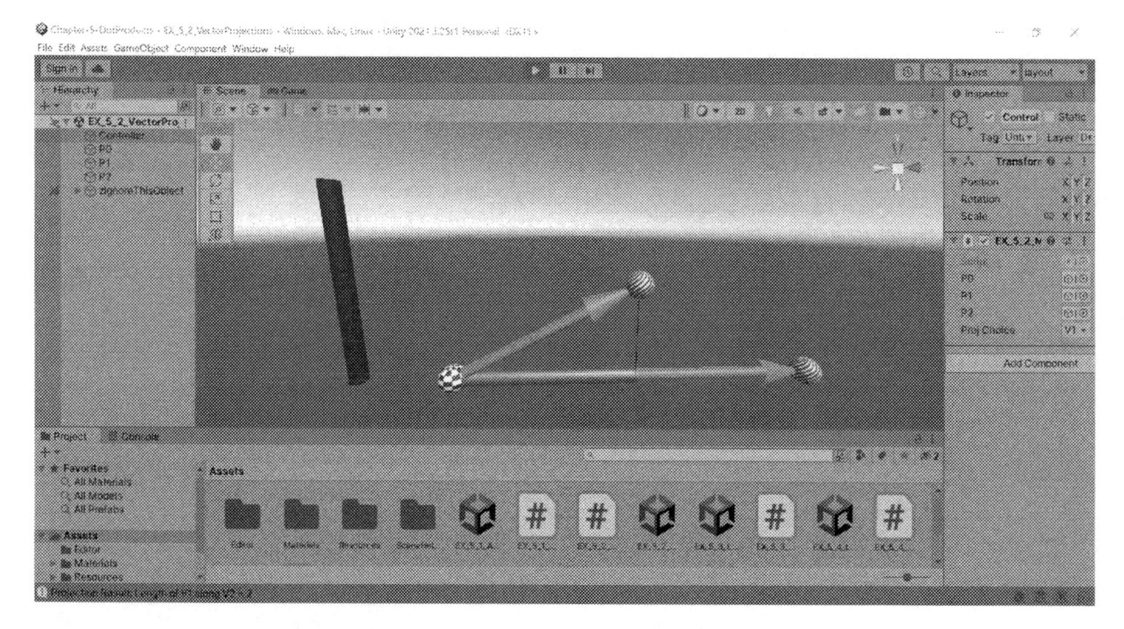

Figure 5-8. *Running the Vector Projections example*

The goals of this example are for you to

- Appreciate the results of normalizing one of the vectors in the dot product operation
- Experience and understand the results of projecting vectors onto each other
- Observe and interact with negative projected distances
- Examine the code that performs vector projection

Examine the Scene

Take a look at the `Example_5_2_VectorProjections` scene and observe the predefined game objects in the Hierarchy Window. In addition to the `Controller`, there are three objects in this scene: `P0`, `P1`, and `P2`. As with the previous example in this chapter, you will manipulate the positions of the three game objects to define two vectors, $\vec{V}_1$ and $\vec{V}_2$,

$$\vec{V}_1 = P_1 - P_0$$

$$\vec{V}_2 = P_2 - P_0$$

and examine the results of projecting these two vectors onto each other.

Analyze Controller MyScript Component

The `MyScript` component on the `Controller` shows the references to the three game objects: the checkered sphere `P0`; two stripped spheres, `P1` and `P2`; and a drop-down menu, `ProjectionChoice`. The drop-down menu allows the following options:

- V1 onto V2: Project $\vec{V}_1$ vector onto $\vec{V}_2$
- V2 onto V1: Project $\vec{V}_2$ vector onto $\vec{V}_1$
- Projection off: Do not perform any projection

Interact with the Example

Click the Play Button to run the example. Take note that by default, the `ProjectionChoice` is set to `V1OntoV2`, and therefore, `MyScript` is computing and displaying the results of projecting $\vec{V}_1$ onto $\vec{V}_2$.

Observe the two vectors, $\vec{V}_1$ and $\vec{V}_2$, that are defined by three positions. $\vec{V}_1$ is cyan and initially is above $\vec{V}_2$, which is magenta. Notice a light, semi-transparent cylinder along the $\vec{V}_2$ vector that is connected with a thin black line to the head of $\vec{V}_1$. The thin black line depicts the projection from the head of $\vec{V}_1$ perpendicularly onto $\vec{V}_2$, where the line intersects $\vec{V}_2$. The semi-transparent cylinder on $\vec{V}_2$ shows the projected length of $\vec{V}_1$ on $\vec{V}_2$. To emphasize the fact that the result of a dot product, or the projected length in this case, is just a floating-point number, this value is used to scale the height of the black bar to the side of the checkered sphere (`P0`). The length of the black bar is always the same as the semi-transparent cylinder. This length is the result of the calculated dot product, and in this scenario is

$$v1LengthOnV2 = \vec{V}_1 \cdot \hat{V}_2 = \left\|\vec{V}_1\right\| \cos\theta$$

Now, select `P1` and manipulate its x-component to change the length of $\vec{V}_1$. Notice that as $\vec{V}_1$ increases in length, the projected length on $\vec{V}_2$, the semi-transparent cylinder, also increases in length resulting in the black bar growing taller. This observation can be explained by the fact that the length of $\vec{V}_1$ is $\left\|\vec{V}_1\right\|$, and as $\left\|\vec{V}_1\right\|$ increases, so does $\left\|\vec{V}_1\right\|\cos\theta$, or `v1LengthOnV2`.

Now, select `P2` and decrease the y-component value to increase the subtended angle. Observe that as the angle increases, the projected length of $\vec{V}_1$ decreases, and when $\vec{V}_1$ and $\vec{V}_2$ become almost perpendicular, the length approaches zero. This observation can be explained by the fact that as the angle θ increases, $\cos\theta$ decreases, and thus `v1LengthOnV2` also decreases. When the two are perpendicular, $\cos\theta$ returns a value of 0, forcing $\|\vec{V_1}\|\cos\theta$ to be 0 as well, which is why no projection is visible when $\vec{V_1}$ and $\vec{V_2}$ are perpendicular. Beyond 90° and to 180°, $\cos\theta$ is negative and thus the dot product result is negative. When this occurs, you will observe the black bar turning red and growing in the negative y-direction. Notice how the semi-transparent projection cylinder is no longer on $\vec{V_2}$, but extending in the opposite direction of $\vec{V_2}$. There are three important observations to make about the value of `v1LengthOnV2`:

- It is a simple floating-point number; this number is a measurement of the length of the projecting vector, $\vec{V_1}$, along the vector being projected onto, $\hat{V_2}$.
- It is the sign of the number that indicates whether $\vec{V_1}$ and $\vec{V_2}$ are within 90° of each other.
- Its magnitude is directly proportional to the length of the projecting vector, $\vec{V_1}$, and the cosine of the subtended angle with $\hat{V_2}$.

It is important to remember the characteristics of the cosine function that its result decreases when the angle increases from 0° to 90°. This means, as you have experienced and observed, that the magnitude of `v1LengthOnV2` is actually inversely proportional to the angle θ for $0° < \theta < 90°$.

Feel free to choose the `V2OntoV1` option for the `ProjectionChoice` variable and to examine and verify the complementary observations for

$$v2LengthOnV1 = \hat{V_1} \cdot \vec{V_2} = \|\vec{V_2}\|\cos\theta$$

Details of MyScript

Open `MyScript` and examine the source code in the IDE. The instance variables are as follows:

```
public enum ProjectionChoice {
    V1OntoV2,
    V2OntoV1,
    ProjectionOff
};
```

```
// Three positions to define two vectors: P0->P1 and P0->P2
public GameObject P0 = null;   // Position P0
public GameObject P1 = null;   // Position P1
public GameObject P2 = null;   // Position P2

public ProjectionChoice ProjChoice = ProjectionChoice.V1OntoV2;

#region For visualizing the vectors
#endregion
```

All the public variables for `MyScript` have been discussed when analyzing the `Controller`'s `MyScript` component. Take note that variables with the enumerated data type show up in the Hierarchy Window as options for a drop-down menu. The `Start()` function for `MyScript` is listed as follows:

```
void Start() {
    Debug.Assert(P0 != null);   // Ensure proper init
    Debug.Assert(P1 != null);
    Debug.Assert(P2 != null);

    #region For visualizing the vectors
    #endregion
}
```

As in all previous examples, the `Debug.Assert()` calls ensure proper setup regarding referencing the appropriate game objects via the Inspector Window. The `Update()` function is listed as follows:

```
void Update() {
    Vector3 v1 = P1.transform.localPosition -
                 P0.transform.localPosition;
    Vector3 v2 = P2.transform.localPosition -
                 P0.transform.localPosition;
    if ((v1.magnitude > float.Epsilon) &&
        (v2.magnitude > float.Epsilon))  {
        // make sure v1 and v2 are not zero vectors
```

```
        switch (ProjChoice) {
            case ProjectionChoice.V1OntoV2:
                float v1LengthOnV2 =
                      Vector3.Dot(v1, v2.normalized);
                Debug.Log("Projection Result:
                      Length of V1 along V2 = " + v1LengthOnV2);
                break;
            case ProjectionChoice.V2OntoV1:
                float v2LengthOnV1 =
                      Vector3.Dot(v1.normalized, v2);
                Debug.Log("Projection Result:
                      Length of V2 along V1 = " + v2LengthOnV1);
                break;
            default:
                Debug.Log("Projection Result: no projection,
                           dot=" + Vector3.Dot(v1, v2));
                break;
        }
    }
    #region  For visualizing the vectors
    #endregion
}
```

The first two lines of the `Update()` function compute

$$\vec{V}_1 = P_1 - P_0$$

$$\vec{V}_2 = P_2 - P_0$$

The `if` condition checks and ensures that the normalization operation will not be performed on zero vectors. When conditions are favorable, the `switch` statement checks the user's projection choice and simply computes and prints the results of one of the following:

$$v1LengthOnV2 = \vec{V}_1 \cdot \hat{V}_2$$

$$v2LengthOnV1 = \hat{V}_1 \cdot \vec{V}_2$$

Takeaway from This Example

This example demonstrates the results of projecting vectors onto each other. Vector projection is computed when one of the two operands of a dot product operation is normalized. Remember, the normalized vector is the one being projected onto. It is important to remember that projection is a simple dot product operation and the result is a signed floating-point number.

Relevant mathematical concepts covered include

- Calculating the dot product with a normalized vector can be interpreted as projecting the length of a vector onto another vector.
- The sign of the projection result indicates if the subtended angle is less than, when positive, or more than, when negative, 90°.
- The projection result is directly proportional to the length of the projecting vector and inversely proportional to the subtended angle when the angle is between 0° and 90°.

Unity tools

- `Enum` data type appears as drop-down menu options in the Hierarchy Window.

EXERCISE

Verify the Vector Projection Formula

When computing v1LengthOnV2 in `MyScript`

```
float v1LengthOnV2 = Vector3.Dot(v1, v2.normalized);
```

verify the projection formula

$$\vec{V}_1 \cdot \hat{V}_2 = \left\|\vec{V}_1\right\| \cos\theta$$

and replace that line with

```
float cosTheta = Vector3.Dot(v1.normalize, v2.normalized);
float v1LengthOnV2 = v1.magnitude * cosTheta;
```

Verify that the runtime results are identical.

Representation of a Line Segment

Figure 5-9 shows two checkered sphere positions, P_0 and P_1, defining a vector, $\vec{V}_1$,

$$\vec{V}_1 = P_1 - P_0$$

Notice that the region bounded by P_0 to P_1 is a segment of a straight line. In this case, the position P_a, when measured along the $\vec{V}_1$ direction, is located before the line segment and position P_b is located after the line segment. In Figure 5-9, positions in between both P_0 and P_1 are described as inside the line segment and thus both P_a and P_b are both outside of the line segment.

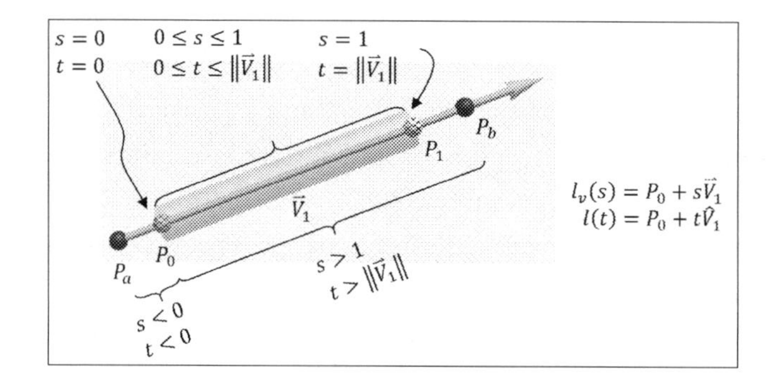

Figure 5-9. *Representing a line segment with a vector*

As you will see in later examples, in many applications it is critical to determine if a position is within the bounds of a line segment defined by two positions. By referencing the vector defined by the two positions, that is, the $\vec{V}_1$ in Figure 5-9, there are two convenient ways to identify a line segment region. The first way is to represent a line segment based on parameterizing the vector $\vec{V}_1$

$$l_v(s) = P_0 + s\vec{V}_1$$

As illustrated in Figure 5-9, the value of s identifies a position along the P_0 and P_1 line segments. For example,

$$l_v(0) = P_0 + 0\vec{V}_1 = P_0$$

$$l_v(0.5) = P_0 + 0.5\vec{V}_1 = \text{midpoint of the line segment}$$

$$l_v(1) = P_0 + 1\vec{V}_1 = P_0 + (P_1 - P_0) = P_1$$

In this way, the value of s is the portion, or percentage, of the line segment covered as measured from P_0 toward P_1 or the portion of the line segment covered along the $\vec{V}_1$ direction starting from P_0. When there is no coverage, or when $s = 0$, the position identified is the beginning position of the line segment, P_0. A complete coverage, or when $s = 1$, is the position identified as the end position of the line segment, P_1. In general, as illustrated in Figure 5-9, a position is within the line segment boundaries when $0 \le s \le 1$. When $s < 0$, for example, P_a, the position is before the beginning position, P_0, and when $s > 1$, for example, P_b, the position is after the end position of the line segment, P_1.

The second way to represent the line segment region bounded by the positions P_0 and P_1 is by parameterizing the normalized $\vec{V}_1$, or $\hat{V}_1$,

$$l(t) = P_0 + t\hat{V}_1$$

In this case, because the vector is normalized, t is the measurement of the actual distance traveled from the beginning position, P_0, toward the end position of the line segment, P_1, or the distance traveled along the $\hat{V}_1$ direction starting at P_0. For this reason, when $t = 0$, or $l(0)$, it signifies that no distance was traveled, and thus the identified position is the beginning of the line segment, P_0. The end position of the line segment is reached when $t = \|\vec{V}_1\|$ or the length of the vector $\vec{V}_1$

$$l(\|\vec{V}_1\|) = P_0 + \|\vec{V}_1\|\hat{V}_1 = P_0 + \vec{V}_1 = P_1$$

As illustrated in Figure 5-9, the range $0 \le t \le \|\vec{V}_1\|$ identifies a position within the line segment boundaries. $t < 0$ and $t > \|\vec{V}_1\|$ identify positions that are before the beginning position and after the end position of the line segment as measured along the $\hat{V}_1$ direction.

The only difference between the two line segments' representations is the normalization of the $\vec{V}_1$ vector

$$l_v(s) = P_0 + s\vec{V}_1$$

$$l(t) = P_0 + t\hat{V}_1$$

When comparing these two representations, the 0 to 1 range of the s parameter in $l_v(s)$ is convenient for determining if a position is within the line segment bounds and the distance measurement of the t parameter in $l(t)$ is advantageous when an actual distance traveled is required in the computation. Note that the s and t parameters are related by a simple scaling factor, $\left\|\vec{V_1}\right\|$,

$$t = s \times \left\|\vec{V_1}\right\|$$

In practice, when serving as part of more elaborate algorithmic computations, line segments are seldom explicitly represented. In these situations, the $l_v(s)$ or $l(t)$ parameterizations are often used interchangeably depending on the needs of the computations.

When represented explicitly, a line segment is often referred to as a **ray**. Rays are always parameterized as $l(t)$ with a normalized direction vector. For this reason, $l(t)$ is often referred to the **vector line equation**, or the ray equation, and is used often in video game development. For example, the Unity `Ray` class, `https://docs.unity3d.com/ScriptReference/Ray.html`, is a straightforward implementation of the line equation.

Inside-Outside Test of a General 1D Interval

Recall from Chapter 2 that a 1D interval is a region that is bounded by a minimum and maximum position along one of the major axes of the Cartesian Coordinate System. With the knowledge of vectors, the definition of an interval can now be relaxed. In general, a 1D interval, or a line segment, is defined as the region bounded by two positions along a direction (instead of just a major axis). In this way, the line segment in Figure 5-9 can be described as a 1D interval with its minimum position at P_0 and its maximum position at P_1 along the $\vec{V_1}$ direction.

Figure 5-10 shows that the inside-outside test for an interval can be based on the comparison of coordinate values or the comparison of distances. Recall that given an interval defined along the Y-axis with min and max positions, a given y-value, v, is inside the interval when

$$min \le v \le max$$

If the value *min* is subtracted from all sides of the equation,

$$min - min \leq v - min \leq max - min$$

then

$$0 \leq (v - min) \leq (max - min)$$

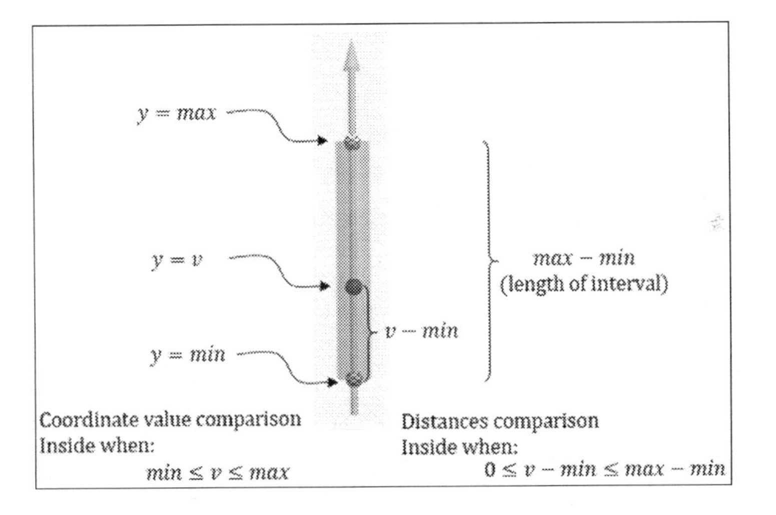

***Figure 5-10.** Inside-outside test based on coordinate values and distances*

This inequality equation says that the inside-outside test can also be determined by examining the distance from the minimum and maximum positions of the interval. For example, a given y-value, v, is inside the Y-axis interval when the distance between v to the minimum position is greater than zero and less than that of the maximum to minimum distance. With this understanding, Figure 5-11 illustrates the case for a general interval, with a direction that may not be aligned with a major axis of the Cartesian Coordinate System, like the Y-axis of Figure 5-10.

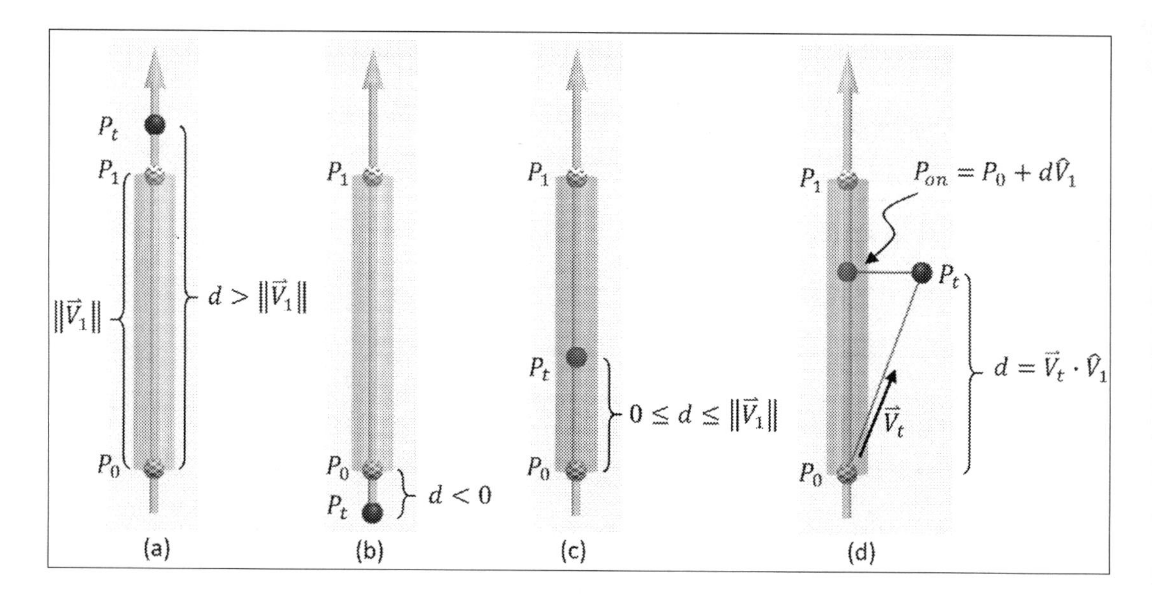

***Figure 5-11.** An interval bounded by P_0 and P_1, or a line segment along the $\vec{V}_1$ direction*

With the knowledge of vectors, you can now define a vector, $\vec{V}_1$, with tail position at P_0, to represent the interval in Figure 5-11, where

$$\vec{V}_1 = P_1 - P_0$$

In this way, the interval is simply the line segment

$$l(t) = P_0 + t\hat{V}_1$$

With the interval being described as a line segment, it should not be surprising that Figures 5-11 (a), (b), and (c) are similar to that of Figure 5-10. Figures 5-11 (a) and (b) illustrate the situation when the position to be tested, P_t, is outside of the line segment interval. Figure 5-11 (a) shows that, d, the symbol representing the distance between P_t and P_0 along the $\vec{V}_1$ direction, is larger than the line segment's magnitude, $d > \|\vec{V}_1\|$, and is thus beyond P_1. Figure 5-11 (b) shows the case when $d < 0$, or when P_t is before P_0. It is obvious that in both Figures 5-11 (a) and (b), P_t is outside of the interval. Figure 5-11 (c) shows that P_t is within the bounds of the interval when $0 \le d \le \|\vec{V}_1\|$. Note the similarities between these three cases with those of Figure 2-2, except instead of the coordinate value comparisons, the inside-outside conditions are restated in Figure 5-11 based on distance measurements.

Figure 5-11 (d) is a more interesting case; here the position of interest, P_t, does not lie on the same line as the interval. You have addressed this type of situation in Chapter 2. You may recall that when working with intervals along the Y-axis, the x- and z-component values are irrelevant when it comes to determining if a position is within a given y-interval. For example, a given position $(-3, 2, 5)$ is inside of the Y-axis interval with a bound of $min = -1$ and $max = 4$ because the y-component value of the position, 2, is bounded by the values of min, −1, and max, 4. In this case, the position $(-3, 2, 5)$ does not lie on the same line as the interval, the Y-axis, and only the coordinate value along the axis direction of interest, the Y-axis value of 2, is considered.

Figure 5-11 (d) translates the interval test knowledge from Chapter 2 using the vector projections you have learned. In this case, $\vec{V_1}$ is the vector from P_0 to P_1 and is the direction that corresponds to the Y-axis where the interval is defined. Given a position of interest, P_t, you can define the vector $\vec{V_t}$ as

$$\vec{V_t} = P_t - P_0$$

then the distance, d in all cases of Figure 5-11, is simply the projected distance of vector $\vec{V_t}$, in the $\vec{V_1}$ direction, or

$$d = \vec{V_t} \cdot \hat{V_1}$$

Note that since $\vec{V_t}$ is projected onto the $\hat{V_1}$ direction, the vector $\vec{V_1}$ must be normalized. Finally, you know that the position, P_{on}, the projection of the P_t position onto $\vec{V_1}$, is $t = d$ along the $l(t)$ line or d distance away from P_0 in the $\vec{V_1}$ direction

$$P_{on} = l(d) = P_0 + d\hat{V_1}$$

Note You can refer back to the initial discussion of vector projection in Figure 5-7. In this case, $\vec{V_t}$ is simply $\vec{V_2}$ and the projected length, d, is $\|\vec{V_2}\|\cos\theta$. When $d > \|\vec{V_1}\|$, the projected length is greater than the size of the vector being projected onto, and when $d < 0$, the subtended angle, θ, is more than 90°.

The Line Interval Bound Example

This example demonstrates the results of the inside-outside test for a general 1D interval (non-axis-aligned interval). This example allows you to interactively define a general 1D interval and manipulate a test position to examine the results of performing the inside-outside test. Figure 5-12 shows a screenshot of running the `EX_5_3_LineIntervalBound` scene from the `Chapter-5-DotProducts` project.

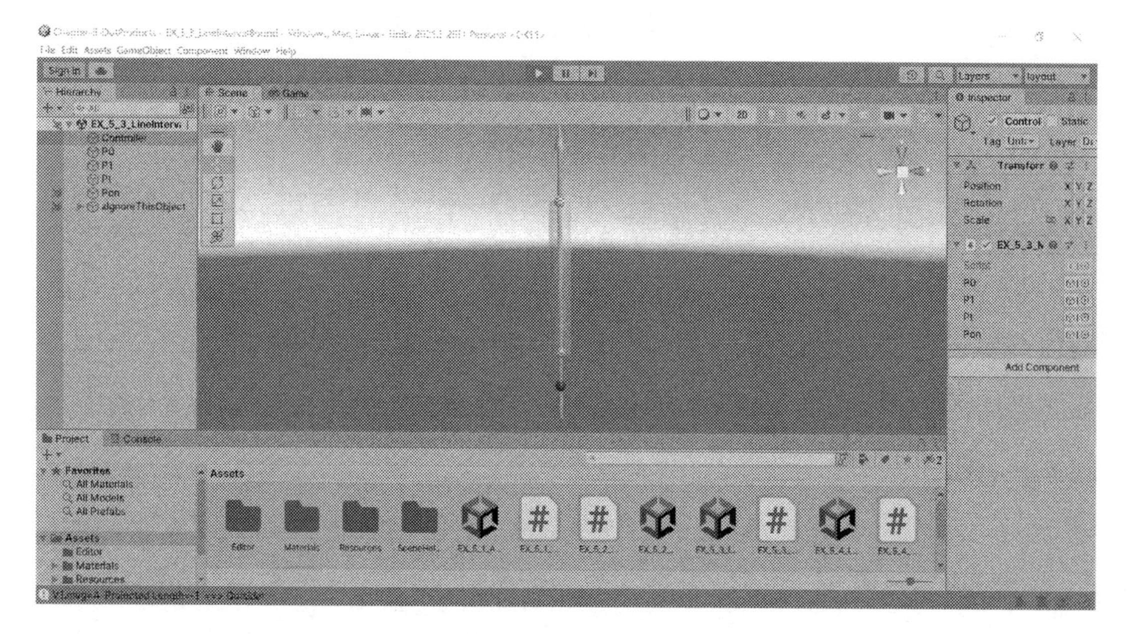

***Figure 5-12.** Running the Line Interval Bound example*

The goals of this example are for you to

- Experience defining and interacting with a general interval
- Examine the projection of a position onto a general interval
- Understand the implementation of an inside-outside test for the general interval

Examine the Scene

Take a look at the `Example_5_3_LineIntervalBound` scene and observe the predefined game objects in the Hierarchy Window. In addition to the `Controller`, there are four objects in this scene: `P0`, `P1`, `Pt`, and `Pon`. Here, `P0` and `P1` are the bounds of the interval, `Pt` is the position to manipulate for the inside-outside test, and `Pon` represents the position when `Pt` is projected onto the interval.

Analyze Controller MyScript Component

The `MyScript` component on the `Controller` shows four variables with names that correspond to the game objects in the scene. For all these variables, the `transform.localPosition` will be used for the manipulation of the corresponding positions.

Interact with the Example

Click the Play Button to run the example. Observe that by default and design, this example is rather similar to the Interval Bounds in 1D example in Chapter 2. Select `Pt` and adjust its y-component value to move the position along the green line that defines the interval. Since `Pt` is on the green line, the projected position, `Pon`, is exactly the same as `Pt`. This is why you do not observe a separate projected position. Notice how the color of the interval changes if `Pt` is inside or outside of the interval. You can compare the interval color change to the debug messages printed in the Console Window and verify that proper inside-outside conditions are being computed. So far, this example has worked in exactly the same manner as the one from Chapter 2.

Now, adjust the x- and z-component values of `Pt` to move the test position away from the green line. Notice that as soon as `Pt` departs from the green line, you begin to observe the position `Pon`. You will also notice that `Pon` is connected to `Pt` by a thin black line that is perpendicular to the green line. Move the camera around to verify that the thin line connecting `Pon` to `Pt` is indeed perpendicular to the green line. You are observing the exact situation illustrated in Figure 5-11 (d).

Now, you can adjust `P0` and `P1` to manipulate the direction and length of the 1D interval. Observe that the perpendicular projection of `Pon` and the inside-outside test results are both consistently updated and correct for any interval you define.

Details of MyScript

Open `MyScript` and examine the source code in the IDE. The instance variables and the `Start()` function are as follows:

```
// Positions: to define the interval, the test, and projected
public GameObject P0 = null;    // Position P0 of interval
public GameObject P1 = null;    // Position P1 of interval
public GameObject Pt = null;    // Pt: test position
public GameObject Pon = null;  // Pon: Pt projected on interval

#region For visualizing the vectors
#endregion

void Start() {
    Debug.Assert(P0 != null);    // Ensure proper init
    Debug.Assert(P1 != null);
    Debug.Assert(Pt != null);
    Debug.Assert(Pon != null);

    #region For visualizing the vectors
    #endregion
}
```

All the public variables for `MyScript` have been discussed when analyzing Controller's `MyScript` component, and as in all previous examples, the `Debug.Assert()` calls in the `Start()` function ensure proper setup regarding referencing the appropriate game objects via the Inspector Window. The `Update()` function is listed as follows:

```
void Update() {
    Vector3 v1 = P1.transform.localPosition -
                 P0.transform.localPosition;

    if (v1.magnitude > float.Epsilon) {
        Vector3 vt = Pt.transform.localPosition -
                     P0.transform.localPosition;
        Vector3 v1n = v1.normalized;
        float d = Vector3.Dot(vt, v1n);
        Pon.transform.localPosition = P0.transform.localPositio
                                      + d * v1.normalized;
```

```
        if ((d >= 0) && (d <= v1.magnitude))
            Debug.Log("V1.mag=" + v1.magnitude +
                  "Projected Length=" + d + "  ==> Inside!");
        else
            Debug.Log("V1.mag=" + v1.magnitude +
                  "Projected Length=" + d + "  ==> Outside!");
    }
    #region  For visualizing the vectors
    #endregion
}
```

The first line of the `Update()` function computes

$$\vec{V}_1 = P_1 - P_0$$

The `if` condition ensures that $\vec{V}_1$ is not a zero vector, which cannot be normalized or projected onto. If $\vec{V}_1$ is not a zero vector, then the four statements within the `if` condition perform the following four computations:

$$\vec{V}_t = P_t - P_0$$

$$\hat{V}_1 = Normalize\left(\vec{V}_1\right)$$

$$d = \vec{V}_t \cdot \hat{V}_1$$

$$P_{on} = P_0 + d\hat{V}_1$$

The `Debug.Log()` function prints the inside-outside status of `Pt` according to $0 \le d \le \left\|\vec{V}_1\right\|$. Note that although the interval is represented by the line equation

$$l(t) = P_0 + t\hat{V}_1$$

this representation is implicit. There is no explicit data structure definition for a specific variable referencing the line equation. This implicit evaluation without explicit representation is rather typical in the application of the line equation.

Takeaway from This Example

This example links the interval discussions in Chapter 2 to the concepts of vectors. At this point, you know how to compute the inside-outside test of a position for a general interval that is not aligned with a major axis. Recall the discussion in Chapter 2, where in Figure 2-7, the point in a bounding area test was derived by applying the one-dimensional interval test twice, once each to two intervals that are defined along two perpendicular directions. The same idea of applying the 1D interval test twice can be used for a general bounding area, and following the same concept once more, you can use the 1D interval test three times for a general bounding box. Now you can perform the inside-outside test of a position for a bounding box with three perpendicular intervals that do not need to be aligned with the major axes!

Though exciting, the non-axis-aligned bounding box has a severe limitation; the collision computation between these boxes is straightforward only when the three corresponding intervals that define the boxes are parallel. In general, given two bounding boxes, each with different interval directions, the collision detection between two such boxes is complex and non-trivial. For this reason, only axis-aligned bounding boxes are typically used in video game development.

Relevant mathematical concepts covered include

- An interval along a direction is a line segment and can be represented by the vector line equation.
- Vector projection can be applied to compute the projected distance of a point along a direction.
- The projected position along a direction can be determined for any given position.

EXERCISES

Verify the Axis-Aligned Interval Discussion with Vectors

Recall that the Y-axis interval is defined by its min and max values. These are actually P_0 with $(0, min, 0)$ and P_1 with $(0, max, 0)$. Now, by computing

$$\vec{V}_t = P_t - P_0$$

$$d = \vec{V} \cdot \hat{V}_1$$

$$P_{on} = P_0 + d\hat{V}_1$$

verify that given a general test position, P_t with (x_t, y_t, z_t), the projected position, P_{on}, is $(0, y_t - min, 0)$, showing that in this case, the x- and z-component values of P_t are indeed irrelevant. You can set up the values of `P0` and `P1` in this example to visualize the described results.

Verify the P_{on} Position

Define $\vec{V}_{on}$ to be

$$\vec{V}_{on} = d\hat{V}_1 - \vec{V}_t$$

and observe that

$$P_{on} = P_t + \vec{V}_{on}$$

Modify `MyScript` to print out P_{on} values based on these equations and then compare them to the `Pon` values currently computed in the script to verify they are identical. Notice that $\vec{V}_{on}$ is also, $P_{on} - P_t$.

Verify that Vector Projection Is Perpendicular

Refer to the previous definition of $\vec{V}_{on}$,

$$\vec{V}_{on} = P_{on} - P_t$$

Since P_{on} is the projection of P_t onto $\hat{V}_1$, it follows that $\vec{V}_{on}$ is perpendicular to $\vec{V}_1$. Recall from the discussion of the dot product that when vectors have a subtended angle of 90°, and because $\cos 90° = 0$, the dot product of two such vectors is zero. Modify `MyScript` to compute and print out the values of $\vec{V}_{on} \cdot \vec{V}_1$ and $\vec{V}_{on} \cdot \hat{V}_1$ and verify that both results are zero.

Line to Point Distance

Imagine an adventure game where hidden treasures are exposed when exploration agents are within their proximity. By now, you know how to define bounding volumes, for example, bounding spheres, for both the treasure and the agent objects, as well as support the detection of collisions between these corresponding bounding volumes. Figure 5-13 illustrates that for a fast-moving agent, the simple bounding sphere collision test may result in missed treasures.

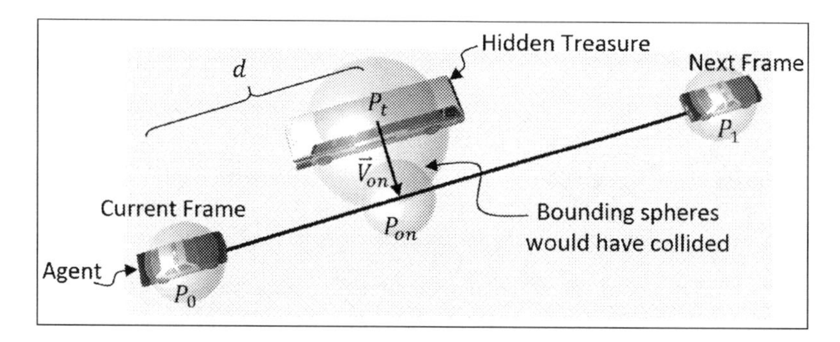

Figure 5-13. *A case where the bounding sphere misses with fast-traveling objects*

In Figure 5-13, both the police car agent and the city bus treasure are bounded by their corresponding bounding spheres. In this case, the police car is traveling at a high speed along the velocity defined by the black line. Here, in one update the car traveled from position P_0 on the left to position P_1 on the right. Notice that the bounding spheres of the car and the bus would have collided around P_{on} if the police car was traveling at a much slower speed. However, at the described high speed, the bounding sphere collisions at both the current frame and the next frame would be false, thereby missing the police car (agent) and the city bus (treasure) collision.

A straightforward approach to address this problem is by modeling this situation as a line to point distance computation. In the case of Figure 5-13, the problem is to find the closest distance between the line segment defined by P_0 and P_1 and the point located at P_t. This distance would be used to compare against the radii of the bounding spheres of the agent and the treasure to determine if a collision should occur during the agent's motion. If this distance is less than the combined radii, then a collision should occur.

From basic geometry, you know that the closest distance between a line segment and a position should be measured along a direction that is perpendicular from the line to the position. Now, you also know that a vector projection projects the head of a vector perpendicularly onto another given vector. In the case of Figure 5-13, these observations can be translated into, defining two vectors,

$$\vec{V}_1 = P_1 - P_0$$

$$\vec{V}_t = P_t - P_0$$

Then you can project $\vec{V}_t$ onto $\vec{V}_1$ to compute P_{on}, the projection of P_t on the vector $\vec{V}_1$. In this case, you know the vector, $\vec{V}_{on}$,

$$\vec{V}_{on} = P_{on} - P_t$$

must be perpendicular to $\vec{V}_1$, and thus the distance between P_t and P_{on}, or $\left\|\vec{V}_{on}\right\|$, is the closest distance between the line segment defined by P_0, P_1 and the position P_t. This distance would be compared with the combined radii of the bounding spheres of the agent and treasure for collision determination.

It is encouraging that this problem and its solution are familiar to those of the line segment and the general interval inside-outside test. Based on the previous discussions, you know that

$$d = \vec{V}_t \cdot \hat{V}_1$$

$$P_{on} = P_0 + d\hat{V}_1$$

You can observe that when the position P_t is within the bounds of the line segment end points, or when $0 \le d \le \left\|\vec{V}_1\right\|$, the closest distance between the line segment and the point is from P_t to P_{on}, or the magnitude of $\vec{V}_{on}$ or $\left\|\vec{V}_{on}\right\|$.

Figures 5-14 (a) and (b) show that P_t can also be outside of the line interval. In these cases, the closest distance measurements are actually between P_t and the end points of the line segment. Figure 5-14 (a) illustrates that when $d < 0$, P_t is located at a position before the line segment and thus the closest distance is actually the distance between P_t and P_0, or simply the magnitude $\vec{V}_t$ or $\vec{V}_t$. Figure 5-14 (b) illustrates that when $d > \left\|\vec{V}_1\right\|$, P_t is located at a position after the line segment and thus the closest distance is the distance between P_t and P_1, or the magnitude of $\left(\overrightarrow{P_t - P_1}\right)$ or $\left\|\overrightarrow{P_t - P_1}\right\|$. Figure 5-14 (c) is the same case as Figure 5-13, when $0 \le d \le \left\|\vec{V}_1\right\|$, and the closest distance is the magnitude of $\vec{V}_{on}$ or $\left\|\vec{V}_{on}\right\|$.

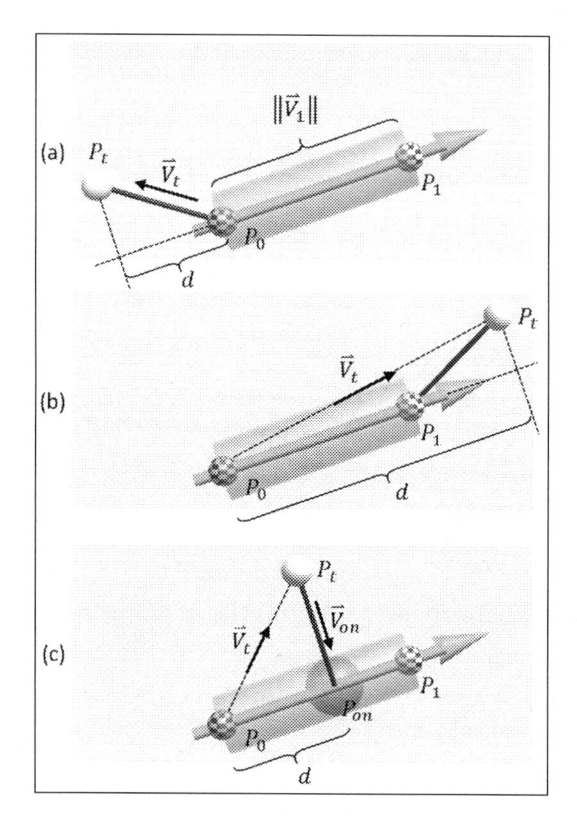

Figure 5-14. *The three conditions of line to point distance calculation*

The Line to Point Distance Example

This example demonstrates the results of the line to point distance computation. This example allows you to interactively define the line segment, manipulate the position of, and examine the results from the line to point distance computation. Figure 5-15 shows a screenshot of running the `EX_5_4_LineToPointDistance` scene from the `Chapter-5-DotProducts` project.

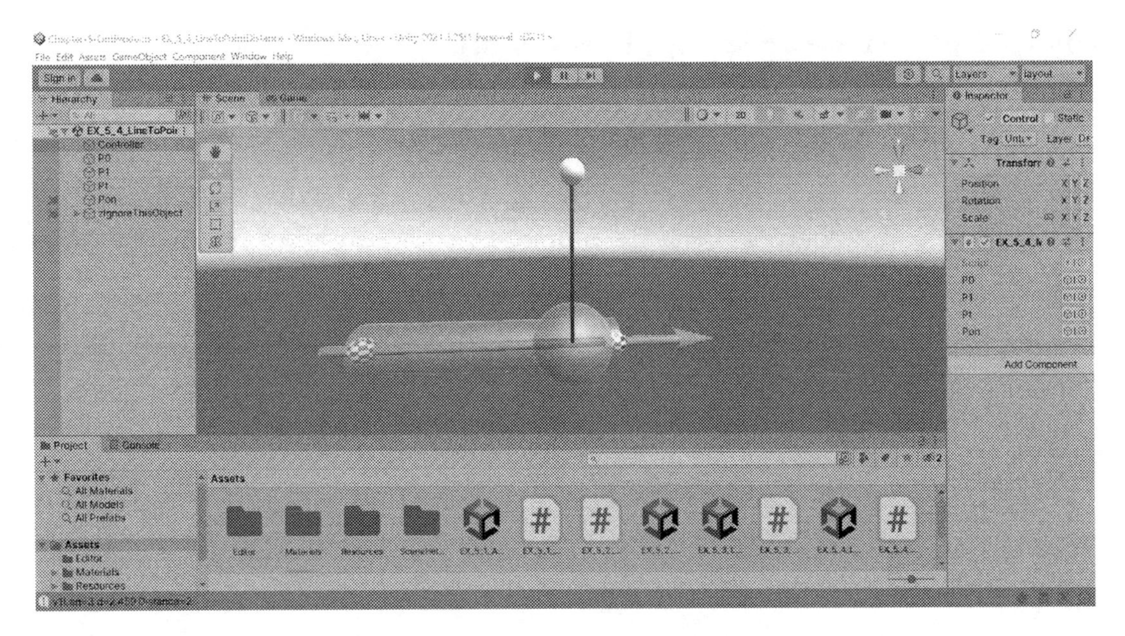

Figure 5-15. *Running the Line to Point Distance example*

The goals of this example are for you to

- Experience working with a straightforward application of the vector dot product concepts
- Interact with and understand the results of line to point distance computation
- Examine the implementation of line to point distance computation

Examine the Scene

Take a look at the `Example_5_4_LineToPointDistance` scene and observe the predefined game objects in the Hierarchy Window. In addition to the `Controller`, there are exact same four objects in this scene as in the previous example: `P0`, `P1`, `Pt`, and `Pon`. Here, `P0` and `P1` are the checkered spheres that identify the line segment. `Pt` is the white sphere and is the position (the point) used for the line to point distance computation. Finally, `Pon`, the red sphere, is the position where `Pt` is projected onto the line.

Analyze Controller MyScript Component

The MyScript component on the Controller shows four variables with names that correspond to the game objects in the scene. For all these variables, the transform.localPosition will be used for the manipulation of the corresponding positions.

Interact with the Example

Click the Play Button to run the example. Observe that P0 and P1 define the green vector direction and a line segment. There is a thin black line connecting Pt, the white sphere, to the projected position, Pon, the red sphere, on the line segment. Select Pt and adjust its y-component value. Try to move Pt away from the line, for example, by increasing the y-component value, and observe the red sphere increase in size. If you move Pt closer instead, you will observe the red sphere shrink. The size of the red sphere, Pon, is directly proportional to the distance between Pt and the line segment. The results of this computation can also be observed in the Console Window.

Now, change the x-component value of Pt to observe the corresponding movement of the projection position, Pon. Notice that when Pt is within the bounds of the line segment, the thin black line connecting Pt to Pon is always perpendicular to the line segment, indicating the projection of Pt onto the line segment. When Pt is moved to outside of the line segment, the thin black line becomes connected to the closest end point of the line segment, either P0 or P1. This signifies that the closest distance in these situations is actually the measurement to one of the end points of the line segment.

You can now select and manipulate P0 and P1 to verify that the distance computation is indeed correct for any line segment, including a line segment defined by the zero vector, which occurs when P0 and P1 are located at the same position.

Details of MyScript

Open MyScript and examine the source code in the IDE. The instance variables and the Start() function are as follows:

```
// Positions: to define the interval, the test, and projected
public GameObject P0 = null;  // Position P0
public GameObject P1 = null;  // Position P1
public GameObject Pt = null;  // For distance computation
public GameObject Pon = null; // closest point on line
```

```
#region For visualizing the line
#endregion

void Start() {
    Debug.Assert(P0 != null);   // Ensure proper init
    Debug.Assert(P1 != null);
    Debug.Assert(Pt != null);

    #region For visualizing the lines
    #endregion
}
```

All the public variables for `MyScript` have been discussed when analyzing `Controller`'s `MyScript` component, and as in all previous examples, the `Debug.Assert()` calls in the `Start()` function ensure proper setup regarding referencing the appropriate game objects via the Inspector Window. The `Update()` function is listed as follows:

```
void Update() {
    float distance = 0;  // closest distance
    Vector3 v1 = P1.transform.localPosition -
                 P0.transform.localPosition;
    float v1Len = v1.magnitude;

    if (v1Len > float.Epsilon) {
        Vector3 vt = Pt.transform.localPosition -
                     P0.transform.localPosition;
        Vector3 v1n = (1f / v1Len) * v1; // <<-- what is this?
        float d = Vector3.Dot(vt, v1n);
        if (d < 0) {
            Pon.transform.localPosition =
                                  P0.transform.localPosition;
            distance = vt.magnitude;
        } else if (d > v1Len) {
            Pon.transform.localPosition =
                                  P1.transform.localPosition;
            distance = (Pt.transform.localPosition -
                          P1.transform.localPosition).magnitude;
        } else {
```

```
            Pon.transform.localPosition =
                        P0.transform.localPosition + d * v1n;
            Vector3 von = Pon.transform.localPosition -
                        Pt.transform.localPosition;
            distance = von.magnitude;
        }
        float s = distance * kScaleFactor;
        Pon.transform.localScale = new Vector3(s, s, s);
        Debug.Log("v1Len=" + v1Len + " d=" + d +
                " Distance=" + distance);
    }

    #region  For visualizing the lines
    #endregion
}
```

The first two lines of the `Update()` function compute

$$\vec{V}_1 = P_1 - P_0$$

$$v1Len = \left\|\vec{V}_1\right\|$$

The `if` condition checks for and avoids performing the normalization operation on a zero vector. When the condition is favorable, the following are computed:

$$\vec{V}_t = P_t - P_0$$

$$\hat{V}_1 = \frac{1}{v1Len}\vec{V}_1 \qquad \text{Note: normalize } \vec{V}_1$$

$$d = \vec{V}_t \cdot \hat{V}_1$$

With P_{on} being the closet point on the line segment and the position being *distance* away from P_t, notice how the computation is governed by the values of the projected length, d:

- When $d < 0$, the condition is as illustrated in Figure 5-14 (a), and

$$P_{on} = P_0$$

$$distance = \left\| \vec{V}_t \right\|$$

- When $d > v1Len$, or $d > \left\| \vec{V}_1 \right\|$, the condition is as illustrated in Figure 5-14 (b), and

$$P_{on} = P_1$$

$$distance = \left\| P_t - P_1 \right\|$$

- The final condition, when $0 \le d \le \left\| \vec{V}_1 \right\|$, is as illustrated in Figure 5-14 (c), and

$$P_{on} = P_0 + d\hat{V}_1$$

$$\vec{V}_{on} = P_{on} - P_t$$

$$distance = \left\| \vec{V}_{on} \right\|$$

The last three lines of code scale the red sphere that represents P_{on} in proportion to the value of *distance* and outputs the computation results to the Console Window.

Takeaway from This Example

This example demonstrates a solution to a fundamental problem in video games and interactive computer graphics. In video games, the closest distance and intersection computations are some of the most straightforward solutions to the problem of missed collisions from fast-moving objects. In graphical interactions, many basic operations depend on the results of line to point distance computation. For example, in a drawing editor, clicking the mouse button to select a line object is typically implemented as determining if the clicked position is sufficiently close to the line object, as clicking perfectly on a one-pixel wide line can be challenging and frustrating!

The solution presented in this example to these types of problems is based on the concepts of vector projection and builds directly on the knowledge gained from the line equation and the general interval inside-outside test discussions. These concepts are some of the most important topics in interactive graphical applications and are widely applied in video game development.

Relevant mathematical concepts covered include

- The distance between a line segment and a point, P_t, can be solved by finding the position, P_{on}, along the line segment that is closest to P_t, and computing the distance between P_{on} and P_t.
- When P_t is outside of the line segment, P_{on} is located at one of the line segment end points.
- When P_t is inside the line segment, P_{on} is the projection of P_t onto the line segment.

EXERCISE

Experience Solving the Missing Collision Problem

Modify `MyScript` to continuously send a fast-moving agent from `P0` to `P1`, for example, traveling at a speed of 20 units per second. You can refer to the `EX_4_3_VelocityAndAiming` scene of `Chapter-4-Vectors` for a sample approach of how to implement this functionality. In your `Update()` function, compute the collision between the agent and the `Pt` sphere. Notice even when the `P0` to `P1` line segment passes right through the `Pt` sphere, you can fail to detect the collision between the agent and the `Pt` sphere. This is because the agent is simply moving too fast for the spheres to overlap. Verify that you can resolve this problem with the line to point distance computation.

Line to Line Distance

Imagine in another adventure game, you want to know if the path of the explorer will come too close to a monster pathway. This is a simple case of determining the distance between two line segments. This problem has a simple and elegant solution that allows you to practice the vector algebra learned. Figure 5-16 illustrates the general case of two line segments, where the problem is how to compute the perpendicular, or the shortest, distance between the lines.

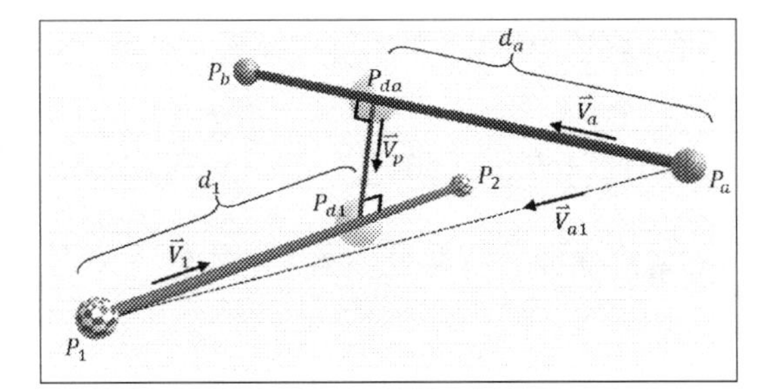

Figure 5-16. *Distance between two line segments*

The problem of finding the closest, or the perpendicular, distance between two given lines is similar to the line to point distance problem. The solution boils down to locating a point on each line where when connected are perpendicular to both of the two given lines. This description is depicted in Figure 5-16, where the two lines are defined by positions P_1 and P_2 and P_a and P_b, respectively. In this figure, the position P_{d1} is d_1 distance away from P_1 and P_{da} is d_a distance away from P_a where the line segment from P_{d1} to P_{da} is perpendicular to both of the other two lines. In this way, the shortest distance between the lines is the length of the vector, $P_{d1} - P_{da}$. In order to find P_{d1} and P_{da}, the task is to find the distances d_1 and d_a. You can begin deriving the solution by defining

$$\vec{V}_1 = P_2 - P_1$$

$$\vec{V}_a = P_b - P_a$$

$$\vec{V}_p = P_{d1} - P_{da}$$

The descriptions of P_{d1} and P_{da} can be formulated as two separate line segments

$$P_{d1} = P_1 + d_1\hat{V}_1$$

$$P_{da} = P_a + d_a\hat{V}_a$$

Since $\vec{V}_p$ is perpendicular to both $\vec{V}_1$ and $\vec{V}_a$, it must be true that both of the following are true:

$$\hat{V}_1 \cdot \vec{V}_p = 0$$

$$\hat{V}_a \cdot \vec{V}_p = 0$$

Now, if you substitute P_{d1} and P_{da} into $\vec{V}_p$, these two equations become

$$\hat{V}_1 \cdot \vec{V}_p = \hat{V}_1 \cdot (P_{d1} - P_{da}) = \hat{V}_1 \cdot \left(P_1 + d_1\hat{V}_1 - P_a - d_a\hat{V}_a\right) = 0$$

$$\hat{V}_a \cdot \vec{V}_p = \hat{V}_a \cdot (P_{d1} - P_{da}) = \hat{V}_a \cdot \left(P_1 + d_1\hat{V}_1 - P_a - d_a\hat{V}_a\right) = 0$$

Note that these are two simultaneous equations with two unknowns, d_1 and d_a. Now, examine the first of the two equations, by following the distributive property of dot product over vector operations, collecting the terms with $\hat{V}_1$, and recognizing $\hat{V}_1 \cdot \hat{V}_1$ is 1.0

$$\hat{V}_1 \cdot \vec{V}_p = \hat{V}_1 \cdot (P_{d1} - P_{da})$$

Substitute the definitions of P_{d1} and P_{da}

$$= \hat{V}_1 \cdot \left(P_1 + d_1\hat{V}_1 - P_a - d_a\hat{V}_a\right)$$

Apply the distributive property of dot product for vector

$$= \hat{V}_1 \cdot P_1 + \hat{V}_1 \cdot d_1\hat{V}_1 - \hat{V}_1 \cdot P_a - \hat{V}_1 \cdot d_a\hat{V}_a$$

Collect the P_1 and P_a terms

$$= \hat{V}_1 \cdot (P_1 - P_a) + \hat{V}_1 \cdot d_1\hat{V}_1 - \hat{V}_1 \cdot d_a\hat{V}_a$$

Apply distributive property over factors d_1 and d_a

$$= \hat{V}_1 \cdot (P_1 - P_a) + d_1\left(\hat{V}_1 \cdot \hat{V}_1\right) - d_a\left(\hat{V}_1 \cdot \hat{V}_a\right)$$

Recognize the fact that $\hat{V}_1$ dot $\hat{V}_1$ is equal to 1

$$= \hat{V}_1 \cdot (P_1 - P_a) + d_1 - d_a\left(\hat{V}_1 \cdot \hat{V}_a\right)$$

Now, let

$$d = \hat{V}_1 \cdot \hat{V}_a$$

$$\vec{V}_{a1} = P_1 - P_a$$

Then

$$\hat{V}_1 \cdot \vec{V}_p = \hat{V}_1 \cdot \hat{V}_{a1} + d_1 - d_a d = 0$$

Following similar simplification steps, left as an exercise, you can show that

$$\hat{V}_a \cdot \vec{V}_p = -\hat{V}_a \cdot \hat{V}_{a1} - d_a + d_1 d = 0$$

In this way, the simultaneous equations become

$$\hat{V}_1 \cdot \hat{V}_{a1} + d_1 - d_a d = 0$$

$$-\hat{V}_a \cdot \hat{V}_{a1} - d_a + d_1 d = 0$$

Recall that dot product results are floating-point numbers; therefore, $\hat{V}_1 \cdot \hat{V}_{a1}$ and $\hat{V}_a \cdot \hat{V}_{a1}$ return simple floating-point numbers. These equations are thus simple algebraic equations that are independent from vector operations, and once again, their simplification and solution derivation are left as an exercise. You can show that the solution to the simultaneous equations is

$$d_1 = \frac{-\left(\hat{V}_1 \cdot \vec{V}_{a1}\right) + d\left(\hat{V}_a \cdot \vec{V}_{a1}\right)}{1 - d^2}$$

$$d_a = \frac{\left(\hat{V}_a \cdot \vec{V}_{a1}\right) - d\left(\hat{V}_1 \cdot \vec{V}_{a1}\right)}{1 - d^2}$$

In this case, to allow easier interpretation of text output, instead of distances you can compute the portion of line segment covered or

$$d_1' = \frac{d_1}{\|\vec{V}\|_1}$$

$$d_a' = \frac{d_a}{\|\vec{V}_a\|}$$

and

$$P_{d1} = P_1 + d_1'\vec{V}_1$$

$$P_{da} = P_a + d_a'\vec{V}_a$$

where you know P_{d1} and P_{da} are within the bounds of their respective line segments only when d_1' and d_a' are both within the range of 0 to 1. Now, the closest distance between the two lines is the distance between P_{d1} and P_{da}, or $\|\overline{P_{d1} - P_{da}}\|$. Note that this is also the length of the vector $\vec{V}_p$, or $\|\vec{V}_p\|$.

The Line to Line Distance Example

This example demonstrates the results of line to line distance computation. This example allows you to interactively define the two line segments and examine the results from the line to line distance computation. Figure 5-17 shows a screenshot of running the `EX_5_5_LineToLineDistance` scene from the `Chapter-5-DotProducts` project.

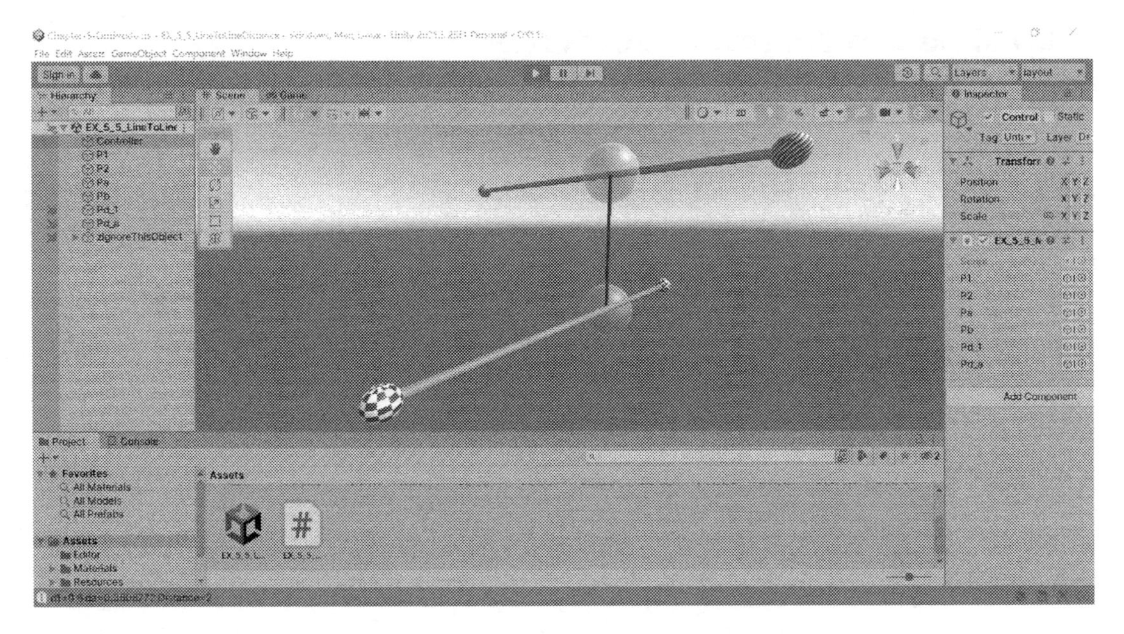

Figure 5-17. Running the Line to Line Distance example

The goals of this example are for you to

- Experience deriving and simplifying non-trivial vector expressions
- Verify solutions to vector equations with a straightforward implementation
- Examine the implementation of line to line distance computation

Examine the Scene

Take a look at the `Example_5_5_LineToLineDistance` scene and observe the predefined game objects in the Hierarchy Window. In addition to the `Controller`, there are three sets of objects defined for the visualization of the two line segments: the two checkered spheres `P1` and `P2`, the two stripped spheres `Pa` and `Pb`, and the two solid color spheres `Pd_1` and `Pd_a`. The `transform.localPosition` of `P1`, `P2` and `Pa`, `Pb` defines the bounding positions of the two line segments. The `transform.localPosition` of `Pd_1` is a position along the line defined by `P1` to `P2` and `Pd_a` a position along the `Pa` to `Pb` line where the distance from `Pd_1` to `Pd_a` is the closest distance between the two lines.

Analyze Controller MyScript Component

The `MyScript` component on the `Controller` shows six variables with names that correspond to the game objects in the scene. These variables are set up to reference the game objects with the corresponding names in the scene.

Interact with the Example

Click the Play Button to run the example. Once running, you will observe two line segments. The first is red and is defined by a pair of checkered spheres, `P1` and `P2`. The second line segment is blue and is defined by a pair of stripped spheres, `Pa` and `Pb`. Along each line segment is a semi-transparent sphere, `Pd_1` on the red line segment and `Pd_a` on the blue line segment. Notice that the two spheres are connected by a thin black line that is perpendicular to both the red and the blue line segments. You are observing the solution to the line to line distance computation.

Now, rotate the Scene View camera to verify that the thin black line is indeed perpendicular to both the red and blue line segments. Feel free to manipulate any of the line segment end points to verify the computation results. Note that when the locations of `Pd_1` or `Pd_a` are outside of their respective line segments, the semi-transparent spheres will turn opaque. You can also observe the text output in the Console Window. There, the values for `d1` and `da` are in the range between 0 and 1, assisting your verification of the corresponding position's inside-outside status on their respective line segment.

Lastly, set both of the line segments to be along the same direction, for example, set `P1` and `P2` to the values $(0,0,0)$ and $(5,0,0)$ and `Pa` and `Pb` to $(0,2,0)$ and $(5,2,0)$. Once done, notice that the results of both `Pd_1` and `Pd_a` are no longer visualized. You can verify in the Console Window that the line segments are in the exact same direction. This is a special case not handled in the derived solution. One of the exercises at the end of this example will tell you what this special case is and allow you to practice handling this special case.

Details of MyScript

Open `MyScript` and examine the source code in the IDE. The instance variables and the `Start()` function are as follows:

```
public GameObject P1, P2;  // define the line V1
public GameObject Pa, Pb;  // define the line Va
public GameObject Pd_1;    // point on V1 closest to Va
public GameObject Pd_a;    // point on va closest to V1

#region For visualizing the line
#endregion

void Start() {
    Debug.Assert(P1 != null);   // Ensure proper init
    Debug.Assert(P2 != null);
    Debug.Assert(Pd_1 != null);
    Debug.Assert(Pa != null);
    Debug.Assert(Pb != null);
    Debug.Assert(Pd_a != null);

    #region For visualizing the line
    #endregion
}
```

All the public variables for `MyScript` have been discussed when analyzing `Controller`'s `MyScript` component, and as in all previous examples, the `Debug.Assert()` calls in the `Start()` function ensure proper setup regarding referencing the appropriate game objects via the Inspector Window. The `Update()` function is listed as follows:

```
void Update() {
    Vector3 v1 = P2.transform.localPosition -
                 P1.transform.localPosition;
    Vector3 va = Pb.transform.localPosition -
                 Pa.transform.localPosition;

    if ((v1.magnitude < float.Epsilon) ||
        (va.magnitude < float.Epsilon))
        return;  // only works with well defined line segments

    Vector3 va1 = P1.transform.localPosition -
                  Pa.transform.localPosition;
    Vector3 v1n = v1.normalized;
    Vector3 van = va.normalized;
```

```
    float d = Vector3.Dot(v1n, van);

    bool almostParallel = (1f - Mathf.Abs(d) < float.Epsilon);

    float d1 = 0f, da = 0f;

    if (!almostParallel) {  // two lines are not parallel
        float dot1A1 = Vector3.Dot(v1n, va1);
        float dotAA1 = Vector3.Dot(van, va1);

        d1 = (-dot1A1 + d * dotAA1) / (1 - (d * d));
        da = (dotAA1 - d * dot1A1) / (1 - (d * d));

        d1 /= v1.magnitude;
        da /= va.magnitude;

        Pd_1.transform.localPosition =
                        P1.transform.localPosition + d1 * v1;
        Pd_a.transform.localPosition =
                        Pa.transform.localPosition + da * va;
        float dist = (Pd_1.transform.localPosition -
                       Pd_a.transform.localPosition).magnitude;
        Debug.Log("d1=" +d1+ " da=" +da+ " Distance=" +  dist);
    } else {
        Debug.Log("Lines are parallel, not handled");
    }

    #region  For visualizing the line
    #endregion
}
```

The first two lines of the `Update()` function compute

$$\vec{V}_1 = P_2 - P_1$$

$$\vec{V}_a = P_b - P_a$$

The code then ensures that both are not zero vectors and continues to compute

$$\vec{V}_{a1} = P_1 - P_a$$

$$\hat{V}_1 \text{ and } \hat{V}_a$$

$$d = \hat{V}_1 \cdot \hat{V}_a$$

Recall that the dot product of two normalized vectors is the cosine of the subtended angle and that the cosine of 0° or 180° is equal to 1 and –1, respectively. For this reason, the `almostParallel` variable is true when $\hat{V}_1$ and $\hat{V}_a$ are almost parallel. In the implementation, the computation only proceeds when the two directions are not almost parallel. This check is necessary because the solutions for both d_1 and d_a involve a division by $1 - d^2$ and when the two directions are almost parallel, $d \approx 1.0$, which means d_1 and d_a will be divided by 0, thus causing neither d_1 nor d_a to be defined. When the two lines are not parallel, the code computes

$$dot1A1 = \hat{V}_1 \cdot \vec{V}_{a1}$$

$$dotAA1 = \hat{V}_a \cdot \vec{V}_{a1}$$

and

$$d_1 = \frac{-dot1A1 + d * dotAA1}{1 - d^2}$$

$$d_a = \frac{dotAA1 - d * dot1A1}{1 - d^2}$$

where notice that both d_1 and d_a are scaled to values between 0 and 1 for positions that are inside the respective line segments, and closest positions are computed accordingly,

$$P_{d1} = P_1 + d_1\vec{V}_1$$

$$P_{da} = P_a + d_a\vec{V}_a$$

And lastly, the closest distance between the two lines is simply the distance between the closest positions

$$dist = \left\| P_{d1} - P_{da} \right\|$$

Takeaway from This Example

Though the presented solution of the line to line distance is interesting, it is incomplete. First of all, the solution does not address the situation when the line segments are parallel. Secondly, the solution does not address the situations when the closest points are outside of the given line segments, that is, when either P_{d1} or P_{da} or both are outside of their corresponding line segments. As in the case of line to point distance, when the closest position is outside of the line segment, the closest distance should be measured to the corresponding end position of the line segment. Although not a complete solution, this example does demonstrate and allow you to practice simplifying vector equations based on the learned vector algebra and serves as a way to illustrate an implementation of a typical solution to vector equations.

Through working with this example, you have observed that the actual vector equations and their solution process may be complex and involved. However, thankfully, as you have also witnessed, the derived solutions are typically elegant and can be implemented in a straightforward fashion with a relatively small number of steps. To ensure proper implementation, it is essential to maintain precise drawings and notes with symbols that correspond to variable names. Lastly, and very importantly, attention must be maintained when working with normalized vs. non-normalized vectors.

Relevant mathematical concepts covered include

- Vector algebra, or the rules governing vector operations, are invaluable in simplifying non-trivial vector equations.

Relevant observations on implementation include

- It is vital to understand and check for situations when mathematical expressions are undefined, for example, normalization of zero vectors, or divisions by 0.
- It is often possible to relate mathematical expressions to real-world geometric orientations. For example, you know that the dot product, $\hat{V}_1 \cdot \hat{V}_a$, computes the cosine of the angle subtended by two vectors; therefore, a value of 1 or -1 means that the vectors are parallel. It is the responsibility of the software developer to understand these implications and ensure all appropriate conditions are considered and supported.

EXERCISES

Verify the Solutions for d_1 and d_a

The derived simultaneous equations for line to line distance are

$$\hat{V}_1 \cdot \left(P_1 + d_1\hat{V}_1 - P_a - d_a\hat{V}_a\right) = 0$$

$$\hat{V}_a \cdot \left(P_1 + d_1\hat{V}_1 - P_a - d_a\hat{V}_a\right) = 0$$

You know

$$\vec{V}_1 = P_2 - P_1$$

$$\vec{V}_a = P_b - P_a$$

$$\vec{V}_{a1} = P_1 - P_a$$

$$d = \hat{V}_1 \cdot \hat{V}_a$$

Now, show that

$$d_1 = \frac{-\left(\hat{V}_1 \cdot \vec{V}_{a1}\right) + d\left(\hat{V}_a \cdot \vec{V}_{a1}\right)}{1 - d^2}$$

$$d_a = \frac{\left(\hat{V}_a \cdot \vec{V}_{a1}\right) - d\left(\hat{V}_1 \cdot \vec{V}_{a1}\right)}{1 - d^2}$$

In your solution derivation process, make sure to pay special attention to normalized and un-normalized vectors.

Handling Parallel Lines

Recall that the solutions for d_1 and d_a are derived based on the observation and simplification of the simultaneous equations

$$\hat{V}_1 \cdot \vec{V}_p = 0$$

$$\hat{V}_a \cdot \vec{V}_p = 0$$

Now, if the two line segments are parallel, then, $\hat{V}_1 = \hat{V}_a$ and thus there is only one equation with two unknowns. For this reason, the derived solution is valid only when $\hat{V}_1 \neq \hat{V}_a$ or when the two lines are not parallel.

In general, the shortest distance between two parallel lines can be determined by computing the shortest distance between one of the lines to the end point on the other line. Now, modify `MyScript` to support distance computation between parallel lines.

Notice your solution assumes both line segments are infinitely long where the closest positions on each line can be outside of their respective line segments. Once again, this is not a complete solution to closest distance between the two finite length line segments. Imagine the explorer and the monster pathways when the closest positions are outside of the line segments, the distance computed would be based on positions that the explorer or the monsters will not move to. The general solution is similar to that of the line to point distance when the closest position is outside, it should be clamp to the corresponding end point.

Summary

This chapter continues with the exploration of vectors by introducing the vector dot product, a tool for analyzing relationships between two vectors. Since a vector is defined by a size and a direction, the tool for analyzing the relationships between two vectors reports on the relative directions and sizes of these vectors.

The definition of the vector dot product is straightforward, the sum of the products of the corresponding components of the two vectors, and the result is a simple signed floating-point number. There are four ways to compute the dot product between two vectors and each offers a unique geometric insight into the resulting floating-point number.

The first way of computing a dot product is by operating on two non-normalized vectors. The resulting floating-point number is the product of the sizes of the two vectors and the cosine of their subtended angle. While the least useful, this floating-point number does provide slight insight into the subtended angle between the two vectors. If the number is positive, then the subtended angle is less than 90°; otherwise, the angle is between 90° and 180°.

The second way of computing a dot product is by operating on two normalized vectors. In this case, the resulting floating-point number is simply the cosine of the subtended angle. This result is invaluable when you need to determine how much two directions differ. In fact, checking the dot product results of two normalized vectors against approximately 0 or 1 for when the two vectors are almost perpendicular or parallel is one of the most frequently encounter test cases in video game development.

The third and fourth way of computing a dot product is to ensure only one of the operands is normalized. In this scenario, you are computing the projected length of the non-normalized vector along the direction of the normalized vector. These forms of computing the dot product have the broadest application. This is because projected sizes, as you have experience with line to point and line to line distance computation, are the basis for computing distances and, as you will learn in the next chapter, for computing intersections.

You have learned about vectors, gained knowledge on how to analyze the relationships between vectors, and applied these concepts in solving some interesting and non-trivial geometric problems. In the next chapter, you will learn about the vector cross product, a tool to relate two vectors to the space that contains those vectors. But before you continue, here are the summaries of the vector dot product definition, rules, and straightforward applications.

Vector Dot Product Definition and Implications

Dot Product Definition	**Remark**
$\vec{V}_1 \cdot \vec{V}_2 = x_1x_2 + y_1y_2 + z_1z_2$	Definition of the dot product, also referred to as the inner product
$\vec{V}_1 \cdot \vec{V}_2 = \|\vec{V}_1\|\|\vec{V}_2\|\cos\theta$	Geometric interpretation of the dot product definition, θ is the angle subtended by the two vectors
$\vec{V}_1 \cdot \vec{V}_1 = \|\vec{V}_1^{\,2}\|$	Dot product of a vector with itself is the squared of its magnitude
$\vec{V}_1 \cdot ZeroVector = ZeroVector$	Dot product with the zero vector is the zero vector

Interpreting the Dot Product Results

Dot Product	Geometric Interpretations
Direction: $\hat{V}_1 \cdot \hat{V}_2 = \cos\theta$	When both operands are normalized, the result of dot product is the cosine of the subtended angle
Projected size: $\hat{V}_1 \cdot \vec{V}_2 = \|\vec{V}_2\|\cos\theta$	Projected size of $\vec{V}_2$ (the un-normalized vector) along the $\hat{V}_1$ (the normalized vector) direction
Projected size: $\vec{V}_1 \cdot \hat{V}_2 = \|\vec{V}_1\|\cos\theta$	Projected size of $\vec{V}_1$ along the $\hat{V}_2$ direction

Insights into the Subtended Angle

Dot Product Results	The Angle θ	Conclusions
$\hat{V}_1 \cdot \hat{V}_2 = \cos\theta = 1$	$\theta = 0°$	The vectors are in the exact same direction, $\hat{V}_1 == \hat{V}_2$
$\hat{V}_1 \cdot \hat{V}_2 = \cos\theta = 0$	$\theta = 90°$	The vector directions are perpendicular to each other
$\hat{V}_1 \cdot \hat{V}_2 = \cos\theta > 0$	$\theta < 90°$	The vectors are pointing along similar directions
$\hat{V}_1 \cdot \hat{V}_2 = \cos\theta < 0$	$\theta > 90°$	The vectors are pointing along similar, but opposite directions
$\hat{V}_1 \cdot \hat{V}_2 = \cos\theta = -1$	$\theta = 180°$	The vectors are pointing in the exact opposite direction $\hat{V}_1 = -\hat{V}_2$

The Line Equations

The line segment bounded by the given two positions, P_0 and P_1, can be expressed as either of the following:

$$lv(s) = P_0 + s\vec{V}_1$$

$$l(t) = P_0 + t\hat{V}_1$$

where

$$\vec{V}_1 = P_1 - P_0$$

and the values of the parameters *s* and *t* provide the following insights into a position on the line segment.

Values of *s*	Values of *t*	Position Identified
$s < 0$	$t < 0$	Measured along the $\vec{V}_1$ direction, a position before the beginning position, P_0
$0 \le s \le 1$	$0 \le t \le \lVert\vec{V}_1\rVert$	A position within the line segment
$s > 1$	$t > \lVert\vec{V}_1\rVert$	Measured along the $\vec{V}_1$ direction, a position after the end position, P_1

CHAPTER 6

Vector Cross Products and 2D Planes

After completing this chapter, you will be able to

- Differentiate between the Left-Handed and Right-Handed 3D Coordinate System
- Discuss the vector cross product definition and the resulting vector direction and magnitude
- Describe the geometric interpretation of the vector cross product
- Relate the 2D plane equation to the vector plane equation and its parameters
- Interpret the geometric implications of the vector plane equation
- Relate the cross product result to 2D plane equations
- Derive an axis frame when given two non-parallel vectors
- Apply the vector concepts learned to solve point to plane distance, point to plane projection, line to plane intersection, and reflecting a vector across a plane

Introduction

In Chapter 4, you learned about vectors—that the relationship between two positions can be defined by a direction and a distance. Vectors and their rules of operation enabled you to precisely describe and analyze object motions. In Chapter 5, you learned about vector dot products—that the relationship between two vectors can be

K. Sung and G. Smith, *Basic Math for Game Development with Unity 3D*,
https://doi.org/10.1007/978-1-4842-9885-5_6

characterized by their subtended angle and projected sizes. The vector dot product and its rules of operation allowed you to accurately represent and analyze arbitrary line segments, including distances between these line segments and other objects. In this chapter, you will learn about how the vector cross product can be used to relate two vectors to the space that defines these vectors and some applications of these concepts.

The result of the vector cross product is a new direction. Interestingly, and as you will learn, this new direction characterizes the space that defines the two vectors as a 2D plane, that is, this new direction defines a plane that both vectors exist on. This new knowledge enables a convenient representation of and the ability to analyze arbitrary 2D planes, including computing distances to, projections onto, and line intersections with any 2D plane. Although these are not direct applications, they are topics that become more comprehensible because of insights gained from the understanding of the vector cross product.

In video games, it is often necessary to process and analyze the relationships between planes and objects or the motion of objects. For example, in a city building game with a top-down view perspective, when a meteoroid is fast approaching the player's city, you may want to project the shadow of the meteoroid as it travels across the city as well as highlight its impact zone to warn players of the impending destruction. Additionally, immediately after the impact, you may want the meteoroid to bounce or slide across the ground. The shadow indicator can be accomplished by projecting the meteoroid onto the city plane, the reflection direction for the bounce is the velocity line reflecting off the ground plane, and the sliding direction would be the reflection direction projected onto the ground plane. As you can see from this brief example, the ability to represent and work with 2D planes is indeed fundamental to video game development.

The chapter begins by introducing conventions for representing a 3D coordinate system so that you can analyze three perpendicular vectors with consistency. The details of the cross products are then described. The application of the cross product results is then showcased in the solution to the inside-outside test of a general 2D region. At this point the chapter takes a slight change in perspective; instead of analyzing problems and solutions based on the results of the cross product, the chapter focuses on applying the insights gained from the vector cross product in the interpretation of the vector plane equation. The remaining of this chapter examines some of the important problems in video game development when working with 2D planes.

3D Coordinate System Convention

Since the analysis of the vector cross product involves understanding the direction of vectors in 3D space, you need to understand the conventions of representing a 3D coordinate system. In 2D space, when referencing the Cartesian Coordinate System, it is a generally agreed upon convention that the origin is on the lower left, the X-axis points toward the right, and the Y-axis points upward. Note that this is a convention and not a mathematical rule or any kind of property. People simply agree to follow these sets of rules.

Unfortunately, there are two sets of generally accepted conventions for 3D space. Although you have been working with 3D vectors, until now, there has not been the need to focus on the specific directions of the major axes. As you will see, unlike the dot product, the vector cross product result is not a simple floating-point number, but a vector that is perpendicular to both of the operand vectors. In this case, it is critical and essential to understand, differentiate, and follow one of the 3D coordinate system conventions. Figure 6-1 illustrates the two different conventions in describing a 3D coordinate system, either according to the left or the right hand. These are referred to as the **Left-** or **Right-Handed Coordinate System**.

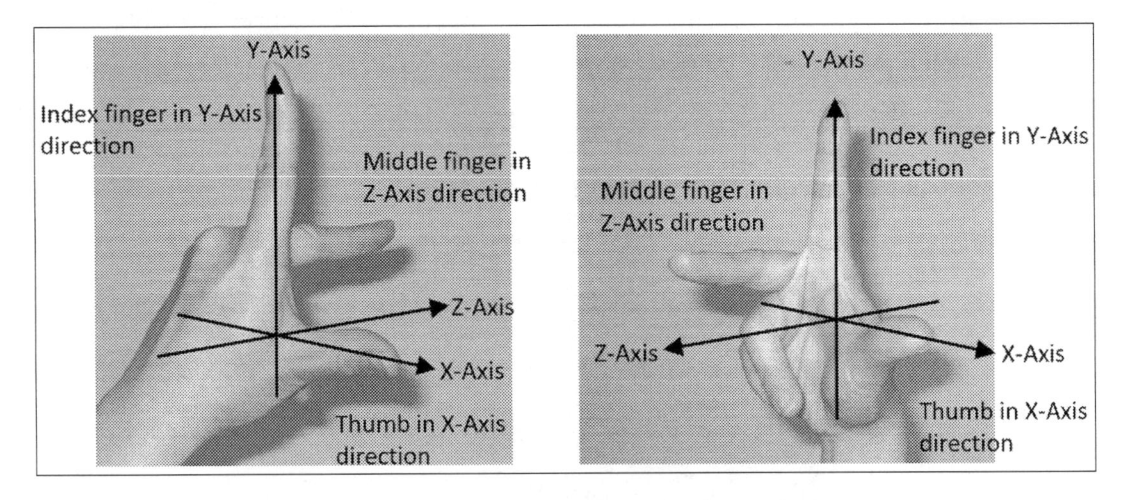

***Figure 6-1.** The directions of the major axes in the Left- and Right-Handed Coordinate System*

In both the Left- and Right-Handed Coordinate Systems, the first three fingers are used to represent and point in the directions of the X-, Y-, and Z-axes. The thumb represents and points in the direction of the X-axis, the index finger the Y-axis, and the middle finger the Z-axis. The left and right images of Figure 6-1 show that under this convention, while the X- and Y-axes still follow the right- and upward directions, the Z-axis directions are opposite. Note that the fingers of the left- and right-hand point toward the directions of the major axes and do not define the location of the axes.

Both the Left- and Right-Handed conventions are accepted in general by the video game and interactive graphics community. These are conventions for analyzing and discussing directions. It is critical to know the reference, the Left- or Right-Handed system, being used and essential to be consistent in following the selected convention. Fortunately, once selected and followed consistently, there are no other consequences or special cases in any of the discussions concerning the fundamentals of vector math. It is simply important to know which convention is used and to be sure to follow that convention consistently throughout.

Unity Follows the Left-Handed Coordinate System

Figure 6-2 shows a screenshot of the Unity Editor Scene View where the top-right coordinate icon is zoomed in upon and shown on the right of the figure. You can verify with your left hand that with your thumb stretching out along the red X-axis, your index finger following the green Y-axis, and your middle finger in the direction of the blue Z-axis, Unity follows the Left-Handed Coordinate System convention. Therefore, this is the convention that will be followed in this book. Once again, all the concepts being discussed are applicable to either 3D coordinate system conventions, as long as you follow the selected convention and maintained consistency.

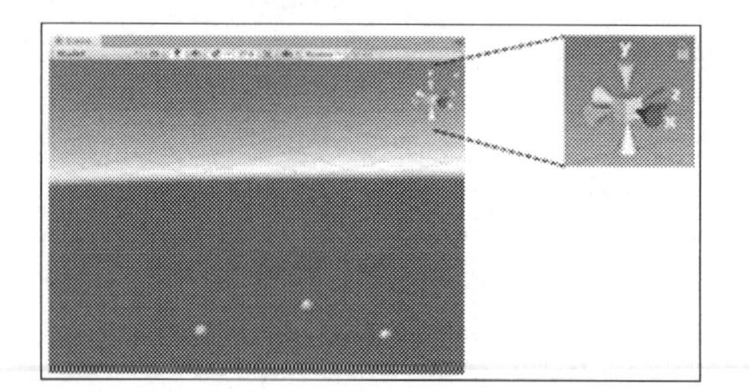

Figure 6-2. *The Unity Editor Scene View Window coordinate icon*

Vector Cross Product: The Perpendicular Direction

Recall in the previous chapter where you verified that a 2D plane can always be derived to draw two non-parallel vectors. This 2D plane is the plane that represents the space or area that defines or contains these two vectors. Through this chapter, you will learn that 2D planes are characterized by a vector that is perpendicular to it and that this perpendicular vector is the result of the cross product between two non-parallel vectors.

Figure 6-3 shows that, in general, there are two directions that are perpendicular to any two non-parallel vectors $\vec{V}_1$ and $\vec{V}_2$. Once again, as discussed previously, these two vectors are depicted at the same tail location for convenient visual analysis. It is important to reiterate that the vector definition is independent of positions and the following discussions are valid even when the two vectors do not share the same tail position.

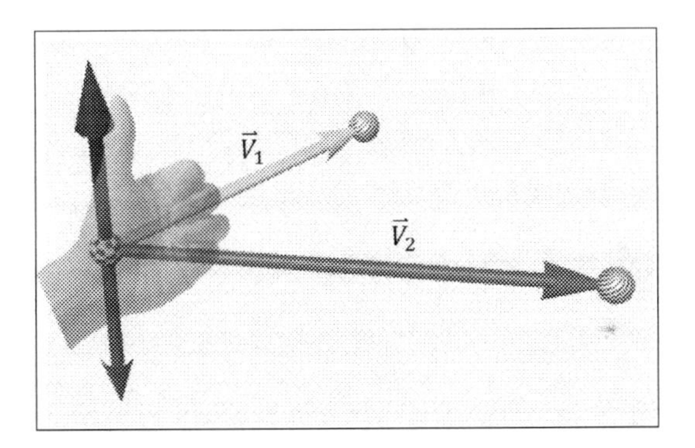

***Figure 6-3.** Vectors that are perpendicular to the two non-parallel vectors, $\vec{V}_1$ and $\vec{V}_2$*

Figure 6-3 shows a left-hand thumb pointing in a direction where the index to little fingers are aligned with the direction of the first vector, $\vec{V}_1$, and then curl toward the second vector, $\vec{V}_2$. The left thumb direction is the one that is perpendicular to the plane that defines $\vec{V}_1$ and $\vec{V}_2$. Of course, the direction opposite to the left thumb is the second direction that is perpendicular to the plane that defines these two vectors.

Note The left hand is used for direction resolution because this book follows Unity's choice of Left-Handed Coordinate System. A Right-Handed Coordinate System would follow the same finger curling process as Figure 6-3 with the right hand and identify a set of directions that seem opposite to that of Figure 6-3. Please do not be concerned. Remember that the left- and right-handed conventions also affect the directions of the major axes. Once again, in the end, both conventions, as long as followed consistently throughout, will produce identical results.

The vector cross product computes the two new directions, along or opposite to the thumb direction in Figure 6-3. These are the two directions that are perpendicular to both of the vectors, $\vec{V}_1$ and $\vec{V}_2$. This chapter will lead you on a journey to examine, understand, and relate these results to 2D planes in 3D space. After which, the problems and solutions associated with 2D planes that are relevant to video game development will be analyzed.

Definition of Vector Cross Product

Given two vectors in 3D space

$$\vec{V}_1 = (x_1, y_1, z_1)$$

$$\vec{V}_2 = (x_2, y_2, z_2)$$

the **cross product**, or vector cross product, between the two vectors is defined as

$$\vec{V}_1 \times \vec{V}_2 = (y_1 z_2 - z_1 y_2,\ z_1 x_2 - x_1 z_2,\ x_1 y_2 - y_1 x_2)$$

Notice that

- Symbol: The symbol for the cross product operation, "×", is literally a "cross".
- Operands: The operation expects two vector operands.
- Result: The result of the operation is a vector with x-, y-, and z-component values.

When compared to the other vector operations you have learned, the cross product also expects two vector operands. Additionally, similar to vector addition and subtraction, and in contrast to the vector dot product, the result of the vector cross product is a vector.

Unlike vector addition and subtraction, the vector cross product result, the x-, y-, and z-component values are not straightforward functions of its operands' corresponding components. Examine these values carefully and you will notice a pattern. For example, the x-component result, $y_1z_2 - z_1y_2$, is the subtraction of the multiplication of operand component values other than their x-components. This pattern is consistent for each of the y- and z-components. Though interesting and important in general, in the context of video game development, these observations do not lead to direct applications.

The left, center, and right tables in Figure 6-4 illustrate an approach that may help you remember the cross product formula. Each of the tables has an x-, y-, and z-heading with two rows consisting of the corresponding component values for the two operand vectors. The left table shows that the x-component cross product result is computed by ignoring the grayed-out x-component values, following the two arrows, and calculating and subtracting the products of the y- and z-components y_1z_2 and z_1y_2. The center table shows a similar computation for the y-component cross product results and the right table for the z-component cross product results. Note that the subtraction order for the y-component is reversed that of the x- and z-components.

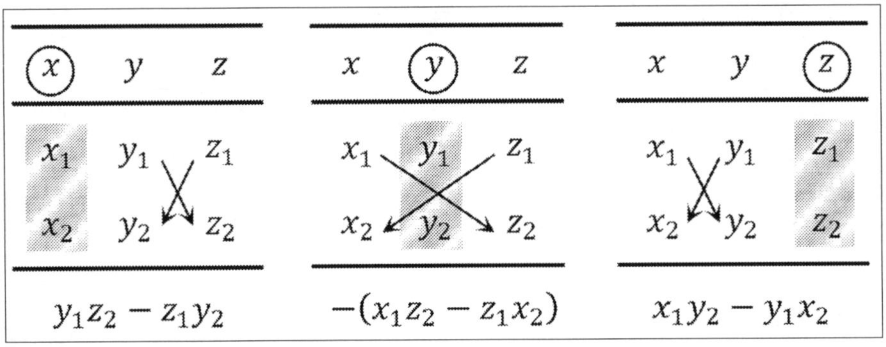

x	y	z
x_1	y_1	z_1
x_2	y_2	z_2
	$y_1z_2 - z_1y_2$	

x	y	z
x_1	y_1	z_1
x_2	y_2	z_2
	$-(x_1z_2 - z_1x_2)$	

x	y	z
x_1	y_1	z_1
x_2	y_2	z_2
	$x_1y_2 - y_1x_2$	

Figure 6-4. *Components of the cross product*

Geometric Interpretation of Vector Cross Products

Figure 6-5 shows the geometric interpretation of the vector cross product. Since Unity follows the Left-Handed Coordinate System, the result of $\vec{V}_1 \times \vec{V}_2$ is a vector in the direction of the thumb on your left hand when following the finger curling process described previously. It follows that for $\vec{V}_2 \times \vec{V}_1$, with the index to little fingers aligned with the first operand, in this case the $\vec{V}_2$ vector, and then curl toward the second operand, or the $\vec{V}_1$ vector, the resulting vector is in the opposite direction (turn your hand so you're giving a thumbs down instead of a thumbs up). The cross product results, $\vec{V}_1 \times \vec{V}_2$ and $\vec{V}_2 \times \vec{V}_1$, are perpendicular to their operand vectors, $\vec{V}_1$ and $\vec{V}_2$, and, as a result, are perpendicular to the plane that defines $\vec{V}_1$ and $\vec{V}_2$.

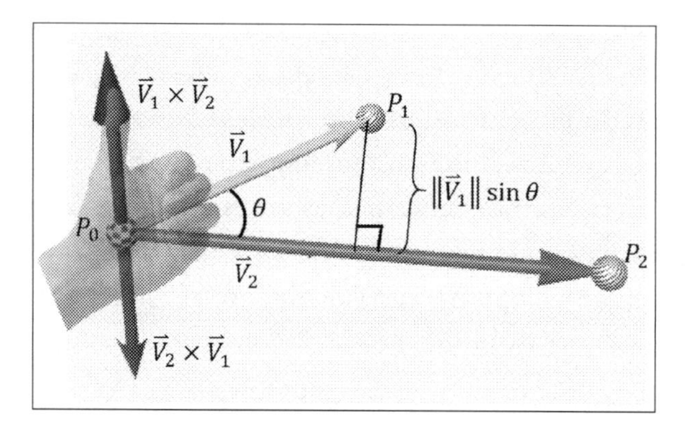

***Figure 6-5.** The directions of vector cross product results*

The magnitude of the vector resulting from the cross product or the magnitude of the perpendicular vector, with details left as an exercise, can be shown to be

$$\|\vec{V}_1 \times \vec{V}\|_2 = \sqrt{(y_1 z_2 - z_1 y_2)^2 + (z_1 x_2 - x_1 z_2)^2 + (x_1 y_2 - y_1 x_2)^2} = \|\vec{V}\|_1 \|\vec{V}_2\| \sin\theta$$

where θ is the subtended angle between $\vec{V}_1$ and $\vec{V}_2$. Notice that when both $\vec{V}_1$ and $\vec{V}_2$ are normalized, thus both with magnitude of 1.0, then

$$\|\hat{V}_1 \times \hat{V}_2\| = \sin\theta$$

Note Although the cross product result encodes the sine of the subtended angle, it is seldom, if ever, used specifically for analyzing subtended angles between vectors. Instead, the dot product is always used. This is because when comparing the two, the cross product operation involves more floating-point operations, and more importantly, the cross product result is a vector and thus a magnitude operation must be performed to convert the vector into a floating-point number for deriving the angle information. In contrast, the dot product is more efficient to compute and the result itself encodes the angle information and thus does not need further processing. For these reasons, the dot product is always used for analyzing angles subtended by vectors, for example, testing for parallel or perpendicular.

In Figure 6-5, notice that $P_0P_1P_2$ is a triangle. Assuming the edge, P_0P_2, is the base, then you know the area of the triangle is the half the length of the base, or $\|\vec{V}_2\|$, multiplied by the height. In this case, the height is the perpendicular distance between P_1 and the edge, P_0P_2, or $\|\vec{V}_1\|\sin\theta$. In this way, the area of the triangle $P_0P_1P_2$ is

$$Area\ of\ Triangle\ P_0P_1P_2 = \frac{1}{2}\|\vec{V}_1\|\|\vec{V}_2\|\sin\theta$$

And the magnitude of the cross product result is twice the area of the triangle

$$\|\vec{V}_1 \times \vec{V}_2\| = 2 \times Area\ of\ Triangle\ P_0P_1P_2 = \|\vec{V}_1\|\|\vec{V}_2\|\sin\theta$$

Though the magnitude of the resulting vector and the sine relationship of the subtended angle are important information to take note of when learning the vector cross product, the analysis presented in the rest of this book only takes advantage of the fact that the resulting vector is perpendicular to the operands and the 2D plane that defines the operand vectors.

Properties of Vector Cross Product

The vector cross product properties of commutative, associative, and distributive over a floating-point scaling factor *s* are summarized in Table 6-1.

Table 6-1. Properties of vector cross product

Properties	Vector Dot Product
Anti-commutative	$\vec{V}_1 \times \vec{V}_2 = -\vec{V}_2 \times \vec{V}_1$
Not Associative	$\left(\vec{V}_1 \times \vec{V}_2\right) \times \vec{V}_3 \neq \vec{V}_1 \times \left(\vec{V}_2 \times \vec{V}_3\right)$
Distributive over scale factor, *s*	$s\left(\vec{V}_1 \times \vec{V}_2\right) = \left(s\vec{V}_1\right) \times \vec{V}_2 = \vec{V}_1 \times \left(s\vec{V}_2\right)$

Table 6-1 shows a set of rather unfamiliar properties. Fortunately, the applications of vector cross products in video game development are often limited to simple operations in the determination of directions. It is seldom for cross product operations to be embedded in complex vector equations. Finally, the definition of the vector cross product states that

$$\vec{V}_1 \times \vec{V}_1 = ZeroVector$$

and that any vector crossed with the zero vector will results in a zero vector

$$\vec{V}_1 \times ZeroVector = ZeroVector \times \vec{V}_1 = ZeroVector$$

The Vector Cross Products Example

This example demonstrates the results of performing the vector cross product between two given vectors. This example allows you to interactively manipulate and define two vectors and then examine the results of performing the cross product between these vectors. Figure 6-6 shows a screenshot of running the `EX_6_1_VectorCrossProducts` scene from the `Chapter-6-CrossProducts` project.

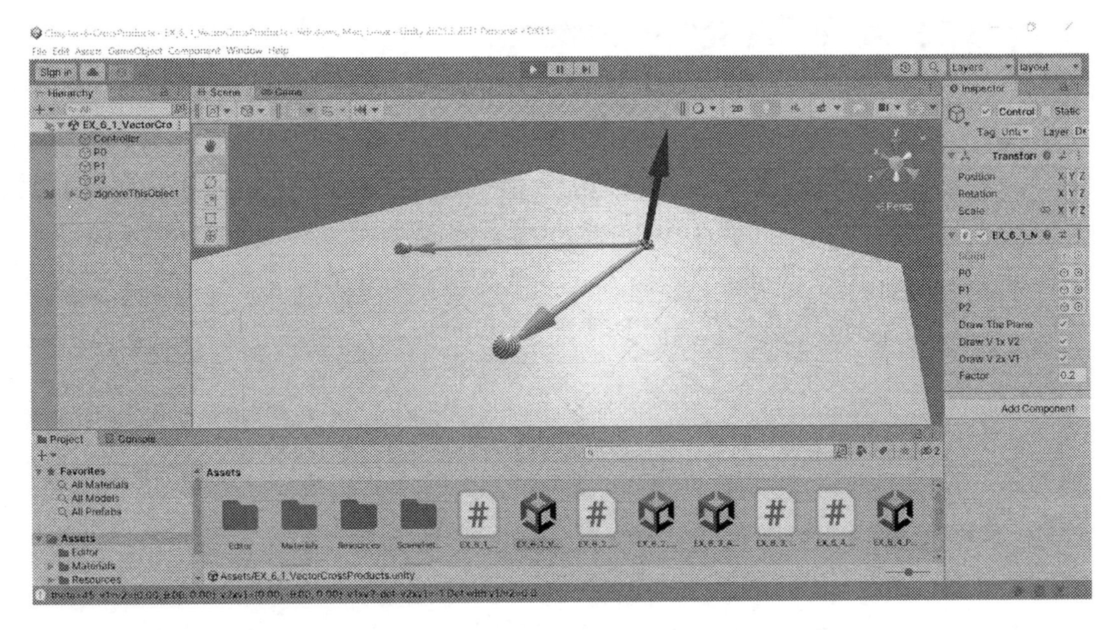

Figure 6-6. Running the Vector Cross Products example

The goals of this example are for you to

- Examine the results of the cross product between two arbitrarily defined vectors
- Verify that the vector resulting from a cross product is perpendicular to both of the operands with a magnitude that is directly proportional to the sine of their subtended angle
- Examine the source code that computes and uses the results of the vector cross product

Examine the Scene

Take a look at the `Example_6_1_VectorCrossProducts` scene and observe the predefined game objects in the Hierarchy Window. In addition to the `Controller`, there are three objects in this scene: a checkered sphere (`P0`) and two striped spheres (`P1` and `P2`). These three game objects will have their corresponding `transform.localPosition` properties referenced to define the two vectors for performing the cross product operations.

Analyze Controller MyScript Component

The `MyScript` component on the `Controller` shows two sets of variables. One set is for defining the two vectors and the other set is for examining the visualization of the cross product between these two vectors and the plane that they define. The first set of variables are `P0`, `P1`, and `P2` and are defined for accessing the game objects with their corresponding names. In this example, you will manipulate the positions of these three game objects to define two vectors, $\vec{V}_1$ and $\vec{V}_2$

$$\vec{V}_1 = P_1 - P_0$$

$$\vec{V}_2 = P_2 - P_0$$

and then examine the result of the cross product between these vectors.

The variables in the second set, `DrawThePlane`, `DrawV1xV2`, and `DrawV2xV1`, are toggles for hiding and showing the plane that defines $\vec{V}_1$ and $\vec{V}_2$ and the corresponding results of the cross products, while the last variable, `Factor`, is the scaling factor applied to the length of the vector from the cross product result, allowing for easier visualization.

Interact with the Example

Click the Play Button to run the example. In the Scene View Window, you will observe two vectors with tail positions located at the checkered sphere, `P0`, and a greenish plane where the two vectors are drawn. The two vectors are $\vec{V}_1$ and $\vec{V}_2$ and are defined by the positions of `P0`, `P1`, and `P2` game objects as previously explained. You will also observe two other vectors in this scene. Both of these vectors are located at the checkered sphere location (`P0`), a black vector that is the result of $\vec{V}_1 \times \vec{V}_2$, and a red vector, the result of $\vec{V}_2 \times \vec{V}_1$. You can confirm that both of these results follow the Left-Handed Coordinate System by extending the index to little fingers of your left hand along the $\vec{V}_1$ direction (the cyan vector) and then curling these fingers toward the $\vec{V}_2$ direction (the magenta vector). In a similar fashion to that of Figure 6-5, your thumb should be pointing along the direction of the black vector which is the result of $\vec{V}_1 \times \vec{V}_2$. You can repeat the left-hand finger curling process to verify that the red vector is indeed pointing in the direction of $\vec{V}_2 \times \vec{V}_1$.

In the Console Window, you can examine the text output where the subtended angle between $\vec{V}_1$ and $\vec{V}_2$ as well as various dot product results are printed for verification purposes. First, you can verify that the printed subtended angles between $\vec{V}_1$ and $\vec{V}_2$

reflect your observations in the Scene View. Next, examine the results of the dot product between the normalized black and red vectors. Since these two vectors are always parallel and pointing in the opposite directions, the angle between them is always 180° and thus the result of the dot product, or the cosine of this angle, is always −1:

$$\left(\hat{V}_1 \times \hat{V}_2\right) \cdot \left(\hat{V}_2 \times \hat{V}_1\right) = -\left(\hat{V}_1 \times \hat{V}_2\right) \cdot \left(\hat{V}_1 \times \hat{V}_2\right) = -1$$

Additionally, the results of the dot product between the cross product result $\left(\hat{V}_1 \times \hat{V}_2\right)$ and the operands, $\hat{V}_1$ and $\hat{V}_2$, are also printed out. You can verify that the cross product result is always perpendicular with its operands by observing that the dot product results between these vectors are always zero, or very close to being zero:

$$\left(\hat{V}_1 \times \hat{V}_2\right) \cdot \hat{V}_1 = \left(\hat{V}_1 \times \hat{V}_2\right) \cdot \hat{V}_2 = 0$$

Note that since the initial values of `P0`, `P1`, and `P2` define the three positions to be on the X-Z plane, the initial $\vec{V}_1$ and $\vec{V}_2$ vectors are also in the X-Z plane. Therefore, the cross product results are vectors pointing in the positive and negative y-directions, perpendicular to both $\vec{V}_1$ and $\vec{V}_2$, and the plane that defines these two vectors is the X-Z plane.

In the following interactions, feel free to toggle and hide any of the components if you find them distracting. You can also adjust the `Factor` value to scale the lengths of the black and red vectors for easier visual examination.

Select `P1` and adjust its z-component value to change the size of $\vec{V}_1$ without changing the subtended angle. Notice that although both are changing, the lengths of the black and red vectors are always the same. This is because both of the vectors vary in direct proportion to the length of $\vec{V}_1$. Now try moving `P1` toward `P2` such that the $\vec{V}_1$ vector approaches $\vec{V}_2$, or move `P1` toward `P0` such that the $\vec{V}_1$ vector approaches the zero vector. Notice that in both cases, the cross product result, the black and the red vectors, both approach a length of zero. You can repeat and verify all these observations by adjusting `P2` or by changing $\vec{V}_2$ in a similar fashion. These manipulations and observations verify that the magnitude of the cross product result is in direct proportion to the magnitude of the operand vectors

$$\left\|\vec{V}_1 \times \vec{V}_2\right\| = \left\|\vec{V}_1\right\| \left\|\vec{V}_2\right\| \sin\theta$$

and that all cross products computed with the zero vector will result in the zero vector.

Now restart the game and adjust the x-component of `P1` to change the subtended angle. Notice that when this angle is between 0° and 90°, the lengths of the black and red vectors vary in direct proportion and then change to vary in the inverse proportion when the angle is beyond 90°. Continue to adjust both the x- and z-component values to increase the subtended angle to beyond 180° and notice the direction swap between the black and red vectors. Recall that a subtended angle is always between 0° and 180°; you can verify with your left hand that after the direction swap, the black vector is still pointing in the direction of $\hat{V}_1 \times \hat{V}_2$.

Notice that until this point, your manipulation has been restricted to the X-Z plane and that the cross product results, the black and red vectors, are always in the positive and negative y-directions. Now, select any of the positions and change the y-component values. As you have observed when investigating the dot product in the previous chapter, the green plane is updated and continues to cut through both $\vec{V}_1$ and $\vec{V}_2$. The interesting observation is that the cross product results, the black and red vectors, are always perpendicular to the green plane. This observation suggests that the green plane is defined by the cross product result. This concept will be explored in the next subsection.

Details of MyScript

Open `MyScript` and examine the source code in the IDE. The instance variables and the `Start()` function are as follows:

```
//Three positions to define two vectors: P0->P1 and P0->P2
public GameObject P0 = null;    // Position P0
public GameObject P1 = null;    // Position P1
public GameObject P2 = null;    // Position P2

public bool DrawThePlane = true;
public bool DrawV1xV2 = true;
public bool DrawV2xV1 = true;
public float Factor = 0.4f;

#region For visualizing the vectors
#endregion
```

```
// Start is called before the first frame update
void Start() {
    Debug.Assert(P0 != null);  // Verify proper editor init
    Debug.Assert(P1 != null);
    Debug.Assert(P2 != null);

    #region For visualizing the vectors
    #endregion
}
```

All the public variables for MyScript have been discussed when analyzing the Controller's MyScript component, and as in all previous examples, the Debug.Assert() calls in the Start() function ensure proper setup regarding referencing the appropriate game objects via the Inspector Window. The Update() function is listed as follows:

```
void Update() {
    Vector3 v1 = P1.transform.localPosition -
                 P0.transform.localPosition;
    Vector3 v2 = P2.transform.localPosition -
                 P0.transform.localPosition;
    Vector3 v1xv2 = Vector3.Cross(v1, v2);
    Vector3 v2xv1 = Vector3.Cross(v2, v1);

    float d = Vector3.Dot(v1.normalized, v2.normalized);
    bool notParallel = (Mathf.Abs(d) < (1.0f - float.Epsilon));
    if (notParallel) {
        float theta = Mathf.Acos(d) * Mathf.Rad2Deg;
        float cd = Vector3.Dot(v1xv2.normalized, v2xv1.normalized);
        float dv1 = Vector3.Dot(v1xv2, v1);
        float dv2 = Vector3.Dot(v1xv2, v2);
        Debug.Log(" theta=" + theta + "  v1xv2=" + v1xv2 +
            "  v2xv1=" + v2xv1 + "  v1xv2-dot-v2xv1=" + cd +
            " Dot with v1/v2=" + dv1 + " " + dv2);
    } else {
        Debug.Log("Two vectors are parallel,
                   cross product is a zero vector");
    }
```

```
    #region  For visualizing the vectors
    #endregion
}
```

The first four lines of the Update() function compute

$$\vec{V}_1 = P_1 - P_0$$

$$\vec{V}_2 = P_2 - P_0$$

$$v1xv2 = \vec{V}_1 \times \vec{V}_2$$

$$v2xv1 = \vec{V}_2 \times \vec{V}_1$$

Next, the cosine of the angle between $\vec{V}_1$ and $\vec{V}_2$ is computed as the dot product of the normalized vectors. This value is examined to ensure that the cross product results will not be zero vectors. The various dot product results are then computed and printed to the Console window.

Note **Collinear** and **collinear test.** In general, given three positions, P_0, P_1, and P_2, that define two vectors, $\vec{V}_1 = P_1 - P_0$ and $\vec{V}_2 = P_2 - P_0$. If $\hat{V}_1 \cdot \hat{V}_2$ is approximately 1 or −1, then you can conclude that the three points are approximately along the same line. In this case, P_0, P_1, and P_2 are referred to as being collinear. The dot product check against approximately 1 or −1 is a convenient collinear test.

Takeaway from This Example

This example demonstrates that the result of the cross product is indeed a vector with a direction that can be derived by curling your left-hand fingers and that the magnitude of the resulting vector is indeed directly proportional to the sizes of the operands and the sine of the subtended angle. You have also confirmed that the cross product of any vector with itself or with the zero vector results in the zero vector. Additionally, you have verified that the cross product is anti-commutative as reversing the operand order results in a vector pointing in the perfectly opposite direction. However, the most interesting observation is that the cross product result is always perpendicular to the operand vectors and thus the 2D plane that contains the two operand vectors.

Relevant mathematical concepts covered include

- The cross product result is a vector that is perpendicular to both of its operands and the 2D plane that contains the operands.
- The magnitude of the vector resulting from a cross product is directly proportional to the magnitude of the operands and the sine of the subtended angle.
- The cross product is not defined when the two operand vectors are derived from three positions that are collinear. This is because three collinear positions can only define one direction and thus one vector, and the cross product of a vector with itself is the zero vector.

EXERCISES

Derive the Magnitude of the Vector Resulting from a Cross Product

Given

$$\vec{V}_1 = (x_1, y_1, z_1)$$

$$\vec{V}_2 = (x_2, y_2, z_2)$$

You know that the cross product is defined as

$$\vec{V}_1 \times \vec{V}_2 = (y_1 z_2 - z_1 y_2,\ z_1 x_2 - x_1 z_2,\ x_1 y_2 - y_1 x_2)$$

where the magnitude of the resulting vector is

$$\|\vec{V}_1 \times \vec{V}_2\| = \|\vec{V}_1\| \|\vec{V}_2\| \sin\theta$$

Recall the trigonometry identity and the dot product definition that

$$\sin^2\theta + \cos^2\theta = 1$$

$$\hat{V}_1 \cdot \hat{V}_2 = \frac{\vec{V}_1 \cdot \vec{V}_2}{\|\vec{V}_1\| \|\vec{V}_2\|} = \cos\theta$$

So

$$\begin{aligned}\left\|\vec{V}_1 \times \vec{V}_2\right\| &= \left\|\vec{V}_1\right\|\left\|\vec{V}_2\right\|\sin\theta \\ &= \left\|\vec{V}_1\right\|\left\|\vec{V}_2\right\|\sqrt{1-\cos^2\theta} \\ &= \left\|\vec{V}_1\right\|\left\|\vec{V}_2\right\|\sqrt{1-\left(\frac{\vec{V}_1\cdot\vec{V}_2}{\left\|\vec{V}_1\right\|\left\|\vec{V}_2\right\|}\right)^2}\end{aligned}$$

Now, simplify the algebra expression and show that

$$\left\|\vec{V}_1 \times \vec{V}\right\|_2 = \sqrt{\left(y_1z_2 - z_1y_2\right)^2 + \left(z_1x_2 - x_1z_2\right)^2 + \left(x_1y_2 - y_1x_2\right)^2}$$

Verify the Cross Product Formula

When computing the cross products in `MyScript`

```
Vector3 v1xv2 = Vector3.Cross(v1, v2);
Vector3 v2xv1 = Vector3.Cross(v2, v1);
```

replace these two lines of code with the explicit cross product definition by creating `v1xv2` and `v2xv1` as new `Vector3` objects with appropriate component values and verify that the runtime results are identical.

The Vector Plane Equation

Throughout the last couple of chapters, you have been working with two vectors defined by three positions and observed that a 2D plane can always be defined when the two vectors are not parallel. Note that both of these observations are identical, and two non-parallel vectors are the same as saying that the three positions that define the two vectors are non-collinear. Intuitively, this should not be surprising because from basic geometry you have learned that three points, as long as they are not all along the same line, define a triangle, and a triangle is the simplest shape in 2D space. For this reason, if a triangle can be formed, as you have observed, then it is always possible to form two non-parallel vectors, and thus a 2D plane can always be defined as well.

Now, you can derive the equation of this 2D plane based on the result of the cross product. Recall from basic geometry that the equation of a 2D plane in 3D space is

$$Ax + By + Cz = E$$

where A, B, C, and E are floating-point constants and x, y, and z are unknowns in 3D space. This equation states that if you gather all the positions (x, y, z) that satisfy the condition where the sum of multiplying x by A, y by B, and z by C is equal to E, then you will find that all these positions are points on the given 2D plane. Interestingly this equation can also be written in vector dot product form, where you can define the vector $\vec{V}$ and a position vector, p, where

$$\vec{V} = (A,\ B,\ C)$$

$$p = (x,\ y,\ z)$$

Then, the 2D plane equation can be written as

$$\vec{V} \cdot p = E$$

Note Recall that a position, *p*, can be interpreted as a position vector, $\vec{V}$, from the origin position, P_0, where

$$\vec{V} = p - P_0 = p$$

Since in this case, P_0 is the origin (0,0,0). To avoid the confusion and nuance of introducing additional symbols, it is a common practice to reuse the symbol of the position (*p*) to represent the corresponding position vector. In the rest of this book, please do not be confused when you encounter language and a symbol such as "following along the position vector *p*." Such statements are always referring to the vector from the origin toward the position, *p*.

If you divide both side of the equation by a nonzero floating-point number, in this case, $\|\vec{V}\|$, the equation becomes

$$\hat{V} \cdot p = \frac{E}{\|\vec{V}\|}$$

Now, let $D = \frac{E}{\|\vec{V}\|}$, then a 2D plane equation can be written as the **vector plane equation** or

$$\hat{V} \cdot p = D$$

This equation may look familiar because it is basically the vector projection equation as illustrated in Figure 5-7. Figure 6-7 shows the geometric interpretation of the vector plane equation.

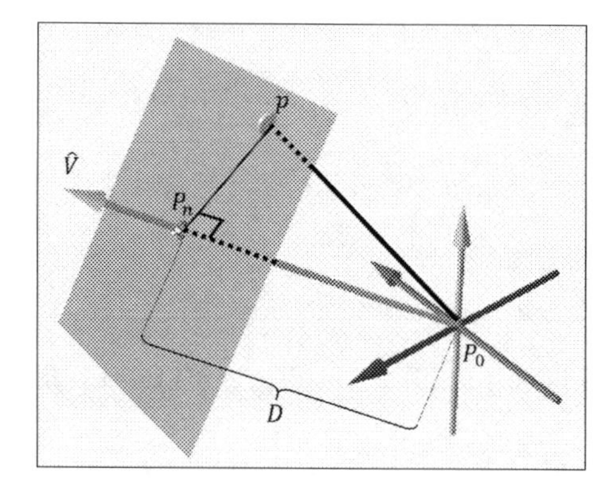

***Figure 6-7.** Geometric interpretation of the vector plane equation*

In Figure 6-7, P_0 is the origin and the vector $\hat{V}$ is the direction from the origin that is perpendicular and passes through a plane at position P_n. The plane is at a distance D from the origin when measured along the direction $\hat{V}$. The vector plane equation states that for any position p on this plane, it is true that the projection of this position vector onto the direction $\hat{V}$ will be of length D. In this way, the vector plane equation identifies all positions that satisfy the projected distance relationship with the $\hat{V}$ vector. As it turns out, these positions define the 2D plane. Notice that you must compute the $\hat{V}$ and D to derive the vector plane equation, $\hat{V} \cdot p = D$:

- Normal vector: $\hat{V}$ is the vector that is perpendicular to the plane; this vector is generally normalized such that the constant D in the equation indicates distance from the origin. As demonstrated in the derivation process, when this vector is not normalized, the magnitude of the vector can be divided through on both sides of the equation to compute the proper value for D.

- Distance to the plane: *D*, when the normal vector is normalized, this is the plane distance from the origin when measured along the $\hat{V}$ direction.

It is important to recognize that the vector plane equation identifies a 2D plane that is of infinite size. Any position in the Cartesian Coordinate System that satisfies the projected distance relationship is part of the solution set of the 2D plane and there are infinitely many positions in the solution set. As will be explored later, a 2D region is a bounded area on a 2D plane. This is analogous to 1D region, or a 1D interval, being a bounded line segment within an infinitely long line that is identified by a line equation.

Note A normal vector is a vector that is perpendicular to a plane. This should not be confused with a normalized vector, which is any vector of size 1. You can compute a normal vector which may not be normalized. You can then decide to normalize the normal vector such that you can work with a normalized normal vector. In the rest of this book, the vector symbol, $\vec{V}_n$, will be used to represent the normal vector of a 2D plane. Once again, a normal vector may or may not be normalized. In this case, $\vec{V}_n$, is a normal vector that is not normalized, and the vector, $\hat{V}_n$, is the normalized plane normal vector.

The Position P_n on a Plane

Notice the position P_n in Figure 6-7; this is the point on the plane that is *D* distance away from the origin when measured along the $\hat{V}_n$ direction. For this reason,

$$P_n = P_0 + D\hat{V}_n = D\hat{V}_n$$

In this case, P_0 is the origin $(0,0,0)$. In the rest of this chapter, the P_n position is computed and displayed on the 2D planes in all examples to provide orientation for and facilitate visualization.

Given a Position on a Plane

If you are given a plane normal vector, $\hat{V}_n$, and a position, P_{on}, that is on the plane, then you know that for any position, p, on the plane, $\overline{p - P_{on}}$ is a vector on the plane and that this vector must be perpendicular to $\hat{V}_n$. This means

$$\hat{V}_n \cdot (p - P_{on}) = 0 \qquad \text{two are perpendicular}$$

This equation can be simplified as follows:

$$\hat{V}_n \cdot p - \hat{V}_n \cdot P_{on} = 0 \qquad \text{distributive property}$$

$$\hat{V}_n \cdot p = \hat{V}_n \cdot P_{on} \qquad \text{move term across equality}$$

$$\hat{V}_n \cdot p = D \qquad P_{on} \text{ is on the plane}$$

which is simply the vector plane equation. This derivation shows that D, the distance from the origin to a plane, can be derived if you know the plane normal and one position on the plane.

Positions on 2D Planes

As a way of verifying the vector plane equation and to provide additional insights, Figure 6-8 shows that it is always possible to compute the point where a position vector intersects a plane.

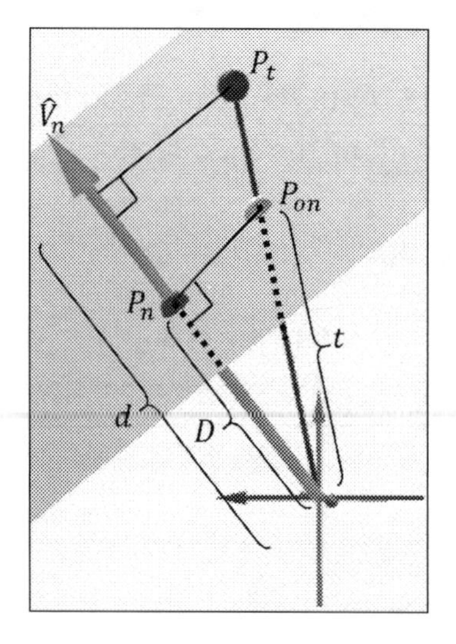

Figure 6-8. *Positions on a given plane*

In Figure 6-8, the given plane is defined by the normalized normal vector, $\hat{V}_n$, and the distance, D, measured along the $\hat{V}_n$ direction from the origin or

$$\hat{V}_n \cdot p = D$$

For any arbitrary position, P_t, it is always possible to compute P_{on}, the point where the position vector P_t intersects the given plane. As illustrated in Figure 6-8, P_{on} is along the position vector P_t and is t distance away from the origin

$$P_{on} = origin + tP_t = tP_t$$

Since P_{on} is on the plane, then it must be true that

$$\hat{V}_n \cdot P_{on} = D$$

or

$\hat{V}_n \cdot tP_t = D$	since $P_{on} = tP_t$
$t\left(\hat{V}_n \cdot P_t\right) = D$	distributive property
$t = \dfrac{D}{\hat{V}_n \cdot P_t}$	divide by $\hat{V}_n \cdot P_t$

With the distance, t, defined, it is now possible to compute the value of P_{on}! In the next example, the plane equation will be examined, especially in relation to the cross product result.

The Vector Plane Equations Example

This example demonstrates the vector plane equation. The example allows you to interactively define a 2D plane, manipulate an arbitrary point, and examine the intersection of this position vector with the 2D plane. Figure 6-9 shows a screenshot of running the `EX_6_2_VectorPlaneEquations` scene from the `Chapter-6-CrossProducts` project.

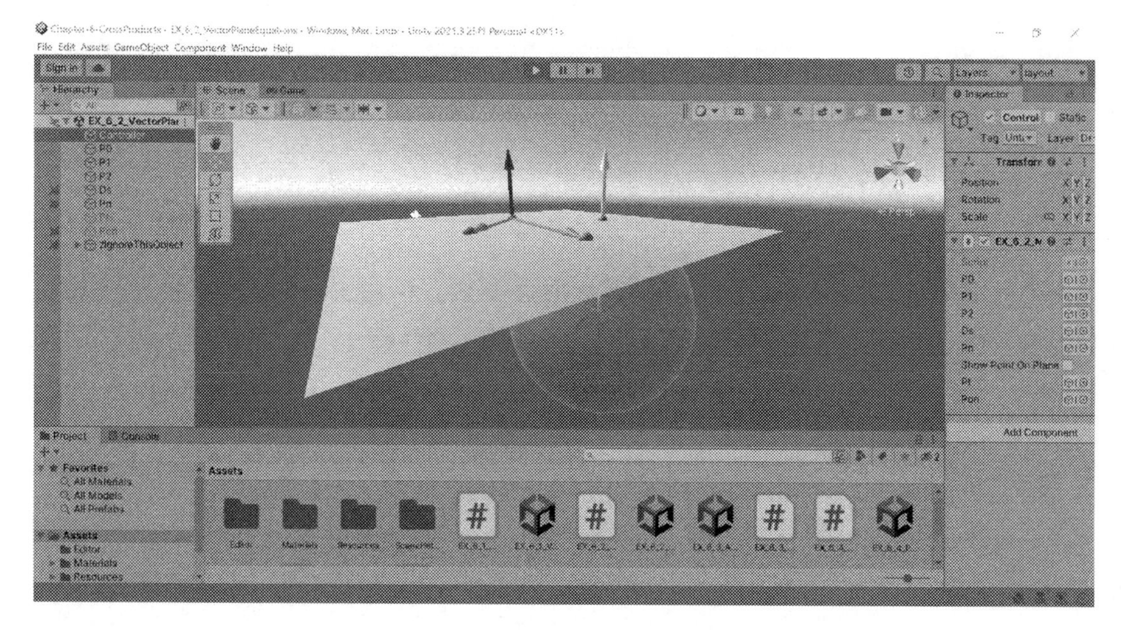

***Figure 6-9.** Running the Vector Plane Equations example*

The goals of this example are for you to

- Understand that the result of the cross product defines a plane normal vector
- Experience working with and gain an understanding of the parameters of the vector plane equation
- Verify the solution to the intersection between a position vector and a 2D plane
- Examine the implementation of working with the vector plane equation

Examine the Scene

Take a look at the `Example_6_2_VectorPlaneEquations` scene and observe the predefined game objects in the Hierarchy Window. In addition to the `Controller`, there are three sets of variables as follows:

- `P0`, `P1`, and `P2`: Game objects for defining two vectors to perform the cross product. The result from the cross product will be used as the plane normal vector.

- `Ds` and `Pn`: `Ds` is a transparent sphere located at the origin for showing the plane distance, *D*, from the origin, and `Pn` is the position where the plane normal vector with tail at the origin intersects the plane. Note, this is the same as saying, `Pn` is the point on the plane with position vector in the plane normal direction.
- `Pt` and `Pon`: `Pt` is a position you can manipulate and `Pon` is the point that the position vector `Pt` intersects with the plane.

Analyze Controller MyScript Component

The `MyScript` component on the `Controller` contains variables with the same name as their referenced game objects in the scene; these variables are used for position manipulations. The only exception is `Ds`, which does not have its position manipulated, instead its radius is set according to the distance, *D*, in the vector plane equation. The variable that doesn't represent any game object, `ShowPointOnPlane`, is a toggle used to control the showing or hiding of `Pt` and `Pon` computation results.

Interact with the Example

Click on Play Button to run the example. Notice that initially the `ShowPointOnPlane` toggle is switched off. You will first focus on examining and understanding the cross product result and its relationship with the plane normal before examining the intersection between a position vector and a plane.

In the initial scene you can observe, similar to the previous example, `P0`, `P1`, and `P2` positions defining the $\vec{V_1}$ (in cyan) and $\vec{V_2}$ (in magenta) vectors. You can also observe the black vector being computed as the result of $\vec{V_1} \times \vec{V_2}$. As with the previous example, the $\vec{V_1}$ and $\vec{V_2}$ vectors are defined on a 2D plane. In this scene, the 2D plane tangents, or touches at a single point, a transparent sphere centered at the origin. Here you will also find a white vector with its tail position at the origin, extending and cutting through the 2D plane perpendicularly at the red position, `Pn`. The white vector is the cross product result and is thus the plane normal vector, $\hat{V}_n$. The transparent sphere mentioned earlier has a radius, *D*, which is defined by projecting position `P0` in the plane normal direction or

$$D = \hat{V}_n \cdot P_0$$

In this way, the 2D plane has a vector plane equation

$$\hat{V}_n \cdot p = D$$

The red sphere on the plane, P_n, is the position vector that is D distance along the $\hat{V}_n$ direction from the origin or

$$P_n = D\hat{V}_n$$

It is worth repeating that this vector plane equation is defined completely by the positions P0, P1, and P2. The plane normal, $\hat{V}_n$, is the cross product of the two vectors defined by those positions, and the plane distance from the origin is the projection of the position vector P_0, in the $\hat{V}_n$ direction. Since the position P0 is referenced in defining both of the parameters of the vector plane equation, adjusting this position causes a profound change in the resulting 2D plane. To verify this, select P0 and adjust its y-component value. Notice the drastic changes to the plane as a result and how the transparent sphere size changes accordingly such that the plane always tangents the sphere. Feel free to adjust any of the P0, P1, and P2 positions to verify that the derived vector plane equation is always correct.

Now that you have verified how the cross product result relates to the plane normal vector and that the plane equation is always correct, you can enable the ShowPointOnPlane toggle. The blue sphere, Pt, is a position that you can manipulate and observe where it would intersect the plane if it followed its direction path to or from the origin or its position vector. The thin black line, extending from the origin to this blue sphere, represents the position vector, P_t. The white sphere, Pon, is the intersection of the position vector P_t with the 2D plane or where the blue sphere would intersect the plane if it followed the black line back to the origin. Feel free to adjust both the 2D plane and the position vector by manipulating the P0, P1, and P2 positions and Pt to verify that the intersection result is always correct. Note that when P_t is perpendicular to $\hat{V}_n$, the position vector will be parallel to the plane and there can be no intersection.

Details of MyScript

Open MyScript and examine the source code in the IDE. The instance variables and the Start() function are as follows:

```
// Defines two vectors: V1 = P1 - P0, V2 = P2 - P0
public GameObject P0 = null;    // The three positions
```

```
public GameObject P1 = null;   //
public GameObject P2 = null;   //

// Plane equation:   P dot vn = D
public GameObject Ds;         // To show the D-value
public GameObject Pn;         // Where Vn crosses the plane

public bool ShowPointOnPlane = true;  // Show or Hide Pt
public GameObject Pt;         // Point to adjust
public GameObject Pon;        // Where Pt intersects the Plane

#region For visualizing the vectors
#endregion

// Start is called before the first frame update
void Start() {
    Debug.Assert(P0 != null);   // Verify proper editor init
    Debug.Assert(P1 != null);
    Debug.Assert(P2 != null);
    Debug.Assert(Ds != null);
    Debug.Assert(Pn != null);
    Debug.Assert(Pt != null);
    Debug.Assert(Pon != null);

    #region For visualizing the vectors
    #endregion
}
```

All the public variables for `MyScript` have been discussed when analyzing the `Controller`'s `MyScript` component, and as in all previous examples, the `Debug.Assert()` calls in the `Start()` function ensure proper setup regarding referencing the appropriate game objects via the Inspector Window. The `Update()` function is listed as follows:

```
void Update() {
    // Computes V1 and V2
    Vector3 v1 = P1.transform.localPosition -
                 P0.transform.localPosition;
    Vector3 v2 = P2.transform.localPosition -
                 P0.transform.localPosition;
```

```
    if ((v1.magnitude < float.Epsilon) ||
        (v2.magnitude < float.Epsilon))
        return;

    // Plane equation parameters
    Vector3 vn = Vector3.Cross(v1, v2);
    vn.Normalize();  // keep this vector normalized
    float D = Vector3.Dot(vn, P0.transform.localPosition);

    // Showing the plane equation is consistent
    Pn.transform.localPosition = D * vn;
    Ds.transform.localScale =
          new Vector3(D * 2f, D * 2f, D * 2f); // diameter

    // Set up for displaying Pt and Pon
    Pt.SetActive(ShowPointOnPlane);
    Pon.SetActive(ShowPointOnPlane);
    float t = 0;
    bool almostParallel = false;
    if (ShowPointOnPlane) {
        float d = Vector3.Dot(vn,
                  Pt.transform.localPosition);  // distance
        almostParallel = (Mathf.Abs(d) < float.Epsilon);
        Pon.SetActive(!almostParallel);
        if (!almostParallel) {
            t = D / d;
            Pon.transform.localPosition =
                t * Pt.transform.localPosition;
        }
    }

    #region  For visualizing the vectors
    #endregion
}
```

The first four lines of the `Update()` function compute the two vectors

$$\vec{V}_1 = P_1 - P_0$$

$$\vec{V}_2 = P_2 - P_0$$

and verify that both are nonzero vectors before continuing. The next three lines compute the vector plane equation parameters

$$\vec{V}_n = \vec{V}_1 \times \vec{V}_2$$

$$\hat{V}_n = \vec{V}_n.\text{Normalized}()$$

$$D = \hat{V}_n \cdot P_0$$

The two lines that follow set the P_n position and the diameter of the transparent sphere, `Ds`, such that you can examine these parameters of the vector plane equation

$$P_n = D\hat{V}_n$$

The `if` condition that follows ensures that `Pt` and `Pon` are computed and displayed only under the command of the user. The two lines in the `if` statement compute

$$d = \hat{V}_n \cdot P_t$$

and verify that d is not close to zero. This check verifies that the plane normal, $\hat{V}_n$, is not almost perpendicular to the position vector, P_t, or that the position vector is not almost parallel to the plane. Recall that in such a case, there can be no intersection and thus `Pon` cannot be computed. When verified that the P_t position vector is not parallel to the plane, the position of `Pon` is computed within the last `if` statement

$$t = \frac{D}{\hat{V}_n \cdot P_t} = \frac{D}{d}$$

$$P_{on} = tP_t$$

Takeaway from This Example

This example demonstrates how three non-collinear positions can define two non-parallel vectors which can define a 2D plane. You have examined and analyzed the parameters of the vector plane equation to develop an understanding for their geometric interpretations. The plane equation

$$\hat{V}_n \cdot p = D$$

describes the plane that is at a distance, D, measured from the origin along the plane normal vector, $\hat{V}_n$. Geometrically, this equation can be interpreted as all positions on this plane have a projected distance, D, when measured from the origin along $\hat{V}_n$. The equation and this interpretation were verified when you manipulated an arbitrary position vector, P_t, and observed the computed intersection position, P_{on}, between the position vector and the plane equation.

By now you have observed quite a few examples of vector value checking, but its importance cannot be overstated. Please do note that the `almostParallel` condition is effectively ensuring that when computing t

$$t = \frac{D}{\hat{V}_n \cdot P_t}$$

the denominator is not a zero value. Once again, it is the responsibility of a video game developer to ensure all mathematical operations performed are well defined and edge cases are checked and handled. Ill-defined conditions for mathematical operations often present themselves as intuitive geometric situations. In this case, when the denominator is close to zero, geometrically, it represents when the position vector, P_t, is almost parallel to the plane and thus an intersection does not exist.

Relevant mathematical concepts covered include

- Three non-collinear positions define two non-parallel vectors which define a 2D plane.
- A 2D plane can be described as being perpendicular to a normal direction and at a fixed distance away from the origin when measured along the normal direction.

- An alternative description of a 2D plane is that it is the collection of all positions with position vectors that have the same projected distance along the plane normal.

EXERCISES

Verify the Vector Plane Equation

The vector plane equation says that all positions on the plane have the same projected distance. Replace P0 with P1 and then P2 in `MyScript` when computing the distance, *D*, and verify that the results are identical.

The Plane at the Negative Distance

Examine the vector plane equation

$$\hat{V}_n \cdot p = D$$

and take note that the distance, *D*, is a projected result and is thus a signed floating-point number. This observation says that there is always a complementary plane that is *D* away in the negative $\hat{V}_n$ direction. Now, modify `MyScript` to compute

$$P_d = -D\hat{V}_n$$

You can visualize this point and begin to imagine the associated plane by defining and using a new sphere game object to represent the position of P_d. This exercise brings home the point that you must be careful with the signs; a simple careless mistake can result in an entirely plausible solution on a completely wrong geometry.

Axis Frames and 2D Regions

Recall that the vector plane equation identifies a 2D plane of infinite size. A 2D region can be defined on this 2D plane for determining if a given position is within the bounds of the region. This functionality is the generalization of the study of interval bounds from Chapter 2. For example, Figure 2-7 illustrated a 2D region on the X-Z plane. Here, the description is a 2D region on any arbitrary plane.

Defining 2D regions on 2D planes is interesting and has some important applications in video game development. However, what is much more important is the implication that given three positions that define two non-parallel vectors, you can actually define a **general axis frame**. Recall that the default axis frame of the Cartesian Coordinate System is the three perpendicular X-, Y-, and Z-axis directions centered at the origin. A general axis frame is three perpendicular directions which need not be aligned with the major axes and can be centered at any position. Figure 6-10 shows such an axis frame centered at the position P_0.

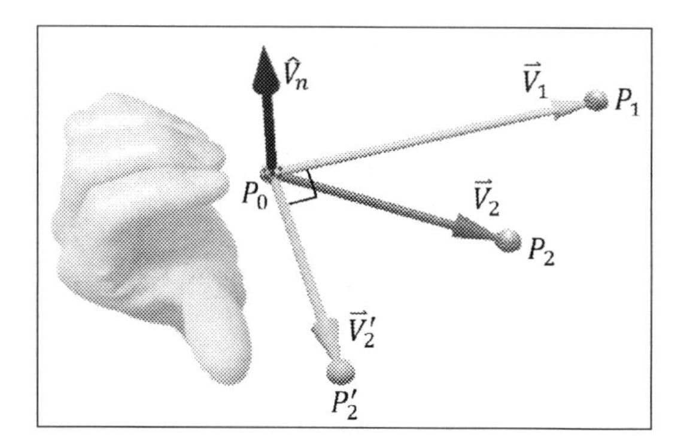

Figure 6-10. *Defining an axis frame*

In Figure 6-10, the three positions, P_0, P_1, and P_2, define two vectors

$$\vec{V}_1 = P_1 - P_0$$

$$\vec{V}_2 = P_2 - P_0$$

When these two vectors are not parallel, a new vector, $\vec{V}_n$, that is perpendicular to both $\vec{V}_1$ and $\vec{V}_2$ can be computed

$$\vec{V}_n = \vec{V}_1 \times \vec{V}_2$$

An important observation is that the cross product of $\vec{V}_n$ with $\vec{V}_1$, as indicated by the curling left hand in Figure 6-10, defines, $\vec{V}_2'$,

$$\vec{V}_2' = \vec{V}_n \times \vec{V}_1$$

a vector perpendicular to both $\vec{V}_n$ with $\vec{V}_1$. Notice that $\vec{V}_1$, $\vec{V}_n$, and $\vec{V}_2'$ are three vectors that are mutually perpendicular and is an axis frame that can be located at any position. In the next chapter you will learn about how this axis frame can serve as the basis for a new coordinate system, for example, serving to define the motion on a navigating spaceship. Here, the focus will be on defining a 2D region and a general bounding box as an exercise.

Bounds on a 2D Plane

Recall from Figure 5-9 that a general 1D interval, or a line segment, is a direction with two positions along that direction defining the beginning and the ending point of that line segment. Also recall from Figure 2-7 that a 2D interval, or a 2D rectangular region, is two 1D intervals along two perpendicular directions. Figure 6-11 shows two perpendicular general 1D intervals. The first interval is along $\vec{V}_1$, with P_0 and P_1, and the second interval is along $\vec{V}_2'$, with P_0 and P_2' as their beginning and ending positions. The two intervals have respective lengths of L_1 and L_2.

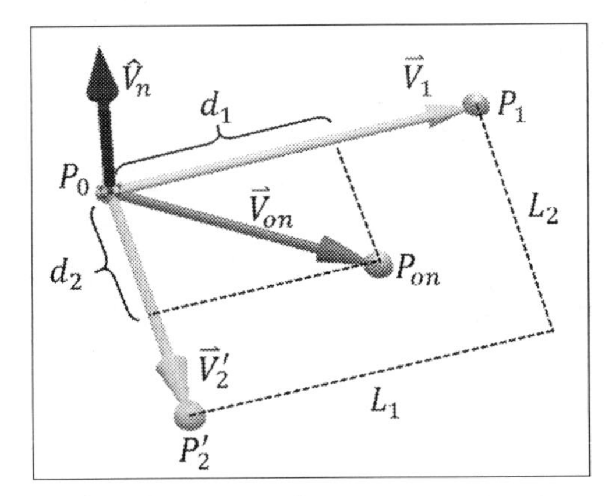

***Figure 6-11.** Inside condition of a general 2D region*

You can follow the exact same logic as in Chapter 2 when generalizing results from a 1D interval to a 2D bounding area and apply the logic to a general axis frame. In this case, instead of 1D intervals along the X- and Z-axes, you are working with general 1D intervals along the $\vec{V}_1$ and $\vec{V}_2'$ directions. The inside-outside status of the 2D region can be determined by applying the general 1D test, as illustrated in Figure 5-11 (d), on each of the two perpendicular general 1D intervals. For example, look at the given position P_{on} in Figure 6-11; this position defines the vector $\vec{V}_{on}$

$$\vec{V}_{on} = P_{on} - P_0$$

The vector, $\vec{V}_{on}$, can be used to determine if the position P_{on} is within the 2D region. In this case, the position P_{on} is within the bounds of the region if the projected size of $\vec{V}_{on}$ along both $\vec{V}_1$ and $\vec{V}_2'$ is positive and smaller than the corresponding interval lengths or

$$d_1 = \vec{V}_{on} \cdot \hat{V}_1 \qquad \vec{V}_{on} \text{ size on } \hat{V}_1$$

$$d_2 = \vec{V}_{on} \cdot \hat{V}_2' \qquad \vec{V}_{on} \text{ size on } \hat{V}_2'$$

With these two projected sizes, the condition for P_{on} being inside the 2D region can be stated by two inequalities: d_1 and d_2 must both be positive and smaller than the length of the corresponding intervals or

$$0 \le d_1 \le L_1 \text{ and } 0 \le d_2 \le L_2$$

Generalization of the Vector Line Equation

Recall the vector line equation that describes all positions located on the line segment which begins from position P_0 and extends in the direction of $\hat{V}_1$ is

$$l(t) = P_0 + t\hat{V}_1$$

In this example, you have observed the corresponding **vector plane equation**, where all positions that are located in the 2D rectangular region begin at position P_0 and extend in the perpendicular directions of $\hat{V}_1$ and $\hat{V}_2'$ as

$$p(d_1, d_2) = P_0 + d_1\hat{V}_1 + d_2\hat{V}_2'$$

Similar to the vector line equation where the range of the value, t, determines the inside-outside status, in 2D region the ranges of the values, d_1 and d_2, determine the inside-outside status of a position. Note the straightforward generalization to the third dimension for a bounding box

$$b(d_1, d_2, d_3) = P_0 + d_1\hat{V}_1 + d_2\hat{V}_2' + d_3\hat{V}_n$$

The Axis Frames and 2D Regions Example

This example builds on the previous example by supporting two additional features. It demonstrates the derivation of axis frames and the determination of the position inside-outside status for a given 2D region. The example allows you to interactively define an axis frame by manipulating three positions while it continuously computes the inside-outside status of the intersection of a position vector with the 2D plane. Figure 6-12 shows a screenshot of running the `EX_6_3_AxisFramesAnd2DRegions` scene from the `Chapter-6-CrossProducts` project.

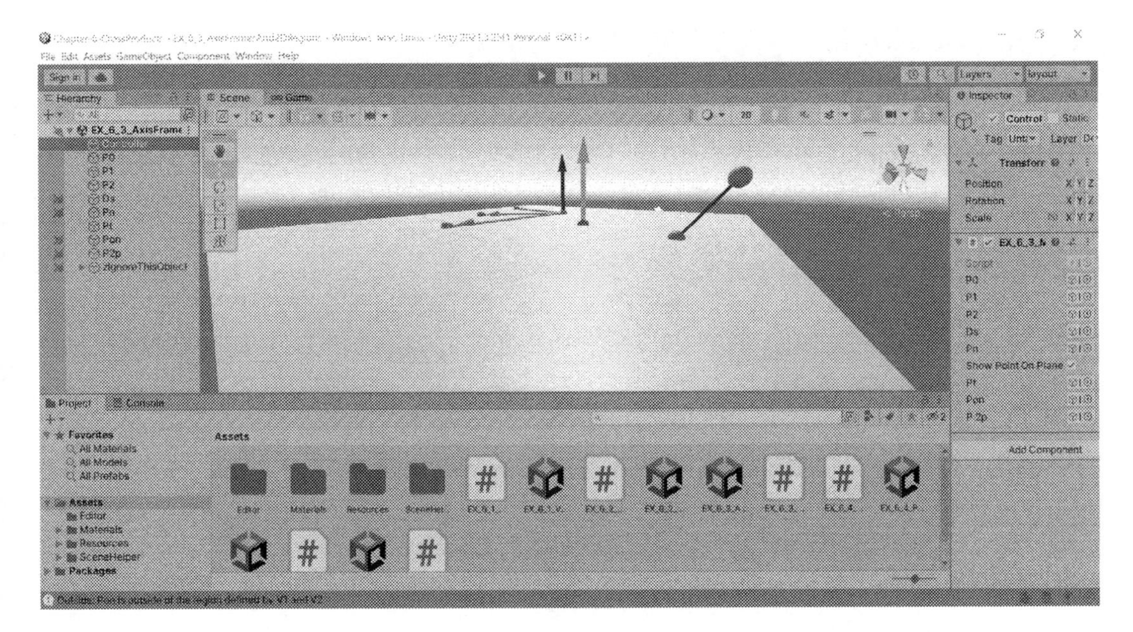

Figure 6-12. *Running the Axis Frames and 2D Regions example*

The goals of this example are for you to

- Observe the creation of axis frames based on three non-collinear positions
- Appreciate the fact that a 2D region on a plane is indeed defined by two perpendicular 1D regions
- Examine the implementation of the axis frame definition and the inside-outside test for the 2D region

Examine the Scene

Take a look at `Example_6_3_AxisFramesAnd2DRegions` scene, observe the predefined game objects in the Hierarchy Window, and note that the only difference between this scene and that of `Example_6_2_VectorPlaneEquations` is a single additional game object, `P2p`. The `transform.localPosition` of this game object will represent the position of $P_2^{'}$ in Figure 6-10, the head position of the $\vec{V}_2^{'}$ vector that is perpendicular to both $\vec{V}_1$ and $\vec{V}_n$. All other game objects serve the same purpose as they did in the previous example.

Analyze Controller MyScript Component

The `MyScript` component on the `Controller` also shows that `P2p` is the only additional variable when compared to the previous example. This new variable is meant to reference the game object with the same name for position manipulation in the script.

Interact with the Example

Click the Play Button to run the example. Notice the almost identical results of this example to that of the previous example. As a quick reminder, pay attention to the checkered sphere, `P0`, and the two striped spheres, `P1` and `P2`. These three positions define the two vectors, $\vec{V}_1$ (in cyan) and $\vec{V}_2$ (in magenta), according to Figure 6-10. The black vector at P_0 is $\vec{V}_n = \vec{V}_1 \times \vec{V}_2$. The blue sphere, P_t, defines the position vector that intersects the plane at P_{on}, the red sphere. The only addition to this scene is the green sphere, `P2p`, identifying the head position of the $\vec{V}_2'$ vector, where this vector has the size of $\vec{V}_2$ and the direction of $\vec{V}_n \times \vec{V}_1$

$$\vec{V}_2' = \left\|\vec{V}_2\right\|\left(\vec{V}_n \times \vec{V}_1\right).normalized$$

Now, select `P2` and manipulate its position. Notice how the green vector, $\vec{V}_2'$, has the exact same length as $\vec{V}_2$ and is always perpendicular to $\vec{V}_1$ and $\vec{V}_n$ and that the three vectors, $\vec{V}_1$, $\vec{V}_2'$, and $\vec{V}_n$, do indeed define a valid axis frame with three perpendicular directions centered at `P0`, independent of where `P0` is located, and as long as `P0`, `P1`, and `P2` are not collinear.

Now restart the scene and select `Pt` and manipulate its position to move `Pon`, the red sphere, into the region bounded by $\vec{V}_1$ and $\vec{V}_2'$ by increasing its x-component value. Notice as soon as `Pon` crosses into the region, its color changes from red to white. As long as `Pon` is located within the 2D region, it will remain white. Feel free to adjust `P0`, `P1`, or `P2` to change the bounds of the region to verify that the inside-outside test is consistent and always correct.

Details of MyScript

Open MyScript and examine the source code in the IDE. The instance variables and the Start() function are as follows:

```
#region Identical to EX_6_2
#endregion
public GameObject P2p;  // The perpendicular version of P2

#region For visualizing the vectors
#endregion

// Start is called before the first frame update
void Start() {
    #region Identical to EX_6_2
    #endregion

    Debug.Assert(P2p != null);

    #region For visualizing the vectors
    #endregion
}
```

As explained, P2p is the only additional variable from an otherwise identical example to the previous subsection. The Update() function is listed as follows:

```
void Update() {
    #region Identical to EX_6_2
    #endregion

    float l1 = v1.magnitude;
    float l2 = v2.magnitude;
    Vector3 v2p = l2 * Vector3.Cross(vn, v1).normalized;
    P2p.transform.localPosition =
                P0.transform.localPosition + v2p;

    bool inside = false;
    if (!almostParallel) {
        Vector3 von = Pon.transform.localPosition -
                      P0.transform.localPosition;
```

```
        float d1 = Vector3.Dot(von, v1.normalized);
        float d2 = Vector3.Dot(von, v2p.normalized);

        inside = ((d1 >= 0) && (d1 <= l1)) &&
                 ((d2 >= 0) && (d2 <= l2));
        if (inside)
            Debug.Log("Inside: Pon is inside of
                       the region defined by V1 and V2");
        else
            Debug.Log("Outside: Pon is outside of
                       the region defined by V1 and V2");
    }
    #region For visualizing the vectors
    #endregion
}
```

The first part of the `Update()` function in the collapsed region contains code that is identical to previous example. Recall that the collapsed code computes $\vec{V}_1$, $\vec{V}_2$, $\vec{V}_n$, and P_{on}. The first four lines of new code derive the vector, $\vec{V}_2'$, of the axis frame and its head position, P_2',

$$L_1 = \|\vec{V}_1\|$$

$$L_2 = \|\vec{V}_2\|$$

$$\vec{V}_2' = L_2\left(\vec{V}_n \times \vec{V}_1\right).normalized$$

$$P_2' = P_0 + \vec{V}_2'$$

When the Pt position vector is not parallel with the plane, Pon is defined, and the inside-outside status is computed by the code in the `if` statement

$$\vec{V}_{on} = P_{on} - P_0$$

$$d_1 = \vec{V}_{on} \cdot \hat{V}_1 \quad \vec{V}_{on} \text{ size on } \hat{V}_1$$
$$d_2 = \vec{V}_{on} \cdot \hat{V}_2' \quad \vec{V}_{on} \text{ size on } \hat{V}_2'$$

And finally, the `inside` condition is computed as

$$inside = (0 \le d_1 \le L_1) \, and \, (0 \le d_2 \le L_2)$$

Takeaway from This Example

This example demonstrates that an axis frame can be defined based on three non-collinear positions. As will be discussed and demonstrated in the next chapter, the ability to derive axis frames is of key importance in supporting many advanced operations in video game development including the support for motion control aboard a navigating spaceship.

The generalization of intervals and bounds is now complete. In Chapter 2, you learned about intervals and bounds that are aligned with the major axes. In Chapter 5, you learned to work with general 1D intervals where the interval does not need to be aligned with any major axis. There, you have also learned that if you were given two general 1D intervals that are perpendicular, then a general 2D region can be defined for inside-outside tests. The challenge was that you did not know how to derive the two perpendicular general 1D intervals. Now, with the knowledge of axis frame derivation, when given three non-collinear positions, you can compute the two perpendicular general 1D intervals and proceed to define a general 2D region.

Following the 2D to 3D generalization logic from Chapter 2, together with the fact that the derived axis frame provides the third perpendicular vector, you can now define and compute the inside-outside status of any position for bounding boxes at any orientation. However, remember that determining the collisions of two bounding boxes based on different axis frames is tedious and non-trivial.

Relevant mathematical concepts covered include

- Three non-collinear positions not only define two non-parallel vectors, they also define an axis frame.
- A general 2D rectangular bound can be defined by two general 1D intervals along perpendicular directions.
- A position can be projected onto any general 1D interval to determine its inside-outside status.

EXERCISES

Implement a General Bounding Box

Modify `MyScript` to include a public floating-point variable, `vnSize`. Initialize it to a reasonable value, for example, 3.0. Use this variable as the size of the third general 1D interval along the $\vec{V}_n$ direction. Notice a general bounding box is now defined with the two intervals identified in Figure 6-11. Now, implement the bounding box inside-outside test for `Pt`. You can print out the status and verify the correctness of your implementation.

Verify the Importance of Cross Product Ordering

Notice that in Figure 6-10, $\vec{V}_2'$ is defined to be $\vec{V}_n \times \vec{V}_1$ and not $\vec{V}_1 \times \vec{V}_n$. This is because a Left-Handed Coordinate System axis frame is followed and thus is required. You can verify with your left hand thumb, index, and middle finger, that the proper third vector to the existing $\vec{V}_n$ and $\vec{V}_1$ must be computed by $\vec{V}_n \times \vec{V}_1$. For example, if you align your index finger with $\vec{V}_n$, then the middle finger is along the $\vec{V}_1$ direction, and your thumb will point in the $\vec{V}_n \times \vec{V}_1$ direction. Alternatively, if your index finger is aligned with $\vec{V}_1$, then, your thumb is in the $\vec{V}_n$ direction, and once again, the middle finger will be in the $\vec{V}_n \times \vec{V}_1$ direction. Now, try reversing the cross product order when computing $\vec{V}_2'$ (the `v2p` variable) and run the game again. Can you explain what you observe?

Projections onto 2D Planes

In video games and many interactive graphical applications, it is a common practice to drop shadows of objects in space to convey hints of relative spatial location. For example, dropping the shadow of an in-flight meteoroid on the grounds of the approaching city or casting the shadow of an amulet tossed by the explorer on the walls of secret chamber to help better track its movement. In these cases, the shadows will convey a clear sense of the actual location of the in-flight objects and will allow the player to strategize their next move and react. Figure 6-13 shows that the shadow casting functionality can be modeled as a point to plane projection problem.

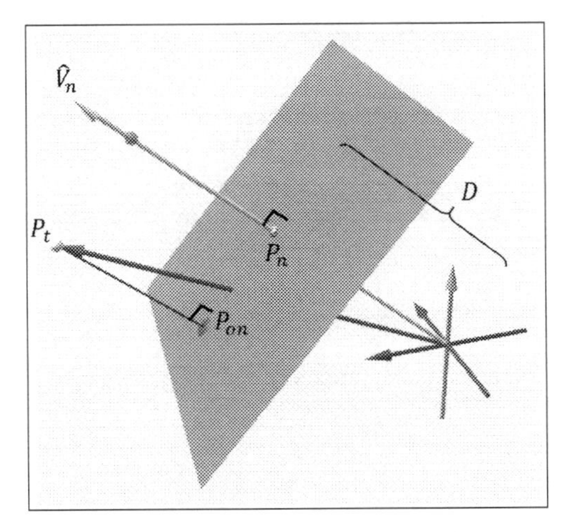

Figure 6-13. *Projection of a point onto a plane or casting shadow onto the plane*

Figure 6-13 shows a plane defined by the plane normal vector, $\hat{V}_n$, located at a distance, *D*, away from the origin. You know that the vector plane equation for this plane is

$$\hat{V}_n \cdot p = D$$

where

$$P_n = D\hat{V}_n$$

In Figure 6-13, P_t is the position of the object in flight and P_{on} is the projection of P_t on the given plane. Note that this projection is along the line connecting P_t to P_{on}, where the projection direction is parallel to the plane normal, $\hat{V}_n$. Figure 6-14 includes the following additional explanation for the derivation of point to plane solution:

$d = P_t \cdot \hat{V}_n$ position vector P_t size on $\hat{V}_n$

$P_l = d\hat{V}_n$ projected position of P_t on $\hat{V}_n$

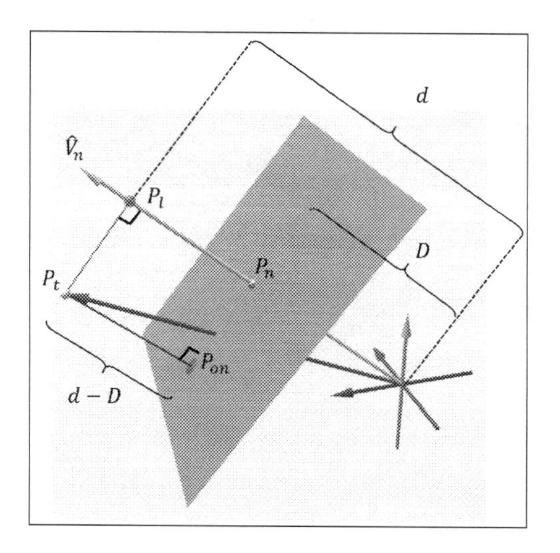

Figure 6-14. *Solving for point to plane projection*

The solution of point to plane projection can be explained by referring to Figure 6-14 and observing the following:

- First, a decision is made that a projection will only occur if position P_t is in front of the plane. This condition is true when the projected length of the P_t position vector in the $\hat{V}_n$ direction is greater than the plane distance, D, or if $d > D$.
- Second, because the projection is along the $\hat{V}_n$ direction, the distance between P_l and P_n is the same as the distance between P_t and P_{on}, and this distance is simply $d - D$.
- Finally, P_{on} is $d - D$ distance away from P_t in the negative $\hat{V}_n$ direction or

$$P_{on} = P_t - (d - D)\hat{V}_n$$

Note The derived solution for the point projection is valid for `Pt` located on either side of the plane. In this case, projection is restricted to one of the sides of the plane to showcase the "in front of" test. Modifying the solution to support proper projections for all locations of `Pt` is left as an exercise for you to complete.

The Point to Plane Projections Example

This example demonstrates the results of point to plane projection computation. The example allows you to interactively define a 2D plane, manipulate the point to be projected, and examine the results of projecting the point onto the plane. Figure 6-15 shows a screenshot of running the `EX_6_4_PointToPlaneProjections` scene from the `Chapter-6-CrossProducts` project.

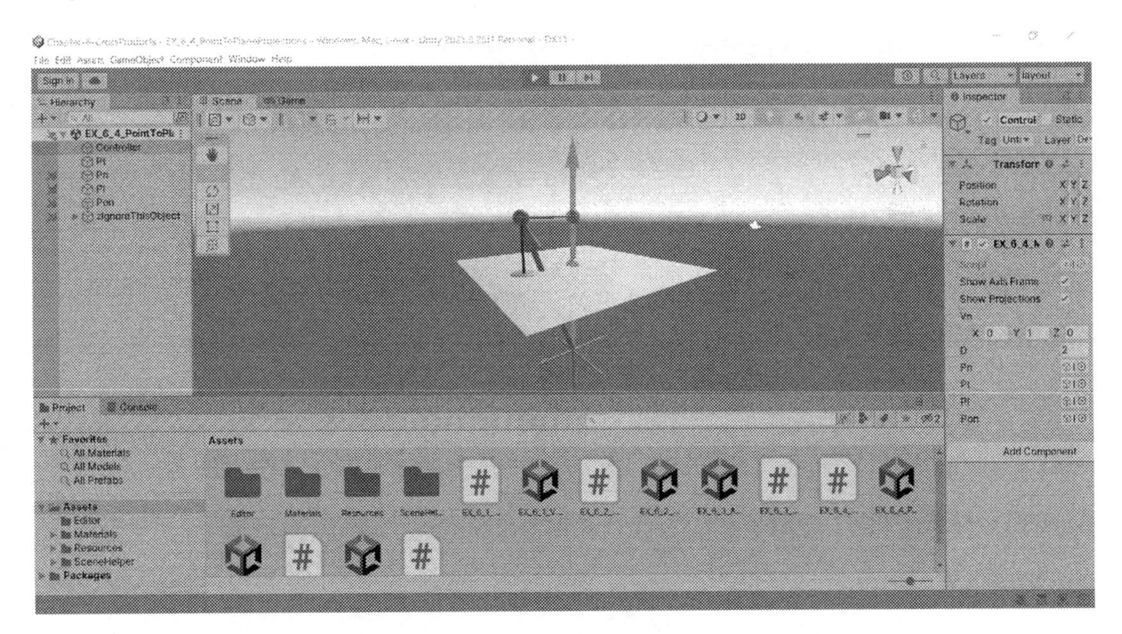

Figure 6-15. Running the Point to Plane Projections example

The goals of this example are for you to

- Gain experience with the "in front of a plane" test
- Verify the solution of point to plane projection

- Examine the implementation of the in front of a plane test and point to plane projection
- Observe the elegance and simplicity of typical implementation of vector solutions

Examine the Scene

Take a look at the `Example_6_4_PointToPlaneProjections` scene and observe the predefined game objects in the Hierarchy Window. In addition to the `Controller`, there are four objects in this scene: `Pn`, `Pt`, `Pl`, and `Pon`. Following the illustration in Figure 6-14, `Pn` is the position vector along the plane normal that intersects the 2D plane, `Pt` is the position to be projected, `Pl` is the projection of `Pt` on the plane normal vector, and `Pon` is the projection of `Pt` on the plane.

Analyze Controller MyScript Component

The `MyScript` component on the `Controller` shows three sets of variables as follows:

- Display toggles: `ShowAxisFrame` and `ShowProjections` will show or hide the axis frame and the projections accordingly. These toggle switches are meant to assist your visualization, allowing you to hide the illustration vectors to avoid screen cluttering.
- Vector plane equation parameters: `Vn` and `D` are the plane normal vector and the distance of the plane from the origin along the normal vector direction and will be used to create and modify the plane.
- Variables for the positions: `Pn`, `Pt`, `Pl`, and `Pon` are variables with names that correspond to the game objects in the scene. For all these game objects, the `transform.localPosition` will be used for the manipulation of their corresponding positions.

Interact with the Example

Click the Play Button to run the example. The white sphere is `Pn`, the white vector is $\hat{V}_n$, the red sphere is `Pt`, and the red vector is the position vector P_t. The semi-transparent black sphere on the white vector or the projected position on the plane normal vector is `Pl`, while the semi-transparent blob on the 2D plane or the projected position on the plane is Pon. Notice the thin green line connecting Pt to Pl; since Pl is the projection of Pt onto the plane normal vector, this line is always perpendicular to the plane normal and parallel to the plane. The thin black line connecting Pt to Pon represents the projection of Pt onto the plane and thus is always perpendicular to the plane and parallel to $\hat{V}_n$. In the following interactions, feel free to toggle off either or both of the display toggles to declutter the Scene View.

With the scene running, first verify the "in front of plane" test by selecting Pt and decreasing its y-component value. Notice that as soon as Pt is below the 2D plane, the projected positions disappear, verifying that the projection computation is only performed when the point, Pt, is in front of the plane. You can also verify this test by manipulating the D or Vn variables to move the plane or rotate the plane normal vector. Notice once again, as soon as Pt drops below the plane, the projected positions will both disappear.

Feel free to manipulate Pt or the plane parameters D or Vn in any way you like. Pay attention to the in front of plane test result and the consistent perpendicular relationships between the green line and the white $\hat{V}_n$ vector and the black line and the 2D plane.

Details of MyScript

Open MyScript and examine the source code in the IDE. The instance variables and the Start() function are as follows:

```
public bool ShowAxisFrame = true;
public bool ShowProjections = true;

// Plane Equation: P dot Vn = D
public Vector3 Vn = Vector3.up;
public float D = 2f;

public GameObject Pn = null;
public GameObject Pt = null;  // Point projected onto the plane
public GameObject Pl = null;  // Projection of Pt on Vn
public GameObject Pon = null; // Projection of Pt on the plane

#region For visualizing the vectors
#endregion

// Start is called before the first frame update
void Start() {
```

```
    Debug.Assert(Pn != null);   // Verify proper editor init
    Debug.Assert(Pt != null);
    Debug.Assert(Pl != null);
    Debug.Assert(Pon != null);

    #region For visualizing the vectors
    #endregion
}
```

All the public variables for `MyScript` have been discussed when analyzing the `Controller`'s `MyScript` component, and as in all previous examples, the `Debug.Assert()` calls in the `Start()` function ensure proper setup regarding referencing the appropriate game objects via the Inspector Window. The `Update()` function is listed as follows:

```
void Update() {
    Vn.Normalize();
    Pn.transform.localPosition = D * this.Vn;
    bool inFront = (Vector3.Dot(Pt.transform.localPosition, Vn) > D);
         // Pt in front of the plane
    Pon.SetActive(inFront);
    Pl.SetActive(inFront);
    float d = 0f;
    if (inFront) {
        d = Vector3.Dot(Pt.transform.localPosition, Vn);
        Pl.transform.localPosition = d * Vn;
        Pon.transform.localPosition =
            Pt.transform.localPosition - (d - D) * Vn;
    }
    #region For visualizing the vectors
    #endregion
}
```

The first three lines of the `Update()` function compute

$\hat{V}_n = \vec{V}_n.Normalize(\ )$	normalize $\vec{V}_n$
$P_n = D\hat{V}_n$	D distance along $\hat{V}_n$
$infront = \left(P_t \cdot \hat{V}_n\right) > D$	P_t is further along $\hat{V}_n$

The `if` condition checks for when P_t is indeed in front of the plane. When the condition is favorable,

$$d = P_t \cdot \hat{V}_n \qquad P_t \text{ size on } \hat{V}_n$$
$$P_l = d \cdot \hat{V}_n \qquad \text{project } P_t \text{ on } \hat{V}_n$$
$$P_{on} = P_t - (d - D)\hat{V}_n \qquad \text{from } P_t \text{ in } -\hat{V}_n$$

Notice the exact one-to-one implementation code when compared with the solution derivation. Once again, the implementation of vector solutions is typically simple and elegant and closely matches the mathematical derivation.

Takeaway from This Example

This example demonstrates an efficient and graceful way to drop shadows which is a commonly encountered situation in video games. The example also demonstrates that the vector solution to projecting along a 2D plane normal is straightforward and stable and involves a small number of lines of code. Additionally, the example shows how dot product results can be used to determine the in front of or behind relationship between an object position and a given 2D plane.

Relevant mathematical concepts covered include

- An object is in front of a given plane when the dot product of the object's position vector with the plane normal is greater than the plane distance from the origin.
- The projection of a position to a given plane is a subtraction of the position vector by a perpendicular distance to the plane, along the plane normal.

EXERCISES

Projection Support for Both Sides of the Plane

Notice that the derivation and the vector solution for projection are valid independent of whether `Pt` is in front of or behind the plane. The analysis of `MyScript` actually demonstrated extra computation to purposefully hide the projection results when `Pt` is not in front of the plane. Modify `MyScript` to disable the in front of check and verify that the projection solution is indeed valid for all positions of `Pt`.

Criteria for Shadow Casting

The result of the "in front of test" is binary—an object is either in front of the plane or not. In this example, an object can either cast shadow or the object cannot cast shadow. Notice that the result from the dot product performed $\left(P_t \cdot \hat{V}_n\right)$ encodes more information than just in front of or not. The result also tells you the projected distance or, if P_t is normalized, the cosine of the subtended angle. This information can be used to refine the criteria of when shadow casting should occur. For example, casting a shadow should only happen when the subtended angle is within a certain range. Now, modify `MyScript` to compute the subtended angle and allow shadows to be casted only when the subtended angle is less than a degree that is under the user's control.

Characteristics of the Shadow Casted

The shadow casted on the 2D plane contains attributes of its object that can also be refined according to the additional information from the projection computation. For example, the projected size on the plane normal $\left(P_t \cdot \hat{V}_n\right)$ carries the height information of the object. This value can be used to scale the size and the transparency of the shadow object. Modify `MyScript` to compute and use the length of the projected size to scale the size of the `Pon` game object.

Let User Manipulate Pn

The very simple relationship between `Pn`, `D`, and `Vn`

$$P_n = D\hat{V}_n$$

states that a user can also define the plane by manipulating `Pn` instead of `D` and `Vn`. In such a case,

$$D = \|P_n\|$$

$$\hat{V}_n = \frac{P_n}{D}$$

Notice that with this approach, instead of the four floating-point numbers, `D`, and the `x-`, `y-`, and `z-`components of `Vn`, the user only has the three floating-point components of `Pn` to manipulate the 2D plane. While this is easier for the user, it also means that the user cannot define planes with `D` of zero. With this caveat in mind, please modify `MyScript` to allow the user the option of defining the 2D plane with either approach.

Projection with 2D Bound Inside-Outside Test

Notice that as you move Pt in the X- and Z-axis directions, the size of the plane adapts and continuously shows the projected position on the plane. In an actual application, a 2D bound would be defined on this plane, and an inside-outside test could be performed and projected positions outside of the 2D bound would simply be ignored. Refer to the previous example where instead of allowing the users to adjust Vn and D to define the plane, three positions, P0, P1, and P2, are used to define both the plane and an axis frame and then a 2D bound. Adapt the solution and support bound testing for the projected position.

Note The last exercise challenges you to replace the Vn and D parameters with three positions to define the 2D plane and an axis frame. In practice, such extra efforts are not necessary. This is because an axis frame is actually conveniently defined by the initial orientation of the 2D plane and the plane normal vector, Vn. This information is available in the rotation matrix of the plane's transform component. However, more advanced knowledge in vector transformations and matrix algebra are required to decode this information. Unfortunately, these are topics beyond the scope of this book. For now, if you want to define an axis frame on a 2D plane, the plane must be defined by three positions that are not collinear. In the rest of the examples in this chapter, 2D plane sizes are always adapting to include the projected or intersected positions as these planes are created using the plane equation which relates better to the math at hand.

Line to Plane Intersection

You may recall that at the end of Chapter 2's discussion of bounds, when comparing what you have learned with the Unity Bounds class, one of the methods whose details were not discussed was

- IntersectRay: Does ray intersect this bounding box?

You are now in a position to closely examine this function. By now, you know that a ray is simply a line segment. The IntersectRay() function computes and returns the closest intersection position between a line segment and the six sides of the bounding box. Note that each side of a bounding box is simply a 2D region as you have previously examined in the Axis Frames and 2D Regions example. The `IntersectRay()` function answers the question of how to intersect a line segment with a 2D plane. This solution is illustrated in Figure 6-16.

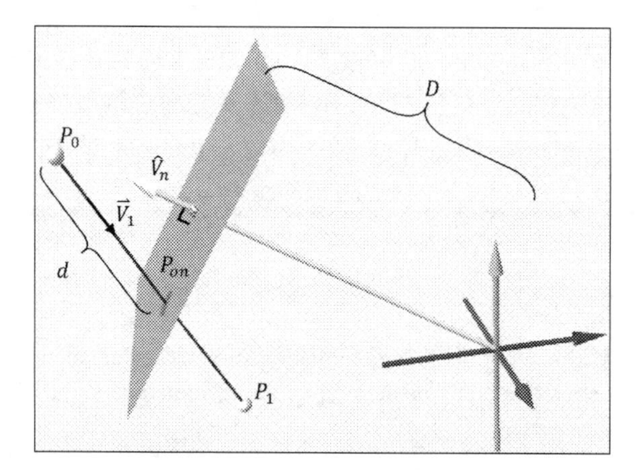

***Figure 6-16.** Solving the line to plane intersection*

Figure 6-16 depicts two positions, P_0 and P_1, that define a vector $\vec{V}_1$

$$\vec{V}_1 = P_1 - P_0$$

where the positions, *p*, along the line segment with parameter *s* can be written as

$$p = P_0 + s\vec{V}_1$$

Notice that in this formulation, since the $\vec{V}_1$ vector is not normalized, *s* values between 0 and 1, or when $0 \le s \le 1$, identify positions that are inside the line segment. In Figure 6-16, the position P_{on} is at a distance, $s = d$, along the $\vec{V}_1$ vector or

$$P_{on} = P_0 + d\vec{V}_1$$

Remember that the vector plane equation states that given a plane defined by normal vector, $\hat{V}_n$, and a distance, *D*, from the origin, all positions, *p*, on the plane satisfy the plane equation

$$p \cdot \hat{V}_n = D$$

In Figure 6-16, the position P_{on} lies on the 2D plane, so

$$P_{on} \cdot \hat{V}_n = D$$

$$\left(P_0 + d\vec{V}_1\right) \cdot \hat{V}_n = D \text{ substitute } P_{on} = P_0 + d\vec{V}_1$$

Note that the only unknown in this equation is *d*, the distance to travel along the line segment. By simplifying this equation, left as an exercise, you can show that

$$d = \frac{D - \left(P_0 \cdot \hat{V}_n\right)}{\left(\vec{V}_1 \cdot \hat{V}_n\right)}$$

With the *d* value computed, you can now find the exact P_{on} position. Note that this solution is not defined when the denominator or $\left(\vec{V}_1 \cdot \hat{V}_n\right)$ is close to zero. Once again, this can be explained by your knowledge of the dot product. A dot product result of zero means that the cosine of the subtended angle is zero, which says the subtended angle is 90° or that the two vectors are perpendicular. These observations indicate that when $\left(\vec{V}_1 \cdot \hat{V}_n\right)$ is close to zero, vectors $\vec{V}_1$ and $\hat{V}_n$ are almost perpendicular, the line segment is almost parallel to the plane, and therefore there can be no intersection between the two.

Note Ray casting is the process of intersecting a line segment or a ray with geometries. For example, if you were told to "cast a ray into a scene," then you would simply intersect geometries in the scene with a given line segment. In this case, you are learning about ray casting with a 2D plane.

The Line Plane Intersections Example

This example demonstrates the results of the line plane intersection solution. The example allows you to interactively define a 2D plane and a line segment and then examine the results of the line plane intersection computation. Figure 6-17 shows a screenshot of running the `EX_6_5_LinePlaneIntersections` scene from the `Chapter-6-CrossProducts` project.

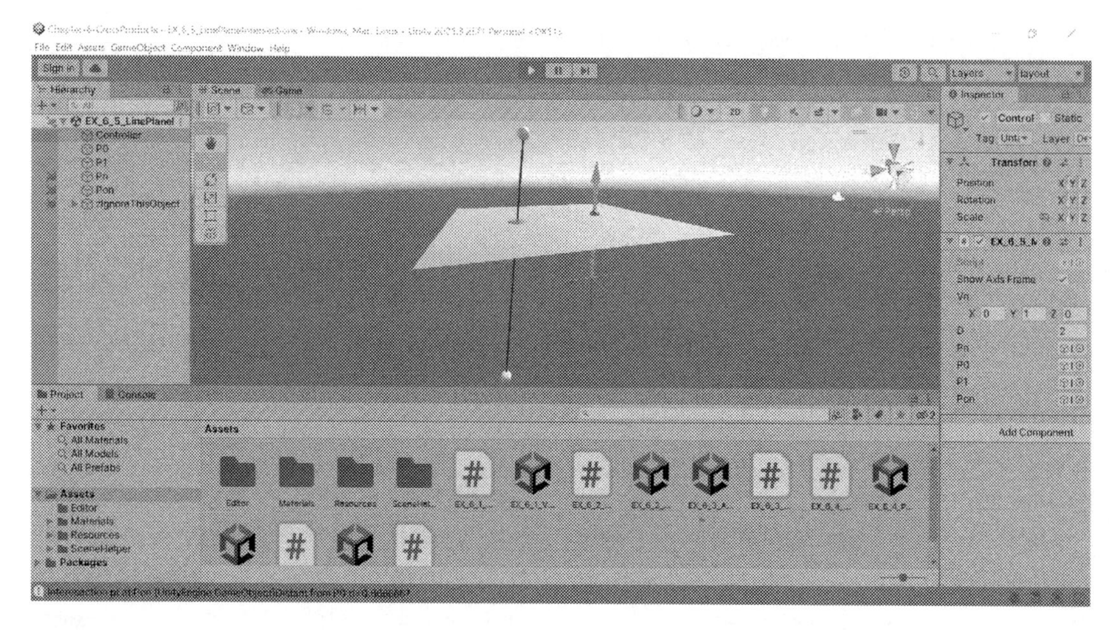

Figure 6-17. *Running the Line Plane Intersections example*

The goals of this example are for you to

- Verify the line plane intersection solution
- Gain experience with the perpendicular vectors test
- Reaffirm that it is important to check for all conditions when a solution is not defined, in this case, when the line segment is parallel to the plane
- Examine the implementation of the line plane intersection solution

Examine the Scene

Take a look at the `Example_6_5_LinePlaneIntersections` scene and observe the predefined game objects in the Hierarchy Window. In addition to the `Controller`, there are four objects in this scene: `Pn`, `P0`, `P1`, and `Pon`. `Pn`, the checkered sphere, is the position on the plane that is at the defined distance, `D`, along the plane normal. This position is displayed to assist in visualizing the 2D plane. The positions `P0` and `P1` define the black line segment, and `Pon` is the intersection position between this line segment and the defined plane.

Analyze Controller MyScript Component

The `MyScript` component on the `Controller` shows three sets of variables as follows:

- Display toggles: `ShowAxisFrame` will show or hide the axis frame to assist your visualization, allowing you to hide the axis frame to avoid screen cluttering.
- Vector plane equation parameters: `Vn` and `D` are the plane normal vector and the distance of the plane from the origin along the normal vector direction. These parameters will be used to create and manipulate the plane.
- Variables for the positions: `Pn`, `P0`, `P1`, and `Pon` are variables with names that correspond to the game objects in the scene. For all these game objects, the `transform.localPosition` will be used for the manipulation of their corresponding positions.

Interact with the Example

Click the Play Button to run the example. You can observe a 2D plane with a white normal vector extending from the origin and passing through the plane at `Pn`. You can also observe a thin black line between the positions `P0` and `P1` that define the line segment. At the intersection of the plane and the line segment is position `Pon`. You should be familiar with the 2D plane and its parameters, `Vn` and `D`.

Select the end points of the line segment, `P0` or `P1`, and adjust its x- and z-component values. Observe that `Pon` changes in response to your manipulation, always locating itself at the line plane intersection. This verifies the solution you have derived for `Pon`. You can verify the intersection computation results by referring to the text output in the Console Window. Remember, the values for the `d` parameterization (see Figure 6-16 for a reminder of what this variable is) are based on a non-normalized vector; therefore, `d` values between 0 and 1 indicate that `Pon` is inside the line segment.

Now, select `P0` and increase its y-component value. When `P0`'s position is above the plane, the `Pon` position is still along the line, but is outside of the line segment, occurring before position `P0`. This fact is reflected by the red line segment between `Pon` and `P0`. Notice that as you continue to increase the `P0` y-component value, as the line segment comes close to being parallel to the plane, the intersection position is located at positions further and further away from `Pn`. Eventually, when `P0` and `P1` y-component values are exactly the same, the line segment and the plane are exactly parallel and therefore there is no intersection between the two. You can verify this condition by referring to the printout in the Console Window. If you continue to increase the P0 y-component value, you will notice the red line segment switching between P0 to Pon to between P1 and Pon. In the case when P0 is above P1, the intersection position is along the line segment and after position P1. When this occurs, the value of d will be greater than 1 which you can verify has happened via the Console Window.

Feel free to manipulate all of the parameters, Vn, D, P0, and P1, and verify that the line plane intersection solution does indeed compute a proper Pon result except when the line is almost parallel to the plane or when the length of the line is very small (when P0 an P1 are located at almost the same position).

Details of MyScript

Open MyScript and examine the source code in the IDE. The instance variables and the Start() function are as follows:

```
public bool ShowAxisFrame = true;

// Plane Equation: P dot Vn = D
public Vector3 Vn = Vector3.up;
public float D = 2f;
public GameObject Pn = null;  // Point on plane along normal

public GameObject P0 = null, P1 = null;  // The line segment
public GameObject Pon = null;  // The intersection position

#region For visualizing the vectors
#endregion

void Start() {
    Debug.Assert(Pn != null);   // Verify proper editor init
    Debug.Assert(P0 != null);
    Debug.Assert(P1 != null);
    Debug.Assert(Pon != null);

    #region For visualizing the vectors
    #endregion
}
```

All the public variables for MyScript have been discussed when analyzing the Controller's MyScript component, and as in all previous examples, the Debug.Assert() calls in the Start() function ensure proper setup regarding referencing the appropriate game objects via the Inspector Window. The Update() function is listed as follows:

```
void Update() {
    Vn.Normalize();
    Pn.transform.localPosition = D * Vn;

    // Compute the line segment direction
    Vector3 v1 = P1.transform.localPosition -
                 P0.transform.localPosition;
    if (v1.magnitude < float.Epsilon) {
        Debug.Log("Ill defined line (magnitude of zero).
                   Not processed");
        return;
    }

    float denom = Vector3.Dot(Vn, v1);
    bool lineNotParallelPlane = (Mathf.Abs(denom) > float.Epsilon);
          // Vn is not perpendicular to V1
    float d = 0;

    Pon.SetActive(lineNotParallelPlane);
    if (lineNotParallelPlane) {
        d = (D - (Vector3.Dot(Vn, P0.transform.localPosition)))
            / denom;
        Pon.transform.localPosition =
            P0.transform.localPosition + d * v1;
        Debug.Log("Intersection pt at:" + Pon +
                  "Distant from P0 d=" + d);
    } else {
        Debug.Log("Line is almost parallel to the plane,
                   no intersection!");
    }
}
```

The first two lines of the `Update()` function normalize the user-specified plane normal vector and compute `Pn`'s position to help the user better visualize the 2D plane. The code that follows computes

$$\vec{V}_1 = P_1 - P_0$$

and checks to ensure that this line segment is well defined and has a nonzero length. When the line is well defined, the denominator for the solution to d, $\vec{V}_1 \cdot \hat{V}_n$, is computed and the condition for the line being parallel to the plane is checked. Note the use of the absolute value function when checking for the perpendicular condition. This is because the subtended angles of 89.99° and 90.01° are both almost perpendicular and the cosine or the dot product results are both close to zero but with different signs. Finally, d is computed and printed out to the Console Window when the line is not almost parallel to the plane.

Takeaway from This Example

This example demonstrates the solution to the line to plane intersection, an important problem that is straightforward to solve based on vector concepts you have learned. The concepts applied include working with the vector plane equation, the sign of the vector dot product, vector projections, and fundamental vector algebra. The line to plane intersection is a core functionality that can be found in typical game engine utility libraries. In the case of Unity, this functionality is presented via the `IntersectRay()` function of the `Bounds` class.

Relevant mathematical concepts covered include

- Two vectors are almost perpendicular when the result of their dot product is close to zero.
- When a line is almost perpendicular to the normal of a plane, it is almost parallel to the plane.
- The intersection point of a line and a plane can be derived based on vector algebra.

Relevant observations on implementation include

- Testing for perpendicular vectors, or when dot product result is close to zero, must be performed via the absolute value function, as very small positive and negative numbers are both close to zero

EXERCISES

Verify the Line Plane Intersection Equation

Recall that in Figure 6-16, the position P_{on} is at a distant, $s = d$, along the $\vec{V}_1$ vector or

$$P_{on} = P_0 + d\vec{V}_1$$

You have observed that since this position is also on the 2D plane

$$\left(P_0 + d\vec{V}_1\right) \cdot \hat{V}_n = D$$

Now, apply the distributive property of the vector dot product over the vector addition operation, and remembering that the result of a dot product is a floating-point number, show that

$$d = \frac{D - \left(P_0 \cdot \hat{V}_n\right)}{\left(\vec{V}_1 \cdot \hat{V}_n\right)}$$

A More General Shadow Casting Solution

One approach to interpret Figure 6-16 is to ignore P_1 and interpret P_{on} as the projection of P_0 on the 2D plane along the $\vec{V}_1$ direction. Given this interpretation, you can now cast shadows of objects onto a 2D plane along any direction specified by the user. Modify `MyScript` to replace `P1` by a 3D projection direction, $\vec{V}_1$, and implement the functionality of casting a shadow of P_0 on the plane along the player-specified $\vec{V}_1$ projection direction.

Ray Casting or Intersecting the General Bounding Box

Refer to your solution from the "Implement a General Bounding Box" exercise from the "Axis Frames and 2D Regions" section. With the results from line plane intersection, you can now implement the `IntersectRay()` function. Modify your solution to this previous exercise by allowing your user to define a line segment and then compute the intersection of the line segment with all six sides of the bounding box. The intersection position between the ray or line segment and the bounding box is simply the closest of all the valid intersection positions.

Mirrored Reflection Across a Plane

The intersection computation from the previous subsection allows you to collide an incoming object with flat planes or walls. In many video games, a typical response to the results of collisions is to reflect the colliding object. For example, when an amulet is tossed by an explorer, it should bounce and reflect off walls or the floor when it collides with them to convey some sense of realism. This reflection is depicted in Figure 6-18 and can be described as reflecting the velocity of an incoming object in the mirrored reflection direction.

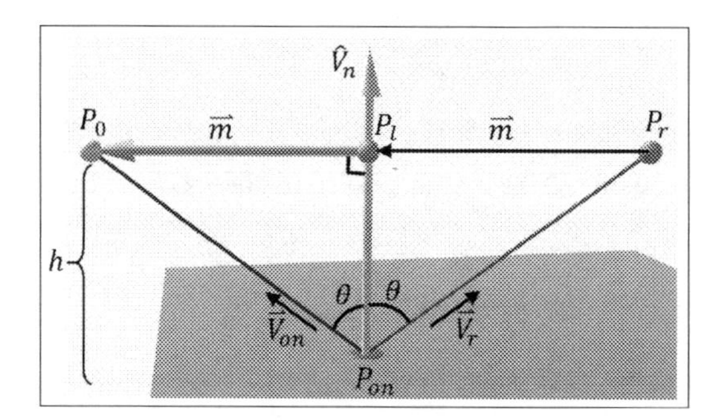

Figure 6-18. *Mirrored reflection across a plane*

In Figure 6-18, P_0, on the left, is the incoming object approaching the plane with normal vector $\hat{V}_n$ and is about to collide with the plane at position P_{on}. P_r is the mirrored reflection of P_0 across the plane normal $\hat{V}_n$ and is the unknown that must be computed.

Since this is a mirrored reflection, the right-angle triangle formed by the incoming object, $P_0P_{on}P_l$, is identical to the one formed by the reflected position, $P_rP_{on}P_l$, where P_l is the position that both P_0 and P_r would project onto in the $\hat{V}_n$ direction. Additionally, the vector, m, from P_l to P_0 is identical to the vector from P_r to P_l. Given these observations, as illustrated in Figure 6-18, you can define the vector $\vec{V}_{on}$ from P_{on} to P_0

$$\vec{V}_{on} = P_0 - P_{on} \qquad \text{vector from } P_{on} \text{ to } P_0$$

Project vector $\vec{V}_{on}$ onto the plane normal direction, $\hat{V}_n$, to compute the length of $\vec{V}_{on}$ when measured along the $\hat{V}_n$ direction

$h = \vec{V}_{on} \cdot \hat{V}_n$ length of $\vec{V}_{on}$ along $\hat{V}_n$

Compute P_l, the projected position of P_0 on the plane normal, $\hat{V}_n$. This position is traveling from P_{on} along the $\hat{V}_n$ direction by the projected distance, h,

$P_l = P_{on} + h\hat{V}_n$ P_{on} along $\hat{V}_n$ by h

With the P_l position, you can compute, $\vec{m}$, the vector from P_l to P_0,

$\vec{m} = P_0 - P_l$ vector from P_l to P_0

And finally, the mirrored reflection position of P_0 across the normal vector $\hat{V}_n$ is simply traveling along the negative $\vec{m}$ vector from P_l

$P_r = P_l - \vec{m}$ traveling by the negative $\vec{m}$

In these steps, you have derived the reflected position, P_r, of the incoming position P_0 with plane normal $\hat{V}_n$ and collision position P_{on}.

The Reflection Direction

The derived solution for P_r can be organized to assist the interpretation of mirrored reflection geometrically:

$P_r = P_l - (P_0 - P_l)$	substitute $\vec{m} = P_0 - P_l$
$= 2P_l - P_0$	collecting the two P_l
$= 2\left(P_{on} + h\hat{V}_n\right) - P_0$	substitute $P_l = P_{on} + h\hat{V}_n$
$= 2P_{on} + 2h\hat{V}_n - P_0$	distributive property
$= P_{on} + 2h\hat{V}_n - \left(P_0 - P_{on}\right)$	group P_{on} with P_0
$= P_{on} + 2h\hat{V}_n - \vec{V}_{on}$	substitute $\vec{V}_{on} = P_0 - P_{on}$
$= P_{on} + 2\left(\vec{V}_{on} \cdot \hat{V}_n\right)\hat{V}_n - \vec{V}_{on}$	substitute $h = \vec{V}_{on} \cdot \hat{V}_n$

Note that this last equation may seem complex; however, it is actually in a simple form. If you define the vector $\vec{V}_r$ to be

$$\vec{V}_r = 2\left(\vec{V}_{on} \cdot \hat{V}_n\right)\hat{V}_n - \vec{V}_{on}$$

Then

$$P_r = P_{on} + \vec{V}_r \qquad \text{from } P_{on} \text{ along } \vec{V}_r$$

Refer to Figure 6-18; this is the exact complement to the incoming position, P_0,

$$P_0 = P_{on} + \vec{V}_{on} \qquad \text{from } P_{on} \text{ along } \vec{V}_{on}$$

In this way, given an incoming direction of $\vec{V}_{on}$ and the normal vector $\hat{V}_n$, the reflected direction, $\vec{V}_r$, is

$$\vec{V}_r = 2\left(\vec{V}_{on} \cdot \hat{V}_n\right)\hat{V}_n - \vec{V}_{on}$$

This is the **reflection direction equation**. Note that this equation says the reflected direction, $\vec{V}_r$, is a function of only two parameters—the incoming direction, $\vec{V}_{on}$, and the normal direction, $\hat{V}_n$, that defines the reflection.

Lastly, it is important to note that in this derivation, the incoming direction, $\vec{V}_{on}$, is defined as a vector pointing away from the intersection position (see the arrow above $\vec{V}_{on}$ in Figure 6-18 for clarification). This convention of defining all vectors to be pointing *away* from the position of interest is a common practice in many video games and computer graphics–related vector solutions.

The Line Reflections Example

This example demonstrates the results of line reflection across a 2D plane. This example allows you to interactively define the line segment and the 2D plane, as well as examine the results of reflecting the line segment across the normal direction of the 2D plane. Figure 6-19 shows a screenshot of running the `EX_6_6_LineReflections` scene from the `Chapter-6-CrossProducts` project.

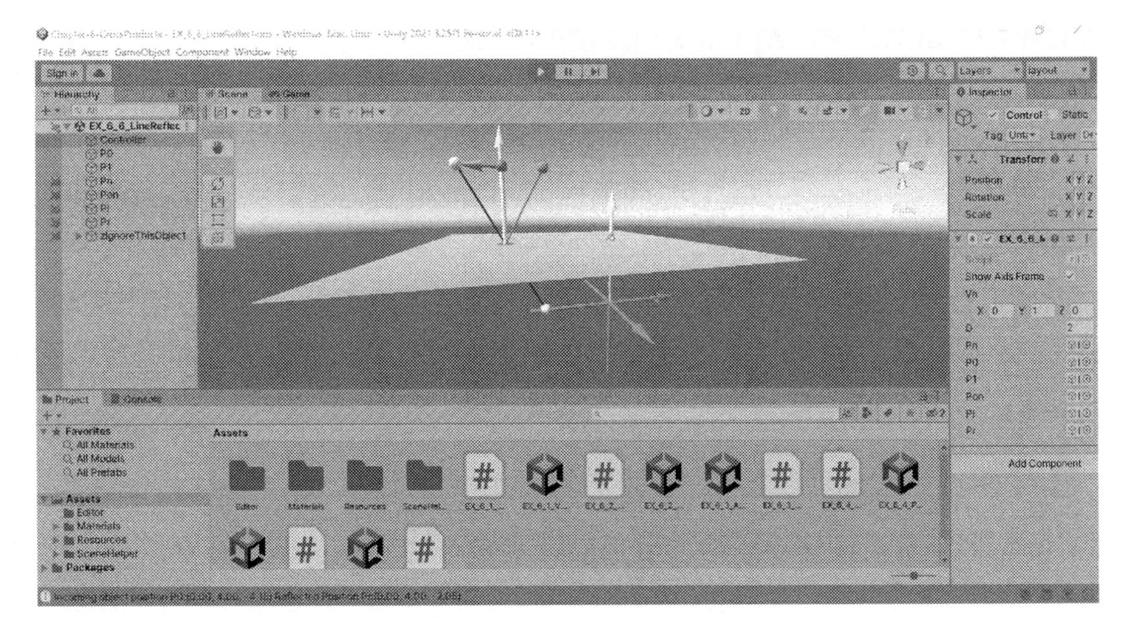

Figure 6-19. *Running the Line Reflections example*

The goals of this example are for you to

- Verify the reflection direction equation
- Examine the reflection of a position across the normal of a plane
- Examine the implementation of the reflection computation

Examine the Scene

Take a look at the `EX_6_6_LineReflections` scene and observe the predefined game objects in the Hierarchy Window. Take note that this example builds directly on the results from the `EX_6_5_LinePlaneIntersects` scene. Similar to the previous example, the parameters, `Vn` and `D`, define the 2D plane where `Pn` is the position on the plane to assist visualization. The parameters `P0` and `P1` define the line segment, and `Pon` is the intersection between the line and the 2D plane.

The two new game objects in this scene are the projection of `P0` on the plane normal vector, `Pl`, and `Pr` the mirrored reflection of `P0` across the plane normal.

Analyze Controller MyScript Component

The MyScript component on the Controller shows that there are two additional public variables with names that correspond to the Pl and Pr game objects. As in previous cases, the transform.localPosition of these variables will be used for the manipulation of the corresponding positions.

Interact with the Example

Click the Play Button to run the example. When compared with the Scene View of EX_6_5_LinePlaneIntersects, you will observe the similar 2D plane defined by Vn and D, the thin black line segment defined by P0 and P1, and their intersection at Pon. Note that the plane normal vector is copied and displayed at Pon to assist in the visualization of reflection. Also note that the green sphere, Pl, is the projection of P0 onto the plane normal, and the green vector is the $\vec{m}$ vector as depicted in Figure 6-18

$$\vec{m} = P_0 - P_l$$

The striped sphere, Pr, connected with a thin red line to Pon, is the mirrored reflection of P0 across the plane normal vector.

Tumble the Scene View camera to examine the running scene from different viewing positions to verify that the red line segment and the black line segment above the plane are indeed mirrored reflections. Notice Pl is the projection of P0 onto the normal vector, and thus, the green $\vec{m}$ vector is always perpendicular to the plane normal vector. You can manipulate the plane, by adjusting Vn and D, and the line segment, by adjusting P0 and P1, to verify that the reflection solution is correct for all cases. Recall from the previous example to be careful when the line segment is almost parallel to the plane as the plane size will increase drastically to accommodate the intersection position that will now be located at a very far distance.

You can set P0 and P1 such that the line segment is in the same direction as the plane normal. Observe that in this case, the reflection direction would be parallel to the normal vector direction and that the projected position, Pl, and the reflected position, Pr, will be located at the same point. In other words, the reflection vector would be exactly the same as in the incoming vector!

Details of MyScript

Open MyScript and examine the source code in the IDE. The instance variables and the Start() function are as follows:

```
#region identical to EX_6_5
#endregion
public GameObject Pl = null;  // Projection of P0 on Vn
public GameObject Pr = null;  // reflected position of P0

#region For visualizing the vectors
#endregion

// Start is called before the first frame update
void Start() {
    #region identical to EX_6_5
    #endregion
    Debug.Assert(Pl != null);
    Debug.Assert(Pr != null);

    #region For visualizing the vectors
    #endregion
}
```

As explained, Pl and Pr are the only additional variables from an otherwise identical example to the previous subsection, and as in all previous examples, the Debug.Assert() calls in the Start() function ensure proper setup regarding referencing these game objects via the Inspector Window. The Update() function is listed as follows:

```
void Update() {
    #region identical to EX_6_5
    #endregion

    float h = 0;
    Vector3 von, m;
    Pr.SetActive(lineNotParallelPlane);
    if (lineNotParallelPlane) {
        von = P0.transform.localPosition -
              Pon.transform.localPosition;
```

```
        h = Vector3.Dot(von, Vn);
        Pl.transform.localPosition =
               Pon.transform.localPosition + h * Vn;
        m = P0.transform.localPosition -
            Pl.transform.localPosition;
        Pr.transform.localPosition =
            Pl.transform.localPosition - m; ;
        Debug.Log("Incoming object position P0:" +
                   P0.transform.localPosition +
                  " Reflected Position Pr:" +
                   Pr.transform.localPosition);
    } else {
        Debug.Log("Line is almost parallel to the plane,
                   no reflection!");
    }

    #region For visualizing the vectors
    #endregion
}
```

Recall that the previous example computes the intersection position, Pon, when the line segment is not almost parallel to the 2D plane. Similar to line plane intersection, a line can only reflect off a plane that it is not parallel with. The `if` condition checks for the parallel condition and outputs a warning message to the Console Window. Otherwise, the five lines inside the `if` condition follow the P_r position derivation exactly and compute

$\vec{V}_{on} = P_0 - P_{on}$	vector from P_{on} to P_0
$h = \vec{V}_{on} \cdot \hat{V}_n$	$\vec{V}_{on}$ size along $\hat{V}_n$
$P_l = P_{on} + h\hat{V}_n$	P_{on} along $\hat{V}_n$ by h
$\vec{m} = P_0 - P_l$	vector from P_l to P_0
$P_r = P_l - \vec{m}$	negative $\vec{m}$ direction

Takeaway from This Example

This example, once again, illustrates a straightforward but important application of vector algebra. Note that the reflection direction equation

$$\vec{V}_r = 2\left(\vec{V}_{on} \cdot \hat{V}_n\right)\hat{V}_n - \vec{V}_{on}$$

is independent of plane to origin distance, D, or the actual incoming object position, P_0, or intersection position P_{on}. As depicted in Figure 6-20, this makes intuitive sense.

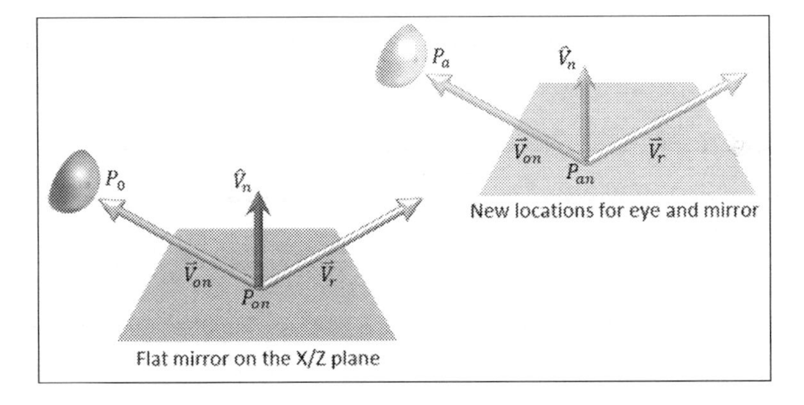

Figure 6-20. *The mirrored reflection direction*

On the left of Figure 6-20, it depicts your eye at an initial position, P_0, looking at a point, P_{on}, on a flat mirror on your desk. The right of Figure 6-20 shows that you have moved your eye and the mirror such that your eye is now located at P_a, and you are looking at a new position, P_{an}, on the mirror. You know that in both of the mirror locations, for the same incoming viewing direction, $\vec{V}_{on}$, as long as the mirror normal, $\hat{V}_n$, is not changed, the reflection direction will always be the same, $\vec{V}_r$. Notice that the reflection direction, $\vec{V}_r$, is only dependent on the incoming direction, $\vec{V}_{on}$, and the mirror normal vector, $\hat{V}_n$. Neither the location of the mirror, which corresponds to the D-value of the plane equation, nor the location of your eye, P_0 and P_a, nor the location of where you are looking at, P_{on} or P_{an}, affects the reflection direction, $\vec{V}_r$. Only your viewing angle and the orientation of the mirror will affect the reflection direction, just as the reflection direction equation states.

Relevant mathematical concepts covered include

- The mirrored reflection direction is a function of the normal vector and incoming direction.
- The mirrored reflection of a position can be found by applying the reflection direction to the impact position.

Relevant observations on implementation include

- In the mirrored reflection implementation, the normal vector must be normalized. Additionally, the vector representing the reflection direction is the same length as the vector representing the incoming direction

EXERCISES

Verify the Reflection Direction

Edit MyScript to replace the implemented solution by first computing the reflection direction, $\vec{V}_r$,

$$\vec{V}_r = 2\left(\vec{V}_{on} \cdot \hat{V}_n\right)\hat{V}_n - \vec{V}_{on}$$

And then compute

$$P_r = P_{on} + \vec{V}_r$$

Verify your results are identical to the existing implementation. How would you modify your solution if $\vec{V}_{on}$ is a normalized vector?

Compare with the Vector3.Reflect() Function

Please refer to https://docs.unity3d.com/ScriptReference/Vector3.Reflect.html; the Unity Vector3 class also supports the reflection function. Edit MyScript to replace the implementation with the Vector3.Reflect() function and verify the results are identical.

Working with the "in Front of" Test

Modify MyScript to reflect the line only when P0 is in front of the 2D plane and P1 is behind the 2D plane.

Support 2D Bound Test

Modify MyScript to remove Vn and D and include three user control positions for defining the plane and a 2D bound where reflection only occurs for intersections that are within the bound.

Summary

This chapter summarizes the discussions on vectors and vector algebra by introducing the vector cross product. You have seen that while the results of the vector dot product relate two vectors via a simple floating-point number, the results of the vector cross product provide information on the space that contains the operand vectors in the form of a new vector in a new direction. This new vector is perpendicular to both operand vectors and has a magnitude that is the product of the sizes of the two vectors and the sine of their subtended angle. You have also learned that the cross product of a vector with itself or with a zero vector is the zero vector. In typical video game-related problems, it is rare to encounter solutions that depend on the result of the cross product of a vector with itself.

You have also learned that an axis frame, or three perpendicular vectors, can be derived from the result of the cross product. This is accomplished by performing one more cross product between the initial cross product result vector and one of the original operand vectors. This newly derived axis frame can serve as a convenient reference for more advanced applications that will be discussed in the next chapter. In this chapter, you experienced working with derived axis frames in 2D space to compute position inside-outside tests for 2D bounds. Remember that it is important to follow the chosen coordinate space convention, left- or right-handed, when computing an axis frame.

You have built on the results of the cross product to gain insights into 2D planes and to relate the algebraic plane equation, $Ax + By + Cz = D$, to the vector plane equation, $P \cdot \hat{V}_n = D$. You have also examined the geometric implications of the vector plane equation where the vector, $\hat{V}_n$, is the plane normal and is perpendicular to the 2D plane and D is the distance between the origin of the Cartesian Coordinate System and the 2D plane measured along the plane normal, $\hat{V}_n$, direction.

These insights into 2D planes allowed the derivation of three important solutions with wide applications in video games and computer graphics applications: projection of a position, intersection with a line segment, and reflection direction. You have interacted with and examined the implementation of these solutions as well as verified that these solutions are general and can work with any input conditions. Lastly, you have observed that the typical implementation of vector solutions match closely with the vector algebraic solution, are elegant, and typically involve a small number of lines of code.

CHAPTER 7

Axis Frames and Vector Components

After completing this chapter, you will be able to

- Understand that the Cartesian Coordinate System is an example of axis frame
- Appreciate that the x-, y-, and z-values of the Cartesian Coordinate System are examples of vector components
- Describe the definition of, and create from three non-collinear positions, an axis frame
- Discuss the components of a vector with respect to any axis frame
- Decompose a vector into the components of any given axis frame
- Define and work with vectors in any axis frame
- Analyze, design, and implement movements of objects in the context of any axis frame

Introduction

You have learned from Chapter 4 that a vector is defined by two nonoverlapping positions. From Chapter 5, you learned that two unique vectors are defined by any three positions that are not collinear and that these two vectors always define a 2D plane. Lastly, from Chapter 6, you have learned that the perpendicular direction to a plane can be derived via a vector cross product, and very importantly, you have also learned that an axis frame can be derived based on this perpendicular direction and the two given

K. Sung and G. Smith, *Basic Math for Game Development with Unity 3D*,
https://doi.org/10.1007/978-1-4842-9885-5_7

vectors. A derived axis frame is a unique 3D coordinate system, just like the Cartesian Coordinate System, that is capable of describing and representing positions and vectors. In this chapter, you will continue to learn about deriving different axis frames and representing and working with positions and vectors in these derived coordinate systems.

Note Recall that a vector points from its **tail** to its **head**.

The Cartesian Coordinate System, with its perpendicular x-, y-, and z-axes, is the most straightforward example of an axis frame. The three axes intersect, with their tails at the position that is referred to as the origin and the axes are directions or unit vectors. In Chapter 6, when you examined the axis frame in Figure 6-10, you saw that in general, the shared tail position of the unit vectors can be located at any arbitrary position, P_0. Thus, an axis frame can be defined simply as three unit vectors that are perpendicular to each other with tails located at the same position, P_0. These three perpendicular unit vectors are referred to as the major axes and the common position that the major axes intersect, P_0, is the origin of the axis frame.

In Chapter 2, you learned that the coordinate values of a position `(x,y,z)` represent distances measured from the origin along their corresponding axes' directions or unit vectors. These coordinate values can be considered the magnitude of vectors in the x-, y-, and z-directions or components of the major axes. In a 3D world, there are exactly three perpendicular unit vectors with exactly three components for each position.

In general, given a position `(x,y,z)` or any vector defined in the Cartesian Coordinate System, it is always possible to compute the corresponding component values for any other axis frame. The converse is also true—that given the component values of any axis frame, it is always possible to compute the corresponding coordinate values in the Cartesian Coordinate System. In other words, it is always possible to represent a vector in the context of any axis frame and to convert the representation to any other axis frames. Among many applications, this ability to represent vectors with respect to any axis frame allows the analysis and manipulation of movements in dynamic environments such as resting and running down a hallway toward the medical bay of a spaceship while that spaceship is actively dodging asteroids.

In video games, there are many applications of representing vectors in different axis frames and working with the resulting components. For example, to continue with the player in a spaceship example, even though resting in the spaceship, the player's position and orientation should be updated as the spaceship navigates in the asteroid field. In this situation, an elegant solution would be to represent the position and orientation of the player in the context of the spaceship's axis frame. In this way, the spaceship's axis frame can be updated as it navigates the asteroid field, while a stationary player in the spaceship can have its particulars remain constant. With these representations, a player resting and facing the front of the spaceship will remain stationary and continue to face the front while the spaceship navigates.

Here are some other examples of working with multiple axis frames in video games:

- Running and swinging a sword in virtual reality where the sword's position is determined by the player's hand position, which is determine by their moving body
- An asteroid mining game where each asteroid spins and has its own gravity system that effects the player
- Riding in a vehicle that has a rotatable mounted turret
- Hopping between a train and horse in a wild west high stakes heist game

In practice, representation and conversion between axis frames are usually integrated as part of and hidden by the scene hierarchy interface. As will be detailed at the end of this chapter, in Unity the functionality of and the transitions between axis frames are delivered via the parent-child relationship that can be created and manipulated in the Hierarchy Window.

This chapter begins by examining the default Cartesian Coordinate System as an axis frame and relates coordinate values to components. The section that follows reviews the definition of general axis frames and derives how to compute the components of positions in these general axis frames. With proper understanding of components, the subsequent section analyzes vectors in general axis frames and discusses the details of representing the same vector in different axis frames. The last section of this chapter simplifies the player in a spaceship example and demonstrates how to achieve independent motion controls for the player moving toward the medical bay while the spaceship navigates.

Note The rest of the book will refer to the axis frame defined by the default Cartesian Coordinate System as the Cartesian axis frame.

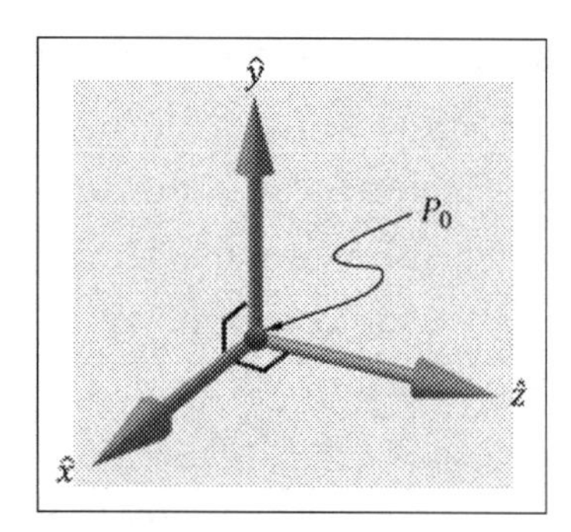

Figure 7-1. *A General axis frame with labels*

Positions in the Cartesian Axis Frame

This section reviews how the Cartesian axis frame, or the Cartesian Coordinate System, represents positions in 3D space. As discussed, in general, an axis frame is defined by three unit vectors, or the major axes, that are perpendicular to each other and intersect at a common position, the origin. Figure 7-1 depicts an example axis frame with labels: P_0 being the origin or the common intersection position and $\hat{x}$, $\hat{y}$, and $\hat{z}$ as the three perpendicular unit vectors.

In the case of the default Cartesian axis frame, the origin, P_0, is simply (0, 0, 0). By convention, the constant x-, y-, and z-directional unit vectors of the Cartesian Coordinate System are referred to as $\hat{i}$, $\hat{j}$, and $\hat{k}$, where

$$\hat{i} = (1,0,0)$$

$$\hat{j} = (0,1,0)$$

$$\hat{k} = (0,0,1)$$

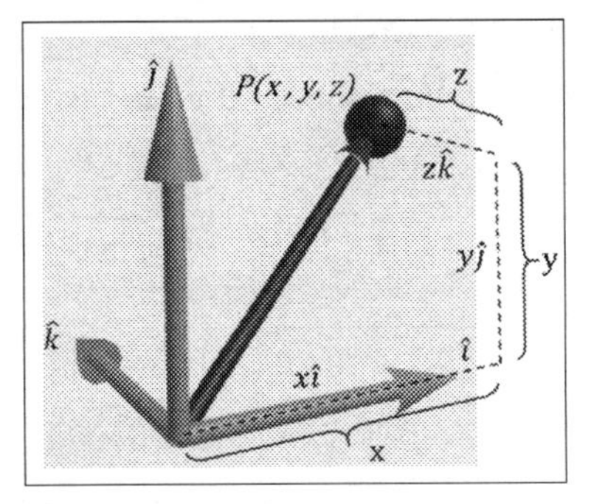

Figure 7-2. *Components of a vector in the default Cartesian axis frame*

Components of a Position Vector

In Chapter 4, when discussing positions, or position vectors, you have learned that

> *position in the Cartesian Coordinate System at P = (x,y,z) can be interpreted as x-, y-, and z-displacements measured along the three major axes from the origin.*

At this point, you have learned enough about vectors to turn this statement into a mathematical expression. As illustrated in Figure 7-2, remembering that x-direction is represented by $\hat{i}$, y-direction is $\hat{j}$, and z-direction is $\hat{k}$, then x-, y-, and z-displacements along the three major axes are simply a vector, $\vec{D}$, that is the sum of the scaled vectors in the $\hat{i}$, $\hat{j}$, and $\hat{k}$ directions or

$$\vec{D} = x\hat{i} + y\hat{j} + z\hat{k}$$

The phrase, "*measured ... from the origin position,*" simply means that the displacement of vector $\vec{D}$ begins from the origin at P_0 or

$$\vec{V} = P_0 + \vec{D} = P_0 + x\hat{i} + y\hat{j} + z\hat{k}$$

$$= (0,0,0) + x(1,0,0) + y(0,1,0) + z(0,0,1)$$

$$= (x,0,0) + (0,y,0) + (0,0,z) = (x,y,z)$$

Notice that in this derivation the coordinate values (x, y, z) are used to scale the corresponding unit vectors of the axis frame, that is, x is used to scale $\hat{i}$, y scaling $\hat{j}$, and z scaled $\hat{k}$. Because the coordinate values scale the corresponding unit vectors of the Cartesian axis frame, these values are referred to as the components of vector $\vec{V}$ in the Cartesian Coordinate System.

Note Vector components are defined with respect to a given axis frame. Coordinate values are components of the Cartesian axis frame. In general, for an axis frame other than the Cartesian axis frame, components of a position are different from the coordinate values of the position.

The Components of Cartesian Axis Frame Example

This example demonstrates that scaling the unit vectors of a Cartesian axis frame with the corresponding coordinate values does indeed compute proper positions. This example allows you to interactively manipulate a position and then examine the corresponding components of the position vector and magnitudes of the unit vectors. Figure 7-3 shows a screenshot of running the `EX_7_1_ComponentsOfCartesianAxisFrame` scene from the `Chapter-7-VectorComponents` project.

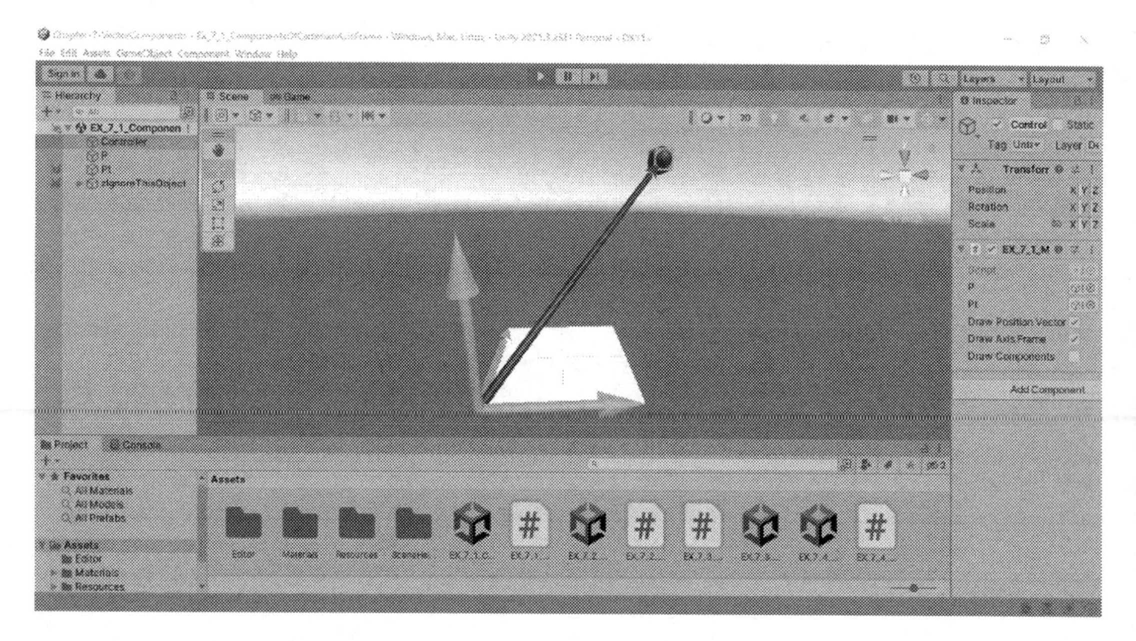

Figure 7-3. *Running the Components of Cartesian Axis Frame example*

The goals of this example are for you to

- Review the significance of coordinate values
- Examine coordinate values as components of a position vector in scaling the corresponding unit vectors of an axis frame
- Verify that the sum of component-scaled unit vectors of an axis frame does indeed compute the proper position

Examine the Scene

Take a look at the `Example_7_1_ComponentsOfCartesianAxisFrame` scene and observe the predefined black sphere, `P`, and red cube, `Pt`, in the Hierarchy Window. In this example, the coordinate values of `P` are used to scale the unit vectors of Cartesian axis frame to compute the position for `Pt`.

Analyze Controller MyScript Component

The `MyScript` component on the `Controller` shows variables with the same name as their corresponding reference game objects in the scene. The toggles draw/hide the position vector of `P`, the default Cartesian axis frame, and the scaled component vectors.

Interact with the Example

Click the Play button to run the example. You can see the game object `Pt` (red cube), overlapping the game object `P` (black sphere). Now, select `P` and manipulate its position. Observe that `Pt` (red cube) always follows and encompasses `P` (black sphere). In this case, the position of `Pt` is computed based on component value-scaled unit vectors of the Cartesian axis frame. This observation verifies that the position P = (x, y, z) is indeed derived by the equation

$$P_0 + x\hat{i} + y\hat{j} + z\hat{k}$$

Examine the scene more closely by selecting the `Controller` and toggling on/off the display of the position vector for the game object `P`, `DrawPositionVector`, and the default Cartesian axis frame, `DrawAxisFrame`. The `DrawComponents` toggle allows you to examine the component-scaled unit vectors: $x\hat{i}$, $y\hat{j}$, and $z\hat{k}$.

Details of MyScript

Open MyScript and examine the source code in the IDE. The instance variables and the Start() function are as follows:

```
public GameObject P = null;  // For manipulation
public GameObject Pt = null; // For computed position
public bool DrawPositionVector = true; // Visualization toggles
public bool DrawAxisFrame = true;
public bool DrawComponents = false;

private Vector3 iV = new Vector3(1f, 0f, 0f);  // unit vectors
private Vector3 jV = new Vector3(0f, 1f, 0f);  //  i, j, and k
private Vector3 kV = new Vector3(0f, 0f, 1f);  //

#region For visualizing the vectors
#endregion

void Start() {
    Debug.Assert(P != null);   // Verify proper setting
    Debug.Assert(Pt != null);

    #region For visualizing the vectors
    #endregion

}
```

All the public variables for MyScript have been discussed when analyzing the Controller's MyScript component, and as in all previous examples, the Debug.Assert() calls in the Start() function ensure proper setup regarding referencing the appropriate game objects via the Inspector Window. The private iV, jV, and kV variables are the corresponding $\hat{i}$, $\hat{j}$, and $\hat{k}$ unit vectors of the Cartesian axis frame. The details of the Update() function are as follows:

```
void Update() {
    // 1. position and  the position vector
    Vector3 Po = Vector3.zero;
    Vector3 v = P.transform.localPosition - Po;
```

```
    // 2. Verify component-scaled unit vector computes position
    Pt.transform.localPosition = Po + v.x*iV + v.y*jV + v.z*kV;

    #region  For visualizing the vectors
    #endregion
}
```

The first two lines of code convert the position of P to a position vector in the Cartesian axis frame by computing the vector from the origin Po to P or

$$\vec{V} = P - P_0$$

Although unnecessary for the Cartesian axis frame because the origin is always (0,0,0), this step is taken explicitly to differentiate and remind you that position vectors are vectors from the origin to the given positions. This seemingly insignificant observation will become important in next sections.

In Step 2, the position of game object Pt is computed by summing the component-scaled unit vectors from the origin following the given equation

$$\text{Pt} = P_0 + x\hat{i} + y\hat{j} + z\hat{k}$$

The two steps in the Update() function follow precisely the given equations and the result is indeed the same position as expected.

Takeaway from This Example

This is a relatively straightforward example demonstrating and verifying the intuitive equations

$$\vec{V} = P - P_0$$

$$P = P_0 + x\hat{i} + y\hat{j} + z\hat{k}$$

The next section will generalize these equations to support derivation of position vector components for different axis frames.

Relevant mathematical concepts covered include

- For the default Cartesian axis frame, the sum of components-scaled unit vectors does indeed compute the proper position.

EXERCISES

Moving the Origin of Cartesian Coordinate

Try the following. Replace the first line of code in the Update() function with

```
Vector3 Po = new Vector3(1f, 1f, 1f);
                    // instead of (0,0,0)
```

and notice that Pt will continue to follow the position of P correctly. In fact, the position of Pt will follow that of P for any Po.

What happened is that a new axis frame is created when Po is set to anything other than (0,0,0). The next section will explore this in depth by deriving and working with components of general axis frames.

Positions in General Axis Frames

In the previous section you interpreted the default Cartesian Coordinate System as an axis frame with the three perpendicular unit vectors, $\hat{i}$, $\hat{j}$, and $\hat{k}$, being the major axes intersecting at the origin $(0,0,0)$. You have also learned to consider the coordinate values of a Cartesian Coordinate position (x,y,z) as the components of its position vector. You will now map these concepts to a general axis frame where the three perpendicular unit vectors may not be aligned with the x-, y-, or z-directions and these vectors may not intersect at $(0,0,0)$.

This section begins with a review of the definition and derivation of a general axis frame. The section then proceeds to analyze positions as position vectors defined in these general axis frames and demonstrates that all positions can be decomposed into components of any given axis frame. You will learn that positions can be expressed and derived based on components from any axis frame.

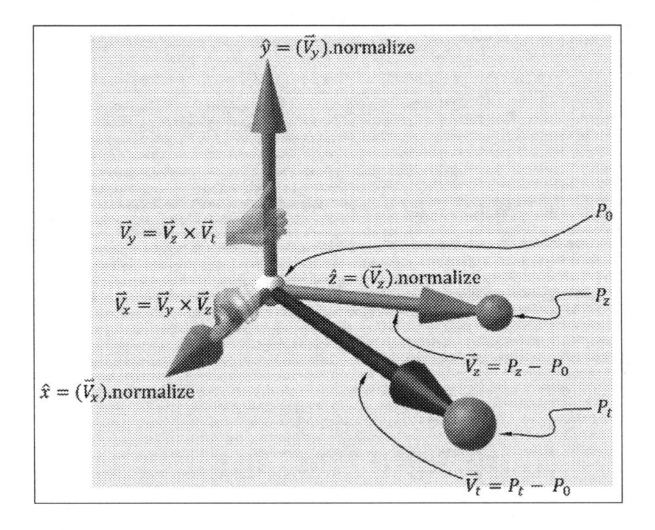

Figure 7-4. *A general axis frame derived from three non-collinear points*

Review of Axis Frame Derivation

As discussed in Chapter 6, an axis frame can be defined by three points that are not collinear. This is reviewed in Figure 7-4. The given three positions labeled on the right of the figure, P_0, P_z, P_t, define two unique vectors, $\vec{V}_z$ and $\vec{V}_t$, with tails located at P_0

$$\vec{V}_z = P_z - P_0$$

$$\vec{V}_t = P_t - P_0$$

Now, let

$$\vec{V}_y = \vec{V}_z \times \vec{V}_t$$

then $\vec{V}_y$ is perpendicular to both $\vec{V}_z$ and $\vec{V}_t$. At this point, $\vec{V}_t$ may not be perpendicular to $\vec{V}_z$. This can be rectified by computing

$$\vec{V}_x = \vec{V}_y \times \vec{V}_z$$

Now, $\vec{V}_x$, $\vec{V}_y$, and $\vec{V}_z$ are three perpendicular vectors which may not be normalized. Let $\hat{x}$, $\hat{y}$, and $\hat{z}$, be the normalized versions of the three vectors and an axis frame is successfully derived, with the three unit vectors intersecting at the origin, P_0.

The default Cartesian Coordinate System is a special example of an axis frame because its $\hat{x}$, $\hat{y}$, and $\hat{z}$ vectors are $\hat{i}$, $\hat{j}$, and $\hat{k}$ with corresponding values of $(1, 0, 0)$, $(0, 1, 0)$, and $(0, 0, 1)$ and that the vectors intersect at the origin with $P_0 = (0, 0, 0)$.

Note An axis frame (in 3D) is defined by three major axes: perpendicular unit vectors, $\hat{x}$, $\hat{y}$, and $\hat{z}$, intersecting at P_0, the origin of the axis frame. It is important to note that P_0 may not be located at $(0, 0, 0)$.

Position Vectors in General Axis Frames

You have been working with positions specified in the default Cartesian axis frame where the origin is conveniently located at $(0, 0, 0)$. For this reason, in the Cartesian axis frame the position, $P = (x, y, z)$, and its position vector, $\vec{V}_p$, always have identical components, where

$$\vec{V}_p = P - (0,0,0) = (x,y,z) - (0,0,0) = (x,y,z) = P$$

This property of having identical components for a position and the corresponding position vector is a special case for the Cartesian axis frame and is not true for any axis frame with origin located at a position other than $(0, 0, 0)$.

In general, the origin of an axis frame, labeled as P_0 in Figure 7-4, can be located at any position in the 3D space. This definition for the origin implies that the general definition of a position vector, $\vec{V}$, for position, P, is

$$\vec{V} = P - P_0$$

Note that since P_0 of an axis frame can be located anywhere, in general, position vectors for the same position may be different across axis frames. Very importantly, given a position, P, its position vector, $\vec{V}$, in an arbitrary axis frame is usually different from the position vector, $\vec{V}_p$,in the Cartesian axis frame.

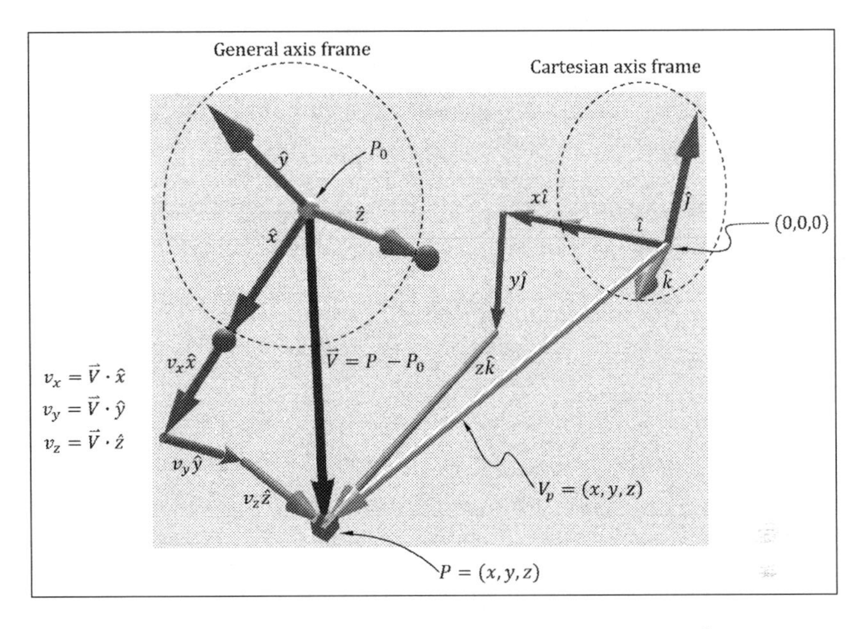

Figure 7-5. *Position vector in general and the Cartesian axis frame*

Figure 7-5 depicts the two position vectors, $\vec{V}$ and $\vec{V}_p$, for the given position, $P = (x, y, z)$, in two axis frames: the $\hat{x}$, $\hat{y}$, and $\hat{z}$ with origin at P_0 in the top left and the default Cartesian axis frame toward the top right of the figure.

Note Each position vector is defined with respect to the origin of the corresponding axis frame.

Components of Position Vectors

It is now possible to derive the components of a position vector by refining the description of a position in the context of a general axis frame:

> *Position P in an axis frame can be interpreted as the displacements measured along the major axes from the origin, P_0.*

In this case, instead of the $\hat{i}$, $\hat{j}$, and $\hat{k}$ of the Cartesian axis frame, a general axis frame has $\hat{x}$, $\hat{y}$, and $\hat{z}$ as the major axes, and the origin, P_0, can be located anywhere. The phrase

from the origin position, P_0

refers to the position vector

$$\vec{V} = P - P_0$$

where

displacements measured along the major axes

are the size of the position vectors measured along the major axes or

$$v_x = \vec{V} \cdot \hat{x}$$

$$v_y = \vec{V} \cdot \hat{y}$$

$$v_z = \vec{V} \cdot \hat{z}$$

Thus, the given description of the position, P, can be formulated as the following equation:

$$P = P_0 + \vec{V}$$

or

$$P = P_0 + (\vec{V} \cdot \hat{x})\hat{x} + (\vec{V} \cdot \hat{y})\hat{y} + (\vec{V} \cdot \hat{z})\hat{z}$$

$$= P_0 + v_x\hat{x} + v_y\hat{y} + v_z\hat{z}$$

Here, v_x, v_y, and v_z are the components of the position vector of P in the axis frame with major axes $\hat{x}$, $\hat{y}$, and $\hat{z}$ and origin P_0. Note the similarity between this equation and the one from the previous section where, in the Cartesian axis frame, with the $\hat{i}$, $\hat{j}$, and $\hat{k}$ as major axes and origin located at $(0, 0, 0)$,

$$P = (0,0,0) + \left(\vec{V}_p \cdot \hat{i}\right)\hat{i} + \left(\vec{V}_p \cdot \hat{j}\right)\hat{j} + \left(\vec{V}_p \cdot \hat{k}\right)\hat{k}$$

$$= (0,0,0) + x\hat{i} + y\hat{j} + z\hat{k}$$

Recall that in this case *x*, *y*, and *z* are components of the position vector in the Cartesian axis frame. Once again, you can observe that in the Cartesian axis frame, and only in the Cartesian axis frame, components are identical to coordinate values.

Note It is important to distinguish between the components (v_x, v_y, v_z) and the coordinate values (x, y, z) of a position. Coordinate values are the results of evaluating components in the context of an axis frame. That is, coordinate values are the results of evaluating $P_0 + v_x\hat{x} + v_y\hat{y} + v_z\hat{z}$.

Figure 7-5 illustrates the preceding derivations where the same position, *P*, is represented by and can be derived based on two different position vectors. On the left shows the accumulation of component-scaled $\hat{x}$, $\hat{y}$, and $\hat{z}$ vectors that resulted in the position vector, $\vec{V}$, while the right side of the figure illustrates the summation of scaled $\hat{i}$, $\hat{j}$, and $\hat{k}$ that resulted in $\vec{V}_p$. Clearly, $\vec{V} \neq \vec{V}_p$, and yet with the two vectors describing offset from the origins along the major axes' directions of their corresponding axis frames, the head of both vectors is located at the same position, *P*. Thus, you can observe that the same position can be expressed and represented as components of different axis frames.

Note Components of a vector are defined with respect to specific axis frames. The process of computing the values for the components, for example, $v_x = \vec{V} \cdot \hat{x}$, is referred to as vector decomposition, or decomposing a vector into its components.

In mathematical terms, axis frames are examples of vector spaces, where the set of three perpendicular unit vectors is an example of a set of basis vectors, and deriving components of a vector to be represented in another axis frame is referred to as changing of basis.

The Components of Any Frame Example

This example demonstrates that for a given position, in addition to the default position vector and components of the Cartesian axis frame, a distinct position vector with a corresponding set of component values can be derived for any axis frame. Figure 7-6 shows a screenshot of running the `EX_7_2_ComponentsOfAnyFrame` scene from the `Chapter-7-VectorComponents` project.

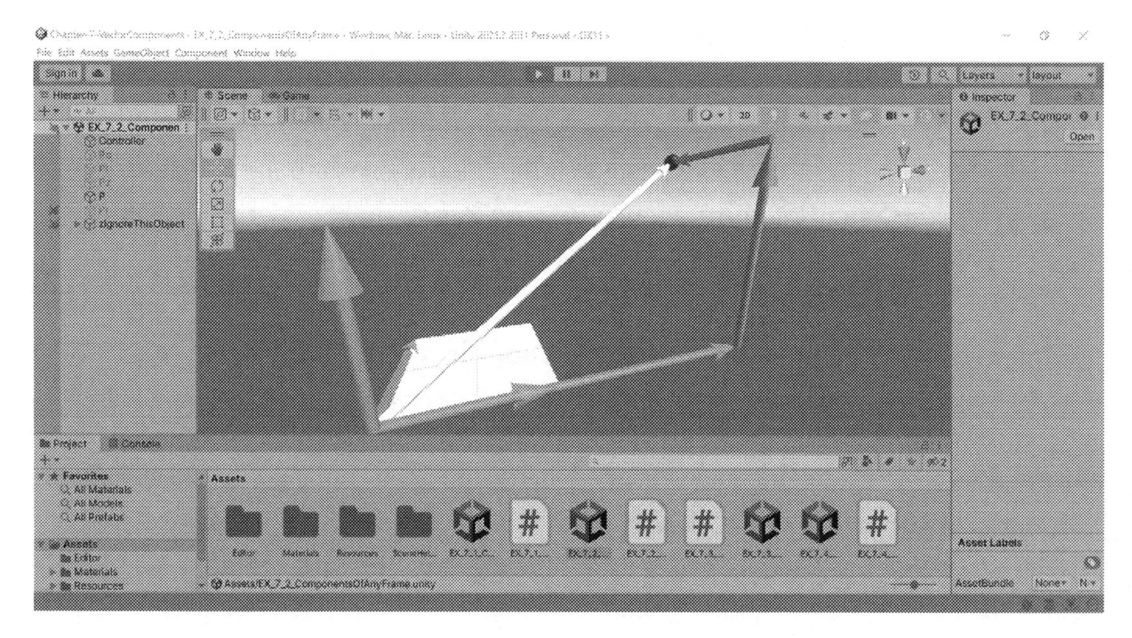

Figure 7-6. Running the Components of Any Frame example

The goals of this example are for you to

- Refamiliarize the steps of deriving an axis frame from three non-collinear positions
- Experience computing and working with vector components
- Examine vector components in any given axis frame
- Appreciate that for the same position, there is a distinct position vector for each different axis frame

Examine the Scene

Take a look at the Example_7_2_ComponentsOfAnyFrame scene and observe the predefined game objects in the Hierarchy Window. In addition to the Controller, there are five objects in this scene: Po (the white sphere), Pt (the red sphere), Pz (the blue sphere), P (the black sphere), and Pr (the green cube). In this case, P is the position of interest; Po, Pt, and Pz are the three non-collinear positions that you can manipulate to define an arbitrary axis frame; and the Pr position is computed based on the components of the corresponding position vector.

Analyze Controller MyScript Component

The MyScript component on the Controller shows that the game objects in the scene are referenced by variables with the same names and that you have the option to show or hide the Cartesian and the derived axis frames.

Note In all examples of this chapter, when attempting to manipulate an axis frame by adjusting the positions of Pt or Pz, you will experience strange constraints and awkwardness. It will appear that the system is fighting against you and often undo or modify your actions. As will be discussed at the end of this chapter, the orientation of an axis frame should be specified and manipulated based on rotation and not independent adjustments of positions. Rotation is a topic that will be discussed in the next chapter. Fortunately, in the context of this chapter, you are focusing on the relative relation of vectors and components to a changing axis frame. Your goal is to manipulate an axis frame, not define or specify a particular axis frame. In all examples of this chapter, simply adjust Pt and Pz to cause changes to the axis frame. Direct your attention on the vectors and components instead of the details of the actual axis frame.

Interact with the Example

Click the Play button to run the example. With the default setting of hiding the details of the derived axis frame, you should observe a scene that is similar to that from the previous example: a position (black sphere) and the corresponding position vector (white vector) with x-, y-, and z-components in the Cartesian axis frame.

Now, select the `Controller` object and flip the axis frame being drawn: disable the showing of Cartesian and enable the derived frame. You should observe a scene that appears to be very similar to the previous. Instead of white, you will observe a position vector in black with components along the $\hat{x}$ (in red), $\hat{y}$ (in green), and $\hat{z}$ (in blue) directions. Notice that the $\hat{x}$, $\hat{y}$, and $\hat{z}$ directions are perpendicular and that the objects `P` (the black sphere) and `Pr` (the green cube) overlap and are located at exactly the same position. The position of `Pr` is computed based on the position vector $\vec{V}$ of position `P` according to

$$\vec{V} = P - P_0$$

and

$$P_r = P_0 + \left(\vec{V} \cdot \hat{x}\right)\hat{x} + \left(\vec{V} \cdot \hat{y}\right)\hat{y} + \left(\vec{V} \cdot \hat{z}\right)\hat{z}$$

You can select and manipulate `Po` (white sphere), `Pt` (red sphere), or `Pz` (blue sphere) to define arbitrary axis frames and observe the changes in major axes' directions and resulting components size, while the position, `P`, always remains stationary. You are observing new sets of component values of the same position for each distinct axis frame defined.

The position of `Pr` cannot be manipulated because this position is computed based on the derived components of position `P`. Take note that `P` and `Pr` always overlap at exactly the same location. This observation verifies that it is always possible to compute coordinate values from components for any given axis frame.

Now select the `Controller` object to re-enable and show the Cartesian axis frame and components. Observe that the position, `P`, is defined by two sets of components: the white position vector of Cartesian axis frame (the $\vec{V}_p$ vector in Figure 7-5) and the black position vector of the defined axis frame (the $\vec{V}$ vector in Figure 7-5). This observation reinforces that any position can be represented and derived by the components of any axis frame. Feel free to manipulate the derived axis frame or `P` to further observe this concept.

Note that the white position vector is $\vec{V}_p$ in Figure 7-5 and it is simply

$$P = \vec{V}_p = (x, y, z)$$

where the sizes of the components are x (in red), y (in green), and z (in blue). The red vector originating from (0, 0, 0) is the $\hat{i}$ vector scaled by x, or $x\hat{i}$, accumulating with $y\hat{j}$ (in green) and then $z\hat{k}$ (in blue). This faithfully implements the equation

$$P = (0,0,0) + x\hat{i} + y\hat{j} + z\hat{k}$$

On the other hand, the black position vector from P_0 (the white sphere) is $\vec{V}$ in Figure 7-5, where

$$\vec{V} = P - P_0$$

In this case, the red vector originating from P_0 is the $\hat{x}$ vector scaled by v_x, or $v_x\hat{x}$, accumulating with $v_y\hat{y}$ (in green) and $v_z\hat{z}$ (in blue), implementing the equation

$$P = P_0 + v_x\hat{x} + v_y\hat{y} + v_z\hat{z}$$

You have now verified that for all positions, in addition to the default position vector of the Cartesian axis frame, a separate position vector can be derived based on the origin and components from any axis frame!

Details of MyScript

Open `MyScript` and examine the source code in the IDE. The instance variables and the `Start()` function are as follows:

```
public GameObject Po = null; // Origin of the reference frame
public GameObject Pt = null; // Position for defining x-dir
public GameObject Pz = null; // Position on z-axis

public GameObject P = null;  // Position to show components
public GameObject Pr = null; // Derived from components

public bool DrawCartesianFrame = true;  // show/hide frames
public bool DrawDerivedFrame = true;

#region For visualizing the vectors
void Start() {
    Debug.Assert(P != null);   // Verify proper editor init
    Debug.Assert(Pr != null);
```

```
    Debug.Assert(Po != null);
    Debug.Assert(Pt != null);
    Debug.Assert(Pz != null);

    #region For visualizing the vectors
    #endregion
}
```

All the public variables for `MyScript` have been discussed when analyzing the `Controller`'s `MyScript` component. The details of the `Update()` function are as follows:

```
void Update() {
    // Step 1: Derive the axis frame
    Vector3 origin = Po.transform.localPosition;
    Vector3 Vt = Pt.transform.localPosition - origin;
    Vector3 zDir = (Pz.transform.localPosition -
                    origin).normalized;
    Vector3 yDir = Vector3.Cross(zDir, Vt).normalized;
    Vector3 xDir = Vector3.Cross(yDir, zDir).normalized;

    // Step 2: Position vector and the components
    Vector3 V = P.transform.localPosition - origin;
    float vx = Vector3.Dot(V, xDir);
    float vy = Vector3.Dot(V, yDir);
    float vz = Vector3.Dot(V, zDir);

    // Step 3: Compute Pr position from the components
    Pr.transform.localPosition = origin +
                                 vx*xDir + vy*yDir + vz*zDir;

    #region  For visualizing the vectors
    #endregion
}
```

Step 1 closely follows the equations for axis frame derivation, where

- `Vt`: $\vec{V}_t = P_x - P_0$
- `zDir`: $\hat{z} = \left(\vec{V}_z\right).normalize = \left(P_x - P_0\right).normalize$

- yDir: $\hat{y} = (\vec{V}_y).normalize = (\vec{V}_z \times \vec{V}_t).normalize$
- xDir: $\hat{x} = (\vec{V}_x).normalize = (\vec{V}_y \times \vec{V}_z).normalize$

Step 2 computes the position vector, $\vec{V}$, for P and the components of the derived axis frame:

- V: $\vec{V} = P - P_0$
- vx, vy, vz: $v_x = \vec{V} \cdot \hat{x}$, $v_y = \vec{V} \cdot \hat{y}$, $v_z = \vec{V} \cdot \hat{z}$

Lastly, Step 3 shows that the coordinate values for the position can be derived based on the axis frame and the components, where

$$\text{Pr: } P = P_0 + v_x\hat{x} + v_y\hat{y} + v_z\hat{z}$$

Note that the same location, *P*, is derived based on a very different computation when compared to that for the Cartesian axis frame.

Takeaway from This Example

Through this example you have verified that the Cartesian Coordinate System is indeed just a special example of axis frame. In general, for any given axis frame, locations of positions can be described by offsets from the origin with three perpendicular vectors scaled by their corresponding components. You have also analyzed and examined the details of deriving the components of a vector for any arbitrary axis frame.

Relevant mathematical concepts covered include

- A general axis frame is defined by three major axes: perpendicular vectors, $\hat{x}$, $\hat{y}$, and $\hat{z}$, with tails intersecting at the origin, P_0.
- For any general axis frame, the position vector, $\vec{V}$, of *P* is

  $$\vec{V} = P - P_0$$

  where the components of vector $\vec{V}$ can be determined by projecting the vector onto each of the three major axis

  $$v_x = \vec{V} \cdot \hat{x},$$

  $$v_y = \vec{V} \cdot \hat{y} \text{, and}$$

  $$v_z = \vec{V} \cdot \hat{z}$$

- Coordinate values of positions in an axis frame can be derived based on the computed components

$$P = P_0 + v_x\hat{x} + v_y\hat{y} + v_z\hat{z}$$

Note the coordinate values are computed based on two separate sets of parameters: those that define the axis frame P_0, $\hat{x}$, $\hat{y}$, and z, and the values of the components v_x, v_y, and v_z.

EXERCISES

Front and Up in an Axis Frame

In the given example, the verification position, `Pr`, is computed according to

$$P_r = P_0 + v_x\hat{x} + v_y\hat{y} + v_z\hat{z}$$

Try changing this expression to

$$P_r = P_0 + v_x\hat{x} + v_y\hat{y} + (v_z + 2)\hat{z}$$

Recall that `Pr` is the green cube, now ensure that the derived axis frame is displayed, manipulate the positions, *P*, `Po`, `Pt`, and `Pz`, and you will observe that the green cube is always a constant offset of 2 units in the $\hat{z}$-axis direction from `P`. If you consider the $\hat{z}$-axis as the front direction, then in this case, `Pr` is always "in front of" `P`. Do you know how to modify the equation for `Pr` such that it is always "on top of" `P`?

Convert Components Between Axis Frames

Given that a position P has components a_d, a_e, and a_f in an axis frame with major axes, $\hat{d}$, $\hat{e}$, $\hat{f}$, and origin at P_a. How can you compute the corresponding components for P in a different axis frame with major axes, $\hat{l}$, $\hat{m}$, $\hat{n}$, and origin at P_b?

The solution process is actually rather straightforward; you would first compute the coordinate values for position P, followed by the position vector and then the new components in the new axis frame. The first step would be to compute the coordinate values of P

$$P = P_a + a_d\hat{d} + a_e\hat{e} + a_f\hat{f}$$

Next, the position vector in the new axis frame

$$\vec{V} = P - P_b$$

Lastly, projecting the position vector to derive the proper components

$$b_l = \vec{V} \cdot \hat{l}$$

$$b_m = \vec{V} \cdot \hat{m}$$

$$b_n = \vec{V} \cdot \hat{n}$$

Vectors in Axis Frames

With the systematic analysis and thorough understanding of positions as position vectors and components in general axis frames, you are now ready to analyze relationships between these positions or vectors in general axis frames. Recall from Chapter 4, a vector is defined by the difference of the corresponding coordinate values between two positions and it encodes the displacements between these two positions. As you will learn in this section, similar to positions, vectors, with all of their elegant properties you learned about in Chapter 4, can also be represented and defined by components in any axis frame. Additionally, just as in the case for positions, there is a distinct set of components describing a vector for each given axis frame and it is always possible to convert between the components of different axis frames.

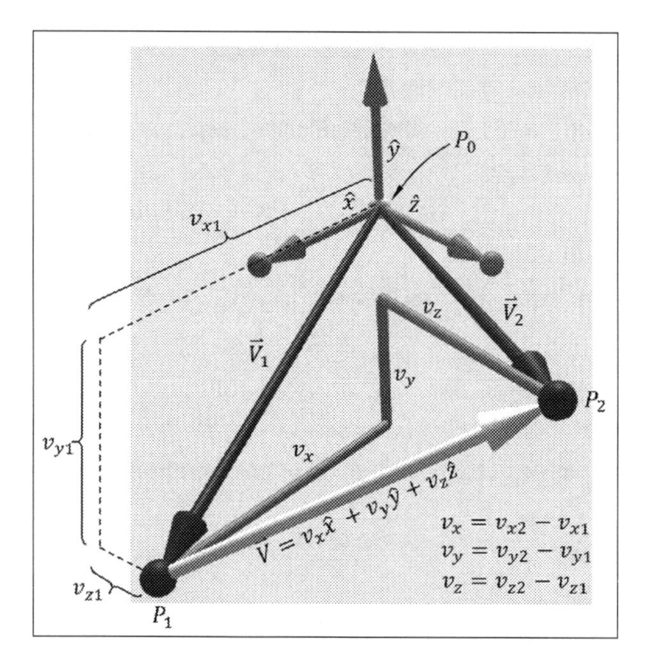

Figure 7-7. *A vector between two positions in an axis frame*

Vector Components

The top portion of Figure 7-7 shows an arbitrary axis frame with $\hat{x}$, $\hat{y}$, $\hat{z}$ as major axes and origin at P_o. A vector, $\vec{V}$, defined by two positions, P_1 and P_2, is illustrated in the lower-center region of the figure.

You have learned that, given the axis frame, the position vector of P_1 is

$$\vec{V}_1 = P_1 - P_0$$

and that the components of P_1 for the given axis frame are v_{x1}, v_{y1}, and v_{z1}, where

$$v_{x1} = \vec{V}_1 \cdot \hat{x}$$

$$v_{y1} = \vec{V}_1 \cdot \hat{y}$$

$$v_{z1} = \vec{V}_1 \cdot \hat{z}$$

and that P_1 is located at

$$P_1 = P_0 + v_{x1}\,\hat{x} + v_{y1}\,\hat{y} + v_{z1}\,\hat{z}$$

In a similar fashion, the location of P_2 can be expressed as follows:

$$P_2 = P_0 + v_{x2}\,\hat{x} + v_{y2}\,\hat{y} + v_{z2}\,\hat{z}$$

Note The details of P_2 and $\vec{V}_2$ are similar to the correspondence of P_1 and $\vec{V}_1$; to avoid excessive cluttering, these are not annotated in Figure 7-7.

The components of a position are derived from the position vector of the position and not the coordinate values of the position.

From Chapter 4, you have learned that the vector, $\vec{V}$, from P_1 to P_2 is defined as

$$\vec{V} = P_2 - P_1$$

$$= \left(P_0 + v_{x2}\,\hat{x} + v_{y2}\,\hat{y} + v_{z2}\,\hat{z}\right) -$$

$$\left(P_0 + v_{x1}\,\hat{x} + v_{y1}\,\hat{y} + v_{z1}\,\hat{z}\right)$$

With P_0 subtracted and collecting terms for each of the major axis,

$$\vec{V} = \left(v_{x2} - v_{x1}\right)\hat{x} + \left(v_{y2} - v_{y1}\right)\hat{y} + \left(v_{z2} - v_{z1}\right)\hat{z}$$

Let

$$v_x = v_{x2} - v_{x1}$$

$$v_y = v_{y2} - v_{y1}$$

$$v_z = v_{z2} - v_{z1}$$

then

$$\vec{V} = v_x \hat{x} + v_y \hat{y} + v_z \hat{z}$$

There are two important observations in this derivation. First, in axis frames, vectors are defined by subtracting the corresponding components of the positions. Second, vectors are always defined by the summation of the major axis directions (the unit vectors) scaled by the difference of the corresponding components from the head and tail positions.

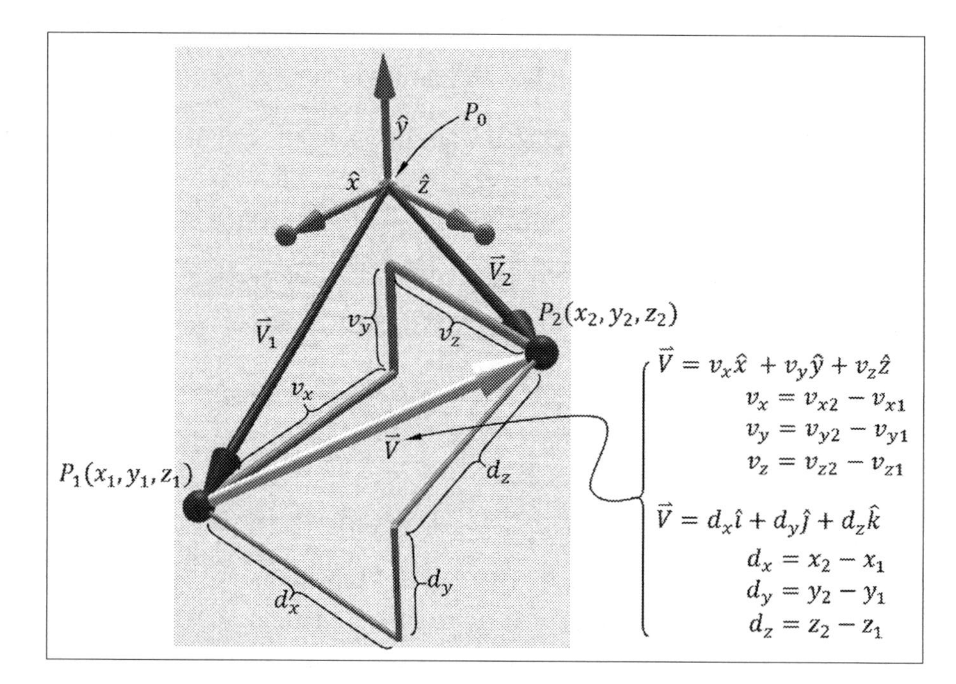

Figure 7-8. *Analyze the vector in Cartesian axis frame*

Analysis in Cartesian Axis Frame

Figure 7-8 shows the details of analyzing the same vector, $\vec{V}$, in the Cartesian axis frame. Assuming the coordinate values of positions P_1 and P_2 to be

$$P_1 = (x_1, y^1, z_1) \text{ and}$$

$$P_2 = (x_2, y_2, z_2)$$

Recall that with origin at (0, 0, 0), the component and coordinate values are identical in the Cartesian axis frame, such that

$$P_1 = (0,0,0) + x_1\,\hat{k} + y_1\,\hat{j} + z_1\,\hat{k} = x_1\,\hat{k} + y_1\,\hat{j} + z_1\,\hat{k}$$

$$P_2 = (0,0,0) + x_2\,\hat{k} + y_2\,\hat{j} + z_2\,\hat{k} = x_2\,\hat{k} + y_2\,\hat{j} + z_2\,\hat{k}$$

In this way, the vector, $\vec{V}$, is defined as

$$\vec{V} = P_2 - P_1 = (x_2 - x_1)\hat{i} + (y_2 - y_1)\hat{j} + (z_2 - z_1)\hat{k}$$

Let

$$d_x = x_2 - x_1$$

$$d_y = y_2 - y_1$$

$$d_z = z_2 - z_1$$

then

$$\vec{V} = d_x\,\hat{i} + d_y\,\hat{j} + d_z\,\hat{k}$$

You have verified that the vector with components, v_x, v_y, and v_z, in the axis frame with $\hat{x}$, $\hat{y}$, and $\hat{z}$ as the major axes and origin at P_o is the same vector with components, d_x, d_y, and d_z in the axis frame with $\hat{i}$, $\hat{j}$, and $\hat{k}$ as the major axes and origin at (0, 0, 0). The key observation is that the same vector is represented by components with distinct values in different axis frames. Lastly, note that since $\hat{i}$, $\hat{j}$, and $\hat{k}$ are constants with values (1, 0, 0), (0, 1, 0), and (0, 0, 1):

$$\vec{V} = d_x\,(1,0,0) + d_y\,(0,1,0) + d_z\,(0,0,1)$$

$$= (d_x,0,0) + (0,d_y,0) + (0,0,d_z) = (d_x,d_y,d_z)$$

This derivation, once again, verifies that for Cartesian axis frame, the values of component and coordinate are identical.

The Vectors in Any Frame Example

This example demonstrates the definition of vectors based on specifying component values and computing the difference in corresponding components from existing positions. Figure 7-9 shows a screenshot of running the `EX_7_3_VectorsInAnyFrame` scene from the `Chapter-7-VectorComponents` project.

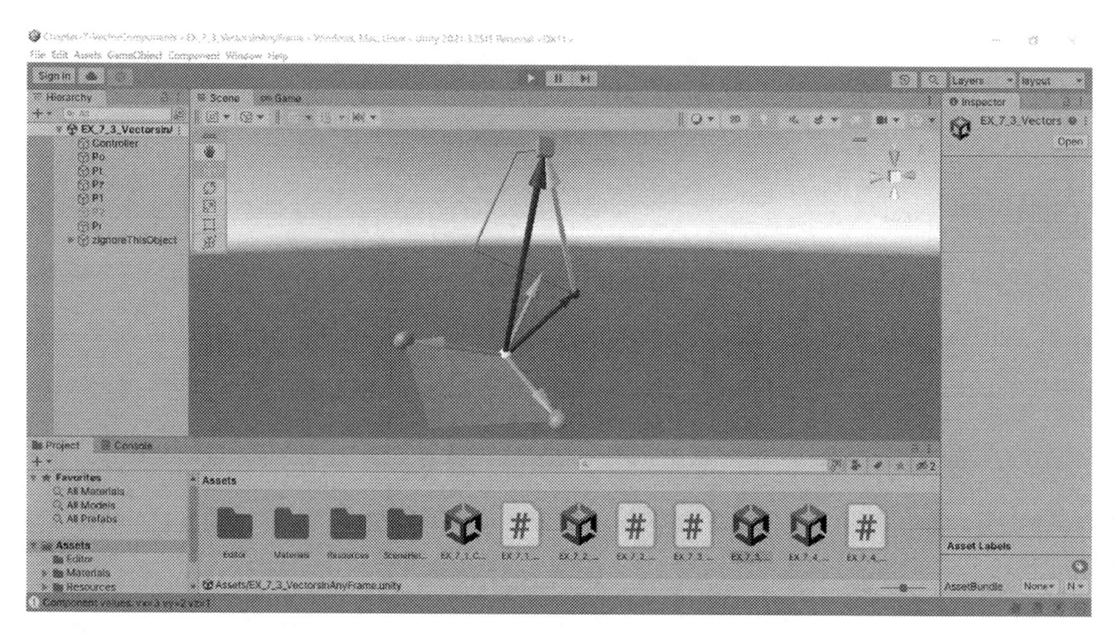

***Figure 7-9.** Running the Vectors in Any Frame example*

The goals of this example are for you to

- Verify that displacements or vectors can be defined explicitly based on component values specified in any axis frame
- Derive vectors based on computing the difference in the corresponding components of two positions
- Examine the defined vectors in the context of and with respect to a changing axis frame

Examine the Scene

Take a look at the `Example_7_3_VectorsInAnyFrame` scene and observe the predefined game objects in the Hierarchy Window. In addition to the `Controller`, there are six objects. Similar to the previous example, `Po` (white sphere), `Pt` (red sphere), and `Pz` (blue sphere) allow the definition and manipulation of an axis frame. The components of P1 and P2 (both black spheres) positions define the vector for examination and the position of Pr (green cube) is derived based on the computed vector.

Analyze Controller MyScript Component

The MyScript component on the Controller shows the six references to the corresponding game objects discussed. Additionally, there are three floating-point controls, vx, vy, and vz. Through these controls, you can specify the values for the components of the $\hat{x}$, $\hat{y}$, and $\hat{z}$ directions to define a vector, $\vec{V}$, for computing the position of Pr

$$\vec{V} = v_x \hat{x} + v_y \hat{y} + v_z \hat{z}$$

$$P_r = P_1 + \vec{V}$$

The four toggles control the showing or hiding of the derived axis frame, the components of the vector, the Cartesian axis frame, and if the vector should be computed based on the positions P1 and P2 (instead of from the specified component values).

Interact with the Example

Click the Play button to run the example. You will observe a white vector from P1 (black sphere) to Pr (green cube). Also visible are the axis frame (with the red plane), the position vectors for P1 and Pr (in black), and the components of the white vector. Pay attention to the components of the white vector: the three perpendicular segments in red, green, and blue showing the displacements along the $\hat{x}$ (in red), $\hat{y}$ (in green), and $\hat{z}$ (in blue) directions. Take note to verify visually that these three components are perpendicular and parallel to their respective axis in the axis frame.

In following this example, your interaction will consist of three categories: examine vectors defined by explicitly specified components, examine vectors in the derived and the Cartesian axis frames simultaneously, and examine vectors computed based on subtracting corresponding components of positions.

Defined by Specified Components

Select the `Controller` and verify the initial values of `vx`, `vy`, and `vz` to be three, two, and one. Notice that these values correspond to the lengths of the displayed components—that the red segment is about three times the length of and the green segment is about two times the length of the blue segment. You can adjust these values to observe the intuitive changes in the corresponding component size that control the white vector and the position of `Pr`. For example, decreasing the value of `vx` will shorten the red component resulting in the position `Pr` moving closer to `P1` along the red component or $\hat{x}$ direction. You have just experienced defining vectors based on specifying component values explicitly.

Now, manipulate the positions of `Po`, `Pt`, and `Pz` to redefine the general axis frame. Observe that when you change the positions of `Pt` and `Pz`, the orientation of the axis frame changes. Since the vector component values are specified explicitly, the lengths of the red, green, and blue component segments do not change when only the directions of $\hat{x}$, $\hat{y}$, and $\hat{z}$ are updated. For this reason, the white vector maintains a constant relative relationship to and follows the axis frame changes. When you change the position of the origin, `Po`, since the vector is defined as a displacement from `P1` and independent from any other positions, the white vector remains constant as expected.

You have interacted with a vector defined by explicitly specified components in a changing axis frame. You have observed that as the orientation of the axis frame changes, such a vector also re-orientates and maintains a constant relative relationship with the axis frame. This can be further understood mathematically. The vector, $\vec{V}$, is defined as

$$\vec{V} = v_x\ \hat{x} + v_y\ \hat{y} + v_z\ \hat{z}$$

With the values of `vx`, `vy`, and `vz` specified and constant, changing the axis frame corresponds to changing $\hat{x}$, $\hat{y}$, and $\hat{z}$. Thus, the constant relative relationship with the underlying axis frame reflects constant displacements with respect to the changing major axes.

Analyze in Derived and Cartesian Axis Frames

Select the `Controller` and toggle `DrawCartesian` to enable the displaying of Cartesian axis frame, position vectors, and components. You will observe an additional and thicker set of red, green, and blue components showing the corresponding $\hat{i}$, $\hat{j}$, and $\hat{k}$ component sizes in the Cartesian axis frame. Try toggling `DrawCartesian` on and off repeatedly to verify and differentiate between the two sets of components.

Now, when you manipulate vx, vy, and vz values, you will continue to observe intuitive changes in the first thinner set of components: only the size of the corresponding component will change! Verify that this may not be the case for the thicker set of components of the Cartesian axis frame. For example, you can adjust the value for vx to observe changes in all three components of the Cartesian axis frame. In this case, you are observing the changes in size along the $\hat{x}$ direction and that the $\hat{x}$ direction is described by the combination of $\hat{i}$, $\hat{j}$, and $\hat{k}$ directions of the Cartesian axis frame. You have just observed the same vector having drastically different component values in two axis frames.

Defined by Positions

Select the `Controller` and toggle off `DrawCurrentFrame`, `DrawComponents`, and `DrawCartesian` for a clean display. Toggle on `VectorFromP1P2` to define the vector $\vec{V}$ by subtracting corresponding components of positions `P1:` v_{x1}, v_{y1}, v_{z1} and `P2:` v_{x2}, v_{y2}, v_{z2},

$$\vec{V} = P_2 - P_2$$

$$= (v_{x2} - v_{x1})\hat{x} + (v_{y2} - v_{y1})\hat{y} + (v_{z2} - v_{z1})\hat{z}$$

Adjust the positions of `P1` and `P2` to update the components of the vector. You can verify the component values are updated by examining either the corresponding fields in the `Controller` or the printouts in the Console Window. Recall that the position for `Pr` is still computed according to

$$P_r = P_1 + \vec{V}$$

In your interactions you will observe that `Pr` position always follows and matches exactly to that of `P2`. You have now verified that vectors can indeed be defined by subtracting the corresponding components of the head and tail positions.

Lastly, and very importantly, select the `Controller` and toggle `DrawCurrentFrame` and `DrawComponents` to re-enable the displaying of the general axis frame and the components. Now, once again manipulate `Pt` and `Pz` to redefine the general axis frame. Since the vector is now defined by two positions that are stationary with respect to the axis frame, as the axis frame changes, the white vector stays constant. However, notice that the components of the vector are defined with respect to the current axis frame and thus are constantly changing when the axis frame is updated. You can observe the printout of the component values in the Console Window. Mathematically, given the vector $\vec{V}$

$$\vec{V} = v_x\,\hat{x} + v_y\,\hat{y} + v_z\,\hat{z}$$

When the axis frame is updated, $\hat{x}$, $\hat{y}$, and $\hat{z}$ are changed, and vx, vy, and vz values are updated to maintain a constant vector.

Details of MyScript

Open `MyScript` and examine the source code in the IDE. The instance variables and the `Start()` function are as follows:

```
public GameObject Po = null;     // Origin of axis frame
public GameObject Pt = null;     // x-direction of frame
public GameObject Pz = null;     // z-direction of frame
public GameObject P1 = null;     // Position for manipulation
public GameObject P2 = null;     // P1P2 defines V
public GameObject Pz = null;     // Position derived from V

public float vx = 3.0f;          // Component values
public float vy = 2.0f;
public float vz = 1.0f;

public bool DrawCurrentFrame = true; // Visualization toggles
public bool DrawComponents = true;
public bool DrawCartesian = false;
public bool VectorFromP1P2 = true;

#region For visualizing the vectors
void Start() {
    Debug.Assert(P1 != null);    // Ensure proper setup
    Debug.Assert(P2 != null);
    Debug.Assert(Pr != null);
    Debug.Assert(Po != null);
```

```
    Debug.Assert(Pt != null);
    Debug.Assert(Pz != null);

    #region For visualizing the vectors
}
```

All the public variables for MyScript have been discussed, and as in all previous examples, the Debug.Assert() calls in the Start() function ensure proper setup regarding referencing the appropriate game objects via the Inspector Window. The details of the Update() function are as follows:

```
void Update() {
   // Step 1: Drive the axis frame
   Vector3 origin = Po.transform.localPosition;
   Vector3 Vt = (Pt.transform.localPosition - origin);
   Vector3 zDir = (Pz.transform. localPosition -
                   origin).normalized;
   Vector3 yDir = Vector3.Cross(zDir, Vt).normalized;
   Vector3 xDir = Vector3.Cross(yDir, zDir).normalized;

   // Step 2: Compute vector components if necessary
   if (VectorFromP1P2) {
       Vector3 V1 = P1.transform.localPosition - origin;
       float vx1 = Vector3.Dot(V1, xDir);
       float vy1 = Vector3.Dot(V1, yDir);
       float vz1 = Vector3.Dot(V1, zDir);

       Vector3 V2 = P2.transform.localPosition - origin;
       float vx2 = Vector3.Dot(V2, xDir);
       float vy2 = Vector3.Dot(V2, yDir);
       float vz2 = Vector3.Dot(V2, zDir);

       // Difference of the P1 and P2 components
       vx = vx2 - vx1;
       vy = vy2 - vy1;
       vz = vz2 - vz1;
   }
```

```
    Debug.Log("Component values: vx="
                + vx + " vy=" + vy + " vz=" + vz);

    // Step 3: compute the vector and position for P2
    Vector3 V = vx * xDir + vy * yDir + vz * zDir;
    // Derive Pr position from computed vector
    Pr.transform.localPosition = P1.transform.localPosition + V;

    // P1.transform.localPosition += 0.001f * V.normalized;
            // What does the above do?

    #region  For visualizing the vectors
}
```

Step 1 is identical to the previous example in deriving the parameters of the axis frame. In Step 2, if user specifies to derive the values of vx, vy, and vz from the P1 and P2 components, then the position vector, V1, for P1 is computed

$$\vec{V}_1 = P_1 - P_0$$

And the components of P1 for the given axis frame, vx1, vy1, and vz1, are derived

$$v_{x1} = \vec{V}_1 \cdot \hat{x}$$

$$v_{y1} = \vec{V}_1 \cdot \hat{y}$$

$$v_{z1} = \vec{V}_1 \cdot \hat{z}$$

The same operations are repeated for P2, and the values for vx, vy, and vz are computed as

$$v_x = v_{x2} - v_{x1}$$

$$v_y = v_{y2} - v_{y1}$$

$$v_z = v_{z2} - v_{z1}$$

Step 3 defines vector, V, and position of Pr to be

$$\vec{V} = v_x \hat{x} + v_y \hat{y} + v_z \hat{z}$$

$$P_r = P_1 + \vec{V}$$

Takeaway from This Example

You have verified that in a general axis frame, vectors can be defined by either specifying component values explicitly or subtracting the corresponding component values of the head and tail positions. You have also verified that given a vector

$$\vec{V} = v_x \hat{x} + v_y \hat{y} + v_z \hat{z}$$

When the component values, v_x, v_y, and, v_z, are specified in a changing axis frame, the vector will update along with the axis frame maintaining a constant relative relationship. On the other hand, to maintain a constant vector in a varying axis frame, the component values must be recomputed.

Relevant mathematical concepts covered include

- Vectors are defined by component-scaled major axes' directions (perpendicular unit vectors) of axis frames.
- It is possible to define a vector to follow and maintain constant relative relationship to a varying axis frame by explicitly specifying the component values.
- It is also possible to define a vector to remain constant in a varying axis frame by continuously updating the component values.

EXERCISE

Velocity in an Axis Frame

Instead of computing the position for Pr, try using the derived vector, V, to update the position of P1 in the Update() function. That is, uncomment the very last line in Step 3 and enable the following:

```
P1.transform.localPosition += 0.001f * V.normalized;
```

The vector is scaled by a small number to avoid drastic position changes. As expected, when running the modified example, you will observe P1 traveling at a constant speed. The constant speed behavior persists even if you manipulate the axis frame. As you have learned, speed is the magnitude of a velocity or the vector; in this case, with the normalized vector, the small number is the actual size of the displacement per update, or the speed.

Now, enable the `VectorFromP1P2` toggle on the `Controller` and try modifying the axis frame. Notice the movement of `P1` is completely independent from the axis frame. This is not surprising as you have observed that the velocity (vector) is derived based on the positions of `P1` and `P2` which are both independent from the axis frame.

A more interesting case is to disable the `VectorFromP1P2` toggle. In this case, notice that the movement is constant with respect to the varying axis frame. Since the velocity (vector) is defined by specified components, as the axis frame changes, velocity follows. This observation suggests a solution for the player in a navigating spaceship example discussed earlier. This will be covered in the next section.

Motion Control in Axis Frames

Recall the example from earlier in the chapter of a player resting but wanting to move toward the medical bay on a navigating spaceship. You know that the position of the player is changing with the navigating spaceship. However, in the context of the spaceship, the player is currently resting with no movement. Additionally, when the player is ready, the movement toward the medical bay is independent of the asteroid dodging maneuvers of the spaceship. That is, the spaceship's turning should not affect the player's pathway of moving toward the medical bay. You are now ready to design a solution to support this scenario.

You have learned that for a general axis frame with origin at P_0 and major axes $\hat{x}$, $\hat{y}$, and $\hat{z}$, a position, P, with components v_x, v_y, and v_z is located at

$$P = P_0 + v_x\hat{x} + v_y\hat{y} + v_z\hat{z}$$

Notice that the location is described by two separate and independent sets of parameters: the axis frame and the components. This observation points to an elegant solution where the spaceship and the player can be described by the two sets of parameters. The first is to describe the location and orientation of the spaceship by the origin and major axes of an axis frame. And the second is to keep track of the player location based on its components with respect to the spaceship axis frame. With this design, as the spaceship navigates, the corresponding axis frame is updated while the components of the position of a resting player stay constant. Then, when the player wants to move, the movement can be represented by updating the components of the player's position independent from the spaceship's axis frame.

For clarity and simplicity of notations, in the following, the superscript c is introduced to represent vectors of components. For example, position P_1 with components v_{x1}, v_{y1}, and v_{z1} and P_2 with components v_{x2}, v_{y2}, and v_{z2} are expressed as

$$P_1^c = \left(v_{x1}, v_{y1}, v_{z1}\right), \text{ and}$$

$$P_2^c = \left(v_{x2}, v_{y2}, v_{z2}\right)$$

The components of the vector, $\vec{V}$, between positions P_1 and P_2 are

$$\vec{V}^c = P_2^c - P_1^c = \left(v_{x2} - v_{x1}, v_{y2} - v_{y1}, v_{z2} - v_{z1}\right)$$

$$= \left(v_x, v_y, v_z\right)$$

It is important to note that, in general, $P_1 \neq P_1^c$; instead,

$$P_1 = P_0 + \left(P_1^c.\text{x}\right)\hat{x} + \left(P_1^c.\text{y}\right)\hat{y} + \left(P_1^c.\text{z}\right)\hat{z}$$

Now, assuming P_1^c is the components of the player's position and P_2^c is that of the medical bay, then the normalized $\vec{V}^c$ or $\hat{V}^c$ is the direction that will lead the player to the medical bay.

Given a speed, *s*, when traveling toward the medical bay, the total traveling of the player at time *t* is $\vec{T}^c$

$$\vec{T}^c = ts\hat{V}^c$$

Now, the components of the player position, P^c, at time *t* are $\vec{T}$ displacements from the initial P_1^c

$$P^c = P_1^c + \vec{T}^c$$

where the actual coordinate values of the player are located at

$$P = P_0 + (P^c.\mathrm{x})\hat{x} + (P^c.\mathrm{y})\hat{y} + (P^c.\mathrm{z})\hat{z}$$

Take note that the preceding derivation is carried out with respect to the components based on the vector notations. Though working on components instead of coordinate values, you are still able to apply all of the vector concepts learned.

This solution defines positions and traveling velocity inside the spaceship by components with respect to the axis frame of the spaceship. In this way, the navigation of the spaceship updates its axis frame while the player location and movements within the spaceship are based on the specifics of the current axis frame at any given instance. The key observation is that while intimately related, the controls of the spaceship and player movements are completely independent. For example, while the spaceship is navigating (axis frame being updated), it is trivial to change the traveling direction of the player to move toward any other position, P_3, in the spaceship, for example, the command deck, and at any other speed, s'. The following example demonstrates the detailed implementation of this design.

The Motion in Axis Frame Example

This example demonstrates the advantage of defining positions and velocities based on components with respect to a changing axis frame. Figure 7-10 shows a screenshot of running the `EX_7_4_MotionInAxisFrame` scene from the `Chapter-7-VectorComponents` project.

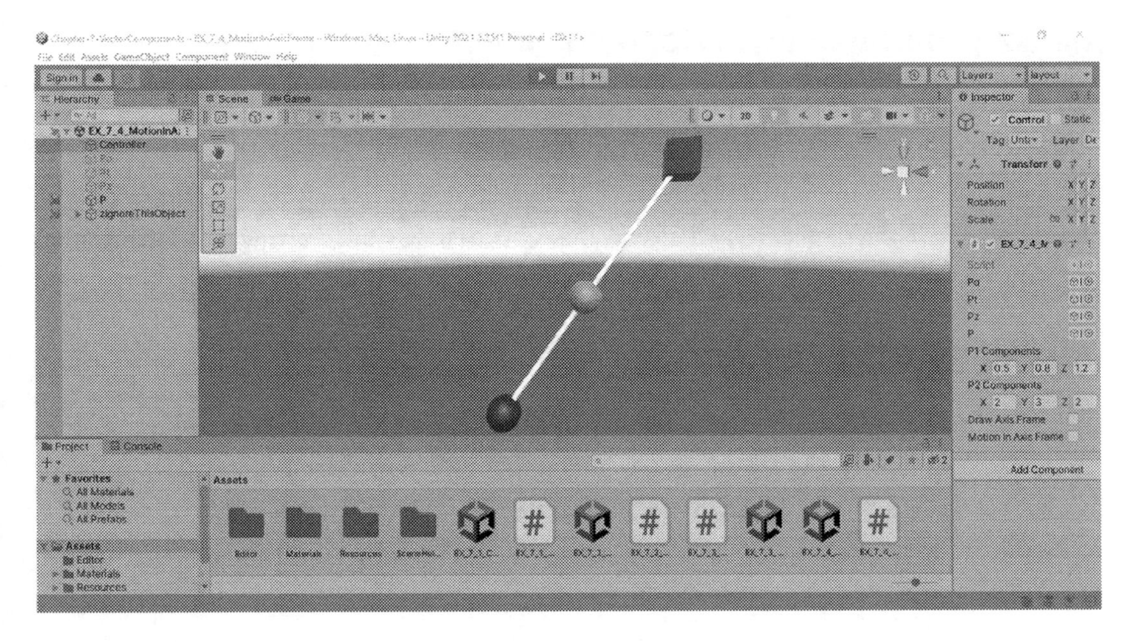

Figure 7-10. *Running the Motion in Axis Frame example*

The goals of this example are for you to

- Explore the application of axis frame concepts covered in this chapter
- Understand the advantage of defining positions and vectors as components with respect to a varying axis frame
- Observe that Cartesian Coordinate axis frame, with origin at (0, 0, 0) and major axes $\hat{i}$, $\hat{j}$, and $\hat{k}$, is indeed a simple example of axis frame and conforms to all of the concepts discussed

Examine the Scene

Take a look at the `Example_7_4_MotionInAxisFrame` scene and observe the predefined game objects in the Hierarchy Window. In addition to the `Controller`, there are four objects. Similar to the previous examples, `Po` (white sphere), `Pt` (red sphere), and `Pz` (blue sphere) allow the definition and manipulation of an axis frame. `P` (green sphere) is the current position of the player within the "spaceship."

Analyze Controller MyScript Component

The MyScript component on the Controller shows the four references to the corresponding game objects discussed. The P1Components and P2Components allow the specification of components for positions P_1 and P_2 representing the initial position of the player and that of the medical bay. The two check boxes toggle the drawing of the axis frame and if the computation should be carried out in the defined or the Cartesian axis frame.

Interact with the Example

Click the Play button to run the example. You will observe the object P (green sphere) continuously travels along a white line from a black sphere to a black cube. Select the Controller object and adjust the values of P1Components and P2Components to observe and verify that the black sphere location is controlled by P1Components and the cube by P2Components. These are the components of P_1 (player location) and P_2 (medical bay location) where the green sphere simulates the continuous motion from P_1 toward P_2.

Now, toggle on DrawAxisFrame to observe the Cartesian axis frame in relation to the objects. You can verify the computation is performed with respect to the Cartesian axis frame by setting P1Components to $(0, 0, 0)$ and P2Components to a location on one of the major axes, for example, $(2, 0, 0)$ or $(0, 2, 0)$. Through these interactions, you have verified that the computation is performed with respect to the Cartesian axis frame.

You can now toggle on MotionInAxisFrame to observe that Po (white sphere), Pt (red sphere), and Pz (blue sphere) are now displayed. At this point, the exact same computations are performed with respect to the defined axis frame. The system behaves in exactly the same manner, except that instead of a constant and static Cartesian axis frame, you can now update the axis frame.

You can simulate the spaceship in motion by selecting and changing the location of Po or manipulating Pt and Pz to rotate the axis frame and simulate asteroid dodging. In all cases, notice how P (green sphere), P_1 (black sphere), and P_2 (black cube) maintain their relative positions to the axis frame as the entire axis frame updates. In addition, note the motion of the green sphere continues as usual and is not affected by the axis frame manipulation.

In this example, P_1 and P_2 are represented by components with respect to a changing axis frame. The position of the traveling object, P, is computed based on velocity (vector) derived from the components of the positions. You have experimented with and observed the independence of axis frame and object motion controls.

Details of MyScript

Open MyScript and examine the source code in the IDE. The instance variables and the Start() function are as follows:

```
public GameObject Po = null;    // Origin of axis frame
public GameObject Pt = null;    // x-direction of frame
public GameObject Pz = null;    // z-direction of frame
public GameObject P = null;     // traveling object
public Vector3 P1Components = Vector3.zero; // P1 Components
public Vector3 P2Components = Vector3.one;  // P2 Components

public bool DrawAxisFrame = true;
public bool MotionInAxisFrame = false;

private const float kSpeed = 0.005f;
private float Traveled = 0f;
#region For visualizing the vectors
void Start() {
    Debug.Assert(P != null);   // Ensure proper setup
    Debug.Assert(Po != null);
    Debug.Assert(Pt != null);
    Debug.Assert(Pz != null);
    #region For visualizing the vectors
 }
```

All the public variables for MyScript have been discussed. The first private constant floating-point variable, kSpeed, defines the speed of the traveling object and the second variable, Traveled, is to accumulate the total distance traveled. As in all previous examples, the Debug.Assert() calls in the Start() function ensure proper setup regarding referencing the appropriate game objects via the Inspector Window. The details of the Update() function are as follows:

```
void Update() {
    // Parameters of an axis frame
    Vector3 origin, xDir, yDir, zDir;

    // Step 1: Set up the axis frame
    if (MotionInAxisFrame) {
```

```
        // Derive the axis frame
        origin = Po.transform.localPosition;
        Vector3 Vt = (Pt.transform.localPosition - origin);
        zDir = (Pz.transform.localPosition -
                origin).normalized;
        yDir = Vector3.Cross(zDir, Vt).normalized;
        xDir = Vector3.Cross(yDir, zDir).normalized;
    } else {
        // Default Cartesian axis frame
        origin = Vector3.zero;  // (0, 0, 0)
        xDir = Vector3.right;   // (1, 0, 0)
        yDir = Vector3.up;      // (0, 1, 0)
        zDir = Vector3.forward; // (0, 0, 1)
    }

    // Step 2: direction and distance traveled
    Vector3 Vc = P2Components - P1Components;
    Traveled += kSpeed * Time.deltaTime; //
    if (Traveled > Vc.magnitude)
        Traveled = 0f; // restart
    Vector3 Tc = Traveled * Vc.normalized;

    // Step 3: components and coordinate of P
    Vector3 Pc = P1Components + Tc;
    P.transform.localPosition = origin +
                   Pc.x * xDir + Pc.y * yDir + Pc.z * zDir;

    #region  For visualizing the vectors
}
```

The first line of the `Update()` function defines the parameters of an axis frame, `origin` (P_0), `xDir` ($\hat{x}$), `yDir` ($\hat{y}$), and `zDir` ($\hat{z}$). The first step is to determine the actual values for the axis frame parameters: either follow the derivation introduced in previous section based on the three manipulatable non-collinear positions `Po`, `Pt`, `Pz` or assign the constant values associated with the Cartesian axis frame. You should take special note of the fact that independent of the values for the axis frame, the rest of the computations are exactly the same. This is the most striking example of the fact that Cartesian axis frame is a specific example of the general axis frame.

Step 2 computes the components of the vector between positions `P1` and `P2` by subtracting the corresponding component values

$$\vec{V}^c = P_2^c - P_1^c$$

The step then accumulates the distance traveled, `Traveled`, where each update results in the coverage of `kSpeed×Time.deltaTime` distance. The implementation checks to ensure that the traveling is always in between `P1` and `P2` and then computes the total traveling

$$\vec{T}^c = ts\hat{V}^c$$

Step 3 computes the components of the green sphere position, `Pc`, by traveling from `P1Components`

$$P^c = P_1^c + \vec{T}^c$$

Lastly, the actual location for the position `P` is computed based on the computed component values, `Pc`

$$P = P_0 + \left(P^c.\text{x}\right)\hat{x} + \left(P^c.\text{y}\right)\hat{y} + \left(P^c.\text{z}\right)\hat{z}$$

The key observation is that the implementation indeed follows the derivation exactly and that independent controls of the motions for the spaceship and the player in the spaceship are accomplished.

Takeaway from This Example

You have observed that it is advantageous to represent locations and velocities of objects by their component values in a constantly changing axis frame.

Relevant mathematical concepts covered include

- Actual locations and velocities of objects can be conveniently represented by components while the reference axis frame varies.

EXERCISES

Vectors from Coordinate vs. Component Values

In this example, the motion vector, $\vec{V}^c$, is computed based on subtracting corresponding component values of the head and tail positions

$$\vec{V}^c = P_2^c - P_1^c$$

An alternative approach is to recognize that P_1 and P_2 locations can be derived based on the specified component values, `P1Components` and `P2Components`

$$P_1 = P_0 + v_{x1}\,\hat{x} + v_{y1}\,\hat{y} + v_{z1}\,\hat{z}$$

$$P_2 = P_0 + v_{x2}\,\hat{x} + v_{y2}\,\hat{y} + v_{z2}\,\hat{z}$$

With the coordinate values computed, the vector, $\vec{V}$, can be computed by subtracting the corresponding coordinate values, just as what you have done in Chapter 4

$$\vec{V} = P_2 - P_1$$

Note that, in this case, the position of P is simply

$$P = P_1 + ts\hat{V}$$

You can modify `MyScript` to implement the preceding cases. This exercise shows you that the same results can be derived based on computations performed with coordinate or component values. When you become familiar with the subject, you are free to choose either to work with.

Axis Frames in Unity

The concepts of axis frame and representing locations as components are crucial and their applications can be found in all interactive graphics software systems, especially in video games. These concepts are applied in all situations when there are interactions and controls involving connected or contained elements of objects, such as player holding on to objects, riding on vehicles, hoping on/off from horses, or a player in a spaceship.

Modern graphical applications typically abstract the detail specifics of axis frame and present the functionality to the end users via the interface to the scene hierarchy: the parent-child relationship between game objects that users can create and manipulate. In the Hierarchy Window of the Unity Editor, when you create a game object as a child of an existing game object, from the perspective of axis frame concepts discussed, you are effectively specifying the child location as components based on the axis frame defined by the parent game object. The actual implementation of the parent-child relationship is abstracted into a more advanced mathematical topic: matrices.

Matrix algebra based on strategically design data structure can encompass and hide the details and the transitions of axis frames. These are interesting topics of discussions for a more advanced book on math for game development.

Summary

This chapter continues with the discussion of positions and vectors by pointing out that the Cartesian Coordinate System is simply an example of the more general concept of axis frames. The chapter analyzes the characteristics of axis frames and explains that coordinate values are component values evaluated in specific axis frames. You have examined the representation and the conversion of components for the same location based on different axis frames. You have also learned to express vectors as components of axis frames and experimented with defining a constant vector with respect to a varying axis frame.

The chapter concludes the coverage with a simplified example of a position (e.g., a player) moving toward a destination (e.g., the medical bay) in a varying axis frame (e.g., a navigating spaceship). You have witnessed the importance and advantage of representing locations as components of an axis frame in accomplishing independent motion controls.

In all of the examples from this chapter, you may have noticed, or felt frustrated by, the awkwardness in manipulating the orientation of the axis frame by adjusting the P_z and P_t objects on the two corresponding major axes. There seem to be strange or arbitrary constraints limiting the interactions where it can be challenging to manipulate these objects to achieve your desired axis frame orientation. This is not surprising as the implicit requirement that P_z and P_t must be on perpendicular axes dictates that the two objects should not be manipulated separately. In this case, what is required is to rotate the entire axis frame as an integral object. This is the topic of study for the next chapter.

CHAPTER 8

Quaternions and Rotations

After completing this chapter, you will be able to

- Appreciate that the rotation of a position is a movement of constant distance around an axis
- Characterize the rotation of a position by an angle and an axis of rotation
- Discuss quaternions as operators for representing rotations
- Implement basic quaternion algebra in rotating positions
- Appreciate that consecutive rotations on objects can be modeled by ordered concatenation of quaternions
- Derive the rotation required to align two arbitrary position vectors
- Describe and model homing and chasing behaviors
- Configure and work with the rotation operator of the `Transform` component on the Unity `GameObject`
- Derive the necessary quaternions to align two axis frames

Introduction

In previous chapters you have analyzed positions, studied intervals, learned to relate two positions via a vector, examined relationships between two vectors via a dot product, and studied the space that contains two vectors via the cross product. In the last chapter, you learned about axis frames and began to understand the convenience of considering

K. Sung and G. Smith, *Basic Math for Game Development with Unity 3D*,
https://doi.org/10.1007/978-1-4842-9885-5_8

multiple coordinate spaces simultaneously in non-trivial situations such as describing motions in a navigating spaceship. You have also encountered awkwardness when trying to manipulate an axis frame by individually adjusting the locations of three non-collinear positions. As discussed, what is desired is a tool for rotating the axis frame as an integrated object. This chapter introduces the quaternion as an operator to rotate positions, or position vectors.

Strategically defined quaternions and the associated algebra are efficient and powerful tools for describing vector rotations. You will learn that rotations can be characterized as angular motions with respect to an axis, where the angle can be derived from the result of a dot product while the axis for the rotation is simply the result of a cross product. Integrated with concepts of interpolation, quaternion rotations are capable of supporting continuous and smooth transitions from an existing direction to a new vector direction. More significantly, quaternions are operators that are independent from any given vector. For this reason, once computed, a quaternion operator can be applied to many instances of vectors, achieving identical rotation operations.

Imagine once more that you are traveling on a spaceship flying through an asteroid field. Now that you know how to apply concepts from axis frames to steer the hero to the medical bay, it is time to learn how to navigate the spaceship to dodge the asteroids. Recall that movement is defined by the changing of position along a vector. Navigating a spaceship generalizes this movement by requiring alignment with an axis frame while moving forward. The spaceship captain would react to the on-coming asteroids by manipulating, or rotating, an axis frame to orientate the spaceship while the spaceship changes its position along the axis that represents the front direction. In other words, the spaceship would continuously move along its forward direction while the captain determines the orientation and forward direction of the spaceship. The knowledge of how to strategically rotate a default axis frame to align with one being manipulated by a user is the key to navigation. Additionally, during navigation, you would want the change of direction to be gradual and smooth as it would in real life. As you can see from this simple example, the ability to effectively and efficiently represent and control rotations is indeed important in video game development.

This chapter begins by introducing quaternions and their rules of operation, or quaternion algebra, that are relevant to describing rotations. Representing rotations with quaternions is then described and analyzed including approaches to aggregate the results from multiple rotations. The second half of this chapter focuses on applying quaternion rotations to align directions and axis frames. To emulate the organic motions of gradual changing from an initial to a final direction, spherical linear interpolation, or SLERP, is introduced. With this knowledge, the actual navigation of a spaceship is left as the last exercise in this chapter.

Note This chapter presents quaternion as an operator, or a tool, from the specific perspective of characterizing and implementing rotations. There is no attempt to cover the fundamental mathematical concept behind quaternions. You can learn more about quaternions in general here: `https://en.wikipedia.org/wiki/Quaternion`.

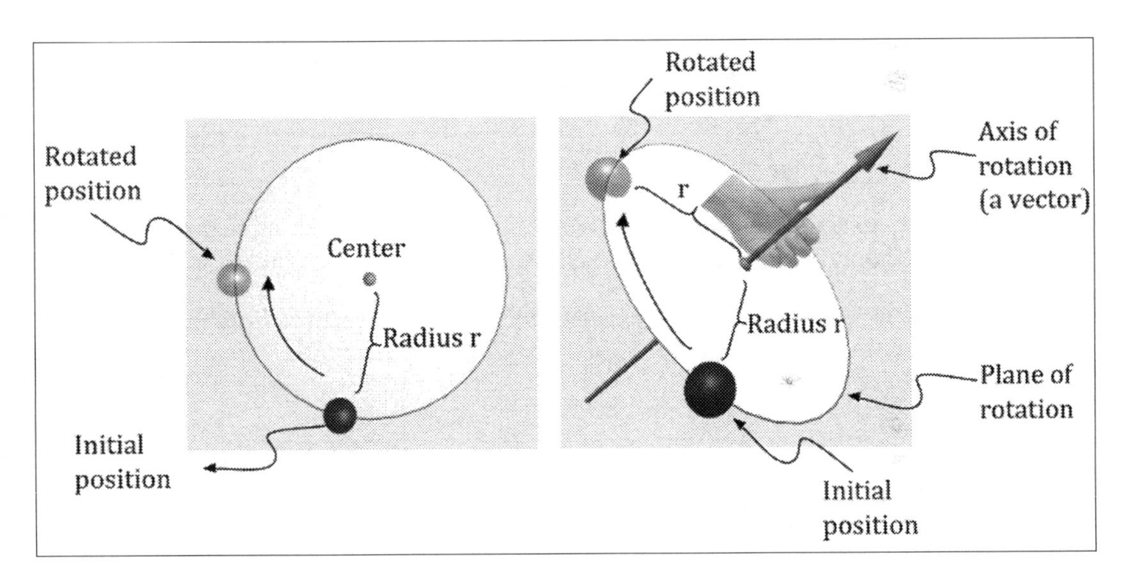

Figure 8-1. *Rotation about an axis in 2D and 3D*

Rotation Terminologies

You may remember when learning about circles that the shape is defined by moving a position while maintaining a constant distance from a second stationary position. As illustrated on the left of Figure 8-1, the stationary position is the center and the constant distance is the radius of the circle. When the movement is less than the circumference of the circle, you may describe that scenario as a rotation that sweeps out an arc or rotating from an initial position to a rotated position.

The right side of Figure 8-1 depicts the exact same rotation, only in 3D. Take note of the following:

- Axis of rotation: A vector that passes through the center of the circle or is the center of the rotation. Rotations are described as rotating with respect to, around, or about the axis of rotation. Note that an axis is simply a direction or a vector.
- Plane of rotation: Both the initial and rotated positions are located on this plane and this plane is always perpendicular to the axis of rotation.
- Direction of rotation: The positive direction of a rotation, in the case of the Left-Handed Coordinate System followed by this book, is pointed to by the thumb when the other four fingers are curled around the axis of rotation. In other words, if the thumb is pointing toward you, the positive direction of a rotation is clockwise.

It is important to note that the preceding terminologies and descriptions are true for any rotation operation. A rotation is a circular movement around the axis of rotation, the initial and rotated positions are always located on the plane of rotation, and the plane of rotation is always perpendicular to the axis of rotation.

Note Rotations in 2D, or the rotation of position (x,y), are always about the Z-axis with the x/y plane being the plane of rotation.

Quaternion: Tuple of Four

Quaternion is a tuple of four floating-point numbers expressed as

$$q = (\mathrm{x,y,z,w})$$

Given two quaternions, q_1 and q_2,

$$q_1 = (x_1, y_1, z_1, w_1)$$

$$q_2 = (x_2, y_2, z_2, w_2)$$

The quaternion multiplication

$$q_r = q_1 q_2 = (x_r, y_r, z_r, w_r)$$

is defined as

$$x_r = x_1 w_2 + y_1 z_2 - z_1 y_2 + w_1 x_2$$

$$y_r = -x_1 z_2 + y_1 w_2 + z_1 x_2 + w_1 y_2$$

$$z_r = x_1 y_2 - y_1 x_2 + z_1 w_2 + w_1 z_2$$

$$w_r = -x_1 x_2 - y_1 y_2 - z_1 z_2 + w_1 w_2$$

Take note that the quaternion multiplication operator takes two quaternions as operands and computes a new quaternion as the result. Given the definition, it is important to recognize that quaternion multiplication is not commutative, that is, in general,

$$q_1 q_2 \neq q_2 q_1$$

However, as you will demonstrate in the exercise at the end of this section, quaternion multiplication is associative; it is always the case that

$$q_1 q_2 q_3 = (q_1 q_2) q_3 = q_1 (q_2 q_3)$$

Lastly, the quaternion identity is

$$q_I = (0,0,0,1)$$

In the exercise at the end of this section, you will show that given any quaternion, q_a, it is always true that

$$q_a = q_I q_a = q_a q_I$$

It will become clear when discussing quaternion concatenation in later sections that the identity quaternion, q_i, plays the important role of serving as the initial value in a concatenation operation.

Encoding of Angle and Axis

A quaternion encodes a rotation of θ degrees along an axis, $\hat{V}_a = (x_a, y_a, z_a)$, as

$$q(\theta, \hat{V}_a) = \left(x_a \sin\frac{\theta}{2},\ y_a \sin\frac{\theta}{2},\ z_a \sin\frac{\theta}{2},\ \cos\frac{\theta}{2} \right)$$

In this encoding, the axis of rotation, $\hat{V}_a$, must be normalized as a unit vector. Notice that with $\hat{V}_a$ being normalized, the magnitude of q, or the sum of the components squared, is one. This magnitude of size one is important to ensure that the size of objects remains the same after a quaternion rotation operation.

The inverse of the q rotation: a rotation of $-\theta$ along the $\hat{V}_a$ axis or a rotation of θ along the negative $\hat{V}_a$ axis is the quaternion

$$q(\theta, \hat{V}_a)^{-1} = \left(-x_a \sin\frac{\theta}{2},\ -y_a \sin\frac{\theta}{2},\ -z_a \sin\frac{\theta}{2},\ \cos\frac{\theta}{2} \right)$$

The derivation for the inversed rotation is left as an exercise at the end of this section.

Rotation Operation

In order to rotate a given position, $P_i = (x_i, y_i, z_i)$, by θ degrees with respect to an axis, $\hat{V}_a$, with a properly encoded quaternion, q, the position must be expressed as a quaternion with the last component being zero

$$P_q = (x_i, y_i, z_i, 0)$$

The rotation operation is then defined by multiplying the rotation quaternion, q, and its inverse, q^{-1}

$$P_r' = q\, P_q\, q^{-1} = (x_r, y_r, z_r, w_r)$$

In an exercise you will show that the w-component of P_r', w_r, is always zero, where the rotated result, P_r, is

$$P_r = (x_r, y_r, z_r)$$

Remember that quaternion multiplication is not commutative and that the order of applying the q-rotation and its inversed is important. The quaternion representing the position to be rotated must be the operand in between q-rotation and its inversed with the q-rotation being on the left-hand side of the position.

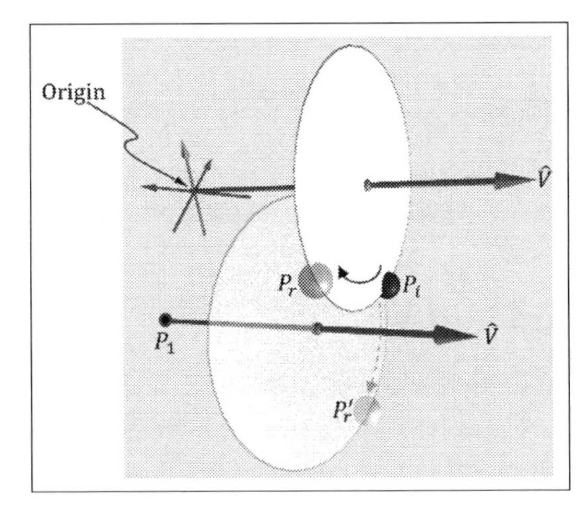

Figure 8-2. *Axis of rotation, $\hat{v}$, that passes through the origin and P_1*

Quaternion Rotation Limitation

Take note that a quaternion is a four floating-point tuple and that all four floating-point numbers are used in the representation of a rotation of θ angle around the $\hat{V} = (x,y,z)$ axis of rotation

$$q(\theta,\hat{V}) = \left(x\sin\frac{\theta}{2},\ y\sin\frac{\theta}{2},\ z\sin\frac{\theta}{2},\ \cos\frac{\theta}{2}\right)$$

Absent from this encoding is the information on the location of the axis of rotation. This is the limitation of quaternion rotation representation: it is a compact and efficient representation of rotations where the axes of rotation are assumed to pass through the origin.

Figure 8-2 explains this limitation by depicting two rotations with identical axes of rotation, $\hat{V}$. The rotation located near the top has the axis, $\hat{V}$, passing through the origin, while the lower rotation axis passes through the position P_1 instead of the origin. The quaternion rotation representation, $q(\theta,\hat{V})$, with no way to encode the P_1 location, is only capable of describing the rotation with the axis $\hat{V}$ passing through the origin. For this reason, applying $q(\theta,\hat{V})$ to rotate P_i will result in P_r. In general, quaternion representation is not capable of describing the rotation from P_i to P_r'.

The discussed quaternion rotation assumes that the axis of rotation passes through the origin. This limitation is not an issue when quaternions are used in concert with other tools, for example, matrices. In such cases, quaternions can support rotations with axes located at any position. However, general rotation with respect to an axis that does not pass through the origin is a subject of coordinate transformation, a more advanced topic not covered in this book. Later in this chapter, you will learn about working with the Unity `Transform` component on `GameObjects` to create rotations with general axes of rotations.

Rotating Positions and Vectors

Recall that vectors are independent of positions. Given a vector, $\vec{V_i} = (x_i, y_i, z_i)$, it is often convenient to depict the vector with tail position located at the origin for visual inspection. When depicted at the origin, the head of vector $\vec{V_i}$ locates at the position $P_i = (x_i, y_i, z_i)$.

In this way, rotating position $P_i = (x_i, y_i, z_i)$ is the same as rotating the head of the position vector for P_i or the vector $\vec{V_i}$. The rotated result $P_r = (x_r, y_r, z_r)$ can also be interpreted as the head of the position vector for P_r or the vector $\vec{V_r} = (x_r, y_r, z_r)$.

This discussion points out that it is equivalent to rotate positions, or head of position vectors, or head of vectors depicted at the origin. When considered in concert with the limitation that quaternions only support rotations with the axis of rotations passing through the origin, in the rest of this chapter, you can interpret coordinate values (x, y, z) as either a position, a position vector, or a vector.

The Rotation with Quaternion Example

This example demonstrates the quaternion rotation operation. It will allow you to interactively manipulate the angle and axis of a rotation and the position to be rotated so that you can observe and verify the quaternion definition, multiplication, and rotation operation. Figure 8-3 shows a screenshot of running the `EX_8_1_QuaternionRotation` scene from the `Chapter-8-Quaternions` project.

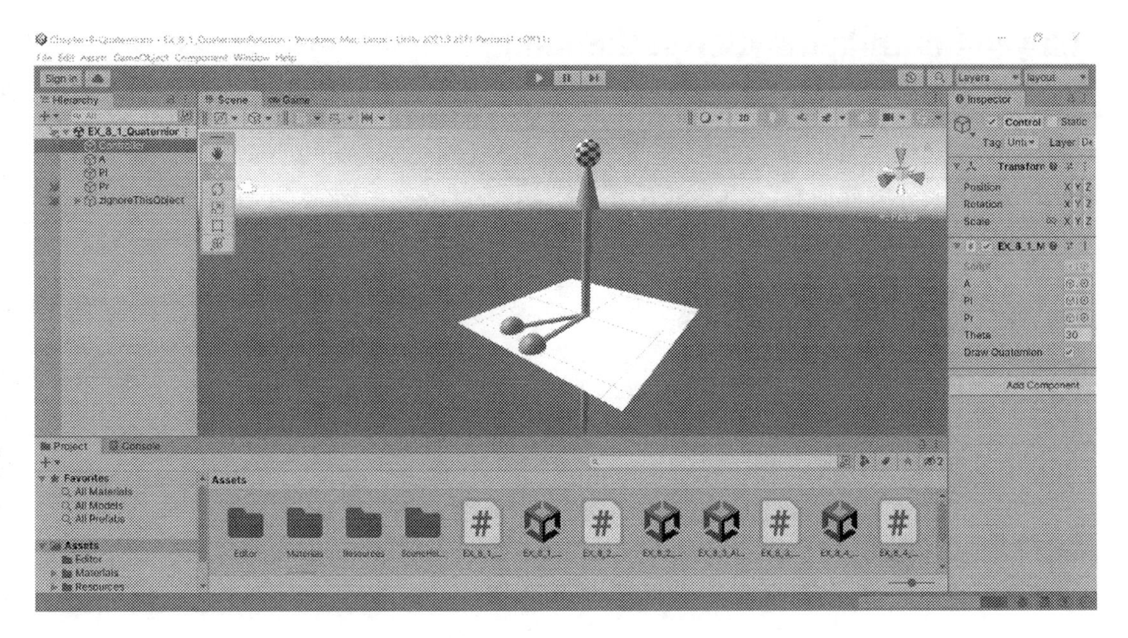

Figure 8-3. *Running the Quaternion Rotation example*

The goals of this example are for you to

- Define quaternion rotations based on specified angles and axes
- Verify the validity of quaternion rotation operation
- Experience and observe the results of quaternion rotations
- Examine the implementation of a quaternion rotation
- Appreciate the limitation of quaternion rotation: the axis of rotation must pass through the origin

Examine the Scene

Take a look at the `Example_8_1_QuaternionRotation` scene and observe the predefined checkered sphere `A`, the green sphere `Pi`, and the red sphere `Pr`. In this example, the rotation quaternion is derived from a user-specified angle and the axis of rotation defined by the position vector to `A`. This quaternion is then used to compute the rotated position `Pr` from the `Pi` position that is under the user control.

Analyze Controller MyScript Component

The `MyScript` component on the `Controller` shows the three variables with the same names as their corresponding reference game objects in the scene. The `Theta` variable is the angle to rotate and the `DrawQuaternion` toggle draws/hides the axis and perpendicular plane that defines the rotation quaternion.

Interact with the Example

Click the Play button to run the example. You can see a red vector that passes through the origin with head located at the position of the `A` sphere. This vector is the axis of rotation, $\hat{V}_a$. You can also observe the green, `Pi`, and red, `Pr`, spheres resting on a white plane that perpendicularly intersects the axis of rotation. These are the user controllable initial (green sphere) and the rotated (red sphere) positions.

The white plane is the plane of rotation where in addition to always intersecting the axis of rotation perpendicularly, the initial and rotated positions, or the green and red spheres, are always resting on this plane. Lastly, the red sphere's location is always a fixed rotation away from the green sphere on the white plane.

Select the `Controller` object and adjust `Theta` to observe that this variable indeed represents the angle between the green and red spheres. Take note to verify that as you increase and decrease the angle of rotation, the red sphere always rotates on the white plane. You can also select and manipulate the green sphere position and observe that the white plane always follows and maintains its perpendicular intersection with the axis of rotation, and that the red sphere is always a constant angular distance away from the green sphere on the white plane. You have observed that a quaternion rotation does indeed always rotate a position by the specified angle and that the rotation is indeed defined with respect to the axis of rotation.

Now select and manipulate the checkered sphere, `A`. As expected, when the checkered sphere position changes, the axis of rotation or the position vector of `A` follows. Take note that as the axis of rotation changes, the green sphere does not move while the white plane follows to maintain its perpendicular intersection with the axis of rotation and always cuts through both the green and red spheres. The location of the rotated red sphere also updates constantly to continue to lie on the white plane and maintains its angular distance from the green sphere. You have now observed and verified that a quaternion rotation always rotates a position perpendicular to the axis of rotation.

Finally, notice that the axis of rotation is defined based on a position vector. This says, the rotation of position Pi is defined with respect to an axis that passes through the origin. Once again, the discussed quaternion rotation only supports rotations with an axis of rotation that passes through the origin.

Details of MyScript

Open MyScript and examine the source code in the IDE. The instance variables and the Start() function are as follows:

```
public GameObject A = null;      // The axis of rotation
public GameObject Pi = null;     // initial position
public GameObject Pr = null;     // rotated position
public float Theta = 30.0f;
public bool DrawQuaternion  = true;

#region For visualizing the vectors
#endregion

void Start() {
    Debug.Assert(A != null);   // Verify proper setting
    Debug.Assert(Pi != null);
    Debug.Assert(Pr != null);
    #region For visualizing the vectors
    #endregion
}
```

All the public variables for MyScript have been discussed when analyzing the Controller's MyScript component, and as in all previous examples, the Debug.Assert() calls in the Start() function ensure proper setup regarding referencing the appropriate game objects via the Inspector Window.

In this example, in addition to Update(), three additional utility functions are defined to support quaternions: definition, QFromAngleAxis(); multiplication, QMultiplication(); and rotation, QRotation(). The details of QFromAngleAxis() are as follows:

```
Vector4 QFromAngleAxis(float angle, Vector3 axis) {
    float useTheta = angle * Mathf.Deg2Rad * 0.5f;
    float sinTheta = Mathf.Sin(useTheta);
    float cosTheta = Mathf.Cos(useTheta);
    axis.Normalize();
    return new Vector4(sinTheta * axis.x,
                       sinTheta * axis.y,
                       sinTheta * axis.z, cosTheta);
}
```

This function receives as input an `angle` θ and `axis` $\hat{V}_a = (x_a, y_a, z_a)$ and encodes the rotation in the returned quaternion

$$q = \left(x_a \sin\frac{\theta}{2},\ y_a \sin\frac{\theta}{2},\ z_a \sin\frac{\theta}{2},\ \cos\frac{\theta}{2} \right)$$

The details of `QMultiplication()` are as follows:

```
Vector4 QMultiplication(Vector4 q1, Vector4 q2) {
    Vector4 r;
    r.x =  q1.x*q2.w + q1.y*q2.z - q1.z*q2.y + q1.w*q2.x;
    r.y = -q1.x*q2.z + q1.y*q2.w + q1.z*q2.x + q1.w*q2.y;
    r.z =  q1.x*q2.y - q1.y*q2.x + q1.z*q2.w + q1.w*q2.z;
    r.w = -q1.x*q2.x - q1.y*q2.y - q1.z*q2.z + q1.w*q2.w;
    return r;
}
```

This function receives two quaternions, q_1 and q_2, where

$$q_1 = (x_1, y_1, z_1, w_1)$$

$$q_2 = (x_2, y_2, z_2, w_2)$$

compute the multiplication

$$q_r = q_1 q_2 = (x_r, y_r, z_r, w_r)$$

and return the resulting quaternion, q_r, where

$$x_r = x_1w_2 + y_1z_2 - z_1y_2 + w_1x_2$$

$$y_r = -x_1z_2 + y_1w_2 + z_1x_2 + w_1y_2$$

$$z_r = x_1y_2 - y_1x_2 + z_1w_2 + w_1z_2$$

$$w_r = -x_1x_2 - y_1y_2 - z_1z_2 + w_1w_2$$

The details of QRotation() are as follows:

```
Vector3 QRotation(Vector4 qr, Vector3 p) {
    Vector4 pq = new Vector4(p.x, p.y, p.z, 0);
    Vector4 qr_inv = new Vector4(-qr.x, -qr.y, -qr.z, qr.w);
    pq = QMultiplication(qr, pq);
    pq = QMultiplication(pq, qr_inv);
    return new Vector3(pq.x, pq.y, pq.z);
}
```

This function receives a quaternion, q_r,

$$q_r = (x_r, y_r, z_r, w_r)$$

and a position, P,

$$P = (x, y, z)$$

computes and returns the result of rotating P by q_r. The first line in this function expresses the input position P as a quaternion, P_q,

$$P_q = (x, y, z, 0)$$

The function then defines the inverse of q_r, q_r^{-1},

$$q_r^{-1} = (-x_r, -y_r, -z_r, w_r)$$

computes the quaternion rotation

$$P_q' = q_r \, P_q \, q_r^{-1} = (x', y', z', w')$$

and returns the resulting position, (x', y', z'). With the utility functions defined, the details of Update() are as follows:

```
void Update() {
    Vector3 axis = A.transform.localPosition;
    Vector4 q = QFromAngleAxis(Theta, axis);
    Pr.transform.localPosition =
                QRotation(q, Pi.transform.localPosition);
    #region  For visualizing the vectors
    #endregion
}
```

The first two lines of the function interpret the location of A as a position vector representing the axis of rotation and construct a rotation quaternion, q, based on the user-specified angle of rotation, Theta. The last line of the function computes the quaternion rotation using the position of Pi and sets the result as the location of Pr.

Takeaway from This Example

This is a straightforward example for verifying the validity of the discussed quaternion definition, multiplication, and rotation.

Relevant mathematical concepts covered include

- Quaternion, a tuple of four floating-point numbers, can be used to represent a rotation.
- Rotating a position by an angle about an axis through the origin can be implemented by multiplying the position with an appropriately defined quaternion and the inverse of that quaternion.
- Quaternion rotation, encoded in four floating-point numbers, is only capable of supporting rotations where the axis of rotation passes through the origin.

EXERCISES

Inverse of a Rotation Quaternion

The rotation quaternion, q, for a rotation with an angle θ along the axis $\hat{V}_a = (x_a, y_a, z_a)$ is defined as

$$q = \left(x_a \sin\frac{\theta}{2},\ y_a \sin\frac{\theta}{2},\ z_a \sin\frac{\theta}{2},\ \cos\frac{\theta}{2} \right)$$

Show that the inverse of q is

$$q^{-1} = \left(-x_a \sin\frac{\theta}{2},\ -y_a \sin\frac{\theta}{2},\ -z_a \sin\frac{\theta}{2},\ \cos\frac{\theta}{2} \right)$$

There are two ways to consider the inverse of a rotation. First, the inverse of a rotation is a rotation by the same angle along the negative rotation axis. In this case, the angle of rotation is still θ and along the negative axis $-\hat{V}_a = (-x_a, -y_a, -z_a)$,

$$q^{-1} = \left(-x_a \sin\frac{\theta}{2},\ -y_a \sin\frac{\theta}{2},\ -z_a \sin\frac{\theta}{2},\ \cos\frac{\theta}{2} \right)$$

Second, an alternative way to consider an inverse of a rotation is a rotation along the same axis by a negative angle. In this approach, the angle of rotation is $-\theta$ and along the same axis $\hat{V}_a$

$$q^{-1} = \left(x_a \sin\frac{-\theta}{2},\ y_a \sin\frac{-\theta}{2},\ z_a \sin\frac{-\theta}{2},\ \cos\frac{-\theta}{2} \right)$$

Since

$$\sin-\alpha = -\sin\alpha$$

$$\cos-\alpha = \cos\alpha$$

The inverse of the rotation is still

$$q^{-1} = \left(-x_a \sin\frac{\theta}{2},\ -y_a \sin\frac{\theta}{2},\ -z_a \sin\frac{\theta}{2},\ \cos\frac{\theta}{2} \right)$$

You have demonstrated that both of the approaches to defining the inverse of a quaternion rotation result in the same expression.

The q_I Identity Quaternion

By following the definition of quaternion inverse and multiplication, show that given the quaternion identity, q_I,

$$q_I = (0,0,0,1)$$

It is always true that

$$q_I^{-1} = q_I$$

And given any quaternion, q_a, it is always true that

$$q_a = q_I q_a = q_a q_I$$

These observations indicate that the quaternion identity is ideal for serving as the initial value when accumulating quaternion multiplication results.

Quaternion Multiplication: Commutative and Associative

It is stated, but without proof, that quaternion multiplication is not commutative and is associative or in general

$$q_1 q_2 \neq q_2 q_1$$

and it is always the case that

$$(q_1 q_2) q_3 = q_1 (q_2 q_3)$$

Knowing the definition of quaternion multiplication, you can now substitute and expand the preceding expressions to demonstrate for yourself that the preceding properties are true in general.

Verify Quaternion Multiplication Is Associative

Notice that in the `QRotation()`, the expression

$$P_q^{'} = q_r \, P_q \, q_r^{-1}$$

is implemented by the following two lines:

```
pq = QMultiplication(qr, pq);
pq = QMultiplication(pq, qr_inv);
```

This two-line implementation corresponds to

$$P_q^{'} = \left(q_r \, P_q\right) q_r^{-1}$$

Since quaternion multiplication is associative, you can switch the order of the two lines of code to implement

$$P_q^{'} = q_r \left(P_q \, q_r^{-1}\right)$$

Now, modify the given code and verify that the example continues to function correctly.

The w-Component of a Quaternion-Rotated Position

Expend the quaternion rotation expression

$$P_q^{'} = q_r \, P_q \, q_r^{-1}$$

and verify that the w-component of $P_q^{'}$ is always zero. You can reconfirm your derivation by making a `Debug.Log()` function call in the `QRotation()` function to print out the value of the w-component of Pq before the return statement.

Verify the Quaternion Rotation Formula

From trigonometry, you know or you can show that the result of rotating a 2D position (*x*, *y*) by θ around the Z-axis is the position

$$x' = x\cos\theta - y\sin\theta$$

$$y' = x\sin\theta + y\cos\theta$$

Note that this rotation can be described by the quaternion rotation $q(\theta,\hat{V})$, where

$$\hat{V} = (0,0,1)$$

or

$$q(\theta,(0,0,1)) = \left(0,0,\sin\frac{\theta}{2},\cos\frac{\theta}{2}\right) \text{ and}$$

$$q^{-1}(\theta,(0,0,1)) = \left(0,0,-\sin\frac{\theta}{2},\cos\frac{\theta}{2}\right)$$

Now, show that the given quaternion rotation formula for the position $P_q = (x, y, 0, 0)$

$$P_r' = q\, P_q\, q^{-1} = (x_r', y_r', z_r', w_r',)$$

is valid for 2D rotation about the Z-axis, where

$$x_r' = x' = x\cos\theta - y\sin\theta$$

$$y_r' = y' = x\sin\theta + y\cos\theta$$

$$z_r' = 0$$

$$w_r' = 0$$

Quaternion Concatenation

You have learned that a quaternion encodes a rotation of θ degrees along an axis, $\hat{V}_a = (x_a, y_a, z_a)$, as

$$q_1(\theta,\hat{V}_a) = \left(x_a \sin\frac{\theta}{2},\ y_a \sin\frac{\theta}{2},\ z_a \sin\frac{\theta}{2},\ \cos\frac{\theta}{2}\right)$$

To rotate a position, $P_i = (x_i, y_i, z_i)$, with the quaternion q_1, you would express the position as a quaternion with the last component being zero

$$P_q = (x_i, y_i, z_i, 0)$$

and compute

$$P_r' = q_1 P_q q_1^{-1} = (x_r, y_r, z_r, w_r)$$

With the w-component, w_r, being zero, the rotated position is

$$P_r = (x_r, y_r, z_r)$$

Now, following the same process, you can continue to rotate the position P_r by another rotation q_2

$$P_r'' = q_2 P_r' q_2^{-1}$$

If you express P_r' as a function of the origin position, P_q,

$$P_r'' = q_2 (q_1 P_q q_1^{-1}) q_2^{-1}$$

Since quaternion multiplication is associative, this same expression can be written as

$$P_r'' = (q_2 q_1) P_q (q_1^{-1} q_2^{-1})$$

In the exercise at the end of this section, you will show that the inverse of $q_2 q_1$, or $(q_2 q_1)^{-1}$, is $q_1^{-1} q_2^{-1}$. If you let

$$q_c = q_2 q_1$$

then

$$P_r'' = q_c P_q q_c^{-1}$$

Note The operation $q_c = q_2 q_1$ combines two rotation quaternions into one and is often referred to as concatenating quaternions. For example, q_c is the concatenated result of q_2 and q_1.

The preceding derivation shows that applying new rotations, q_2, on a q_1 rotated result, P_r', is the same as concatenating q_2 and q_1 and applying the resulting rotation, q_c, on the initial position, P_q. The key observation is that quaternion rotations can be concatenated to capture the combined results of multiple subsequent rotations.

Remember that quaternion multiplication is not commutative and that $q_c = q_2\, q_1$ is in general different from $q_d = q_1 q_2$. The order of rotation is important: the order for q_c is q_1 first than q_2, while the order for q_d is q_2 first than q_1. These two rotations are different in general.

Note The quaternion, $q_c = q_2\, q_1$, encodes a rotation that performs q_1 first followed by q_2. It may be counterintuitive, but although q_1 is on the right-hand side of the concatenation further away from the assignment, the q_1 operation is performed first.

The Quaternion Concatenation Example

This example demonstrates the results of applying multiple quaternions and a single concatenated quaternion in rotating a position. This example allows you to interactively manipulate three individual rotations and examine the results of applying the rotations independently verses the concatenated result as one single quaternion. Figure 8-4 shows a screenshot of running the `EX_8_2_QuaternionConcatenation` scene from the `Chapter-8-Quaternions` project.

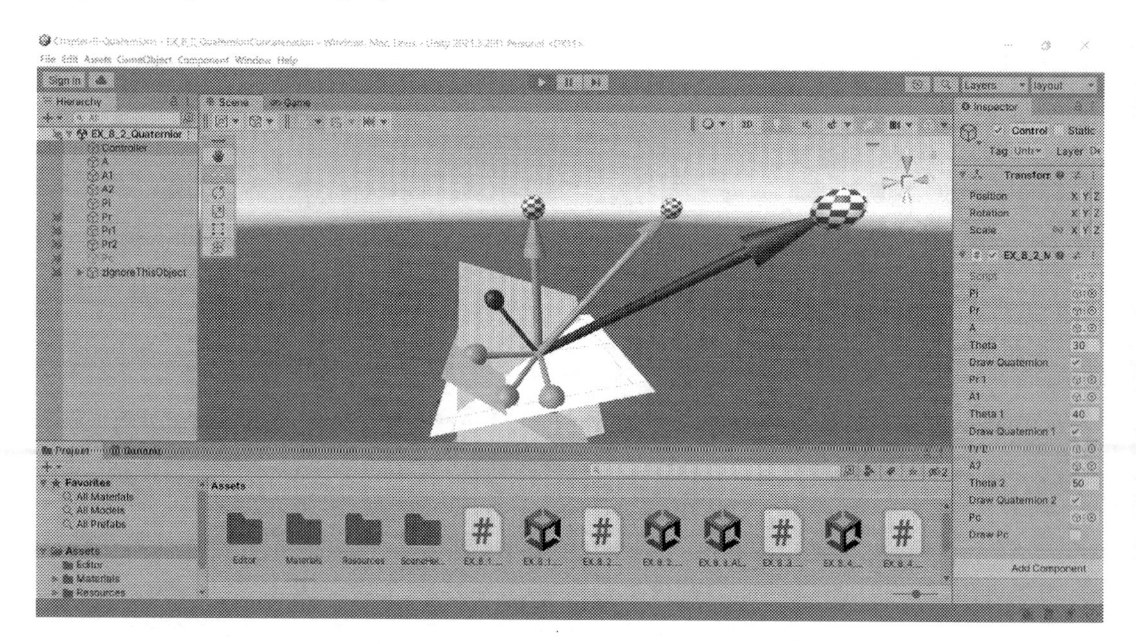

Figure 8-4. Running the Quaternion Concatenation example

The goals of this example are for you to

- Examine the results of applying multiple quaternion rotations to a position
- Gain experience with concatenation of quaternion rotations
- Verify that the concatenated quaternion delivers identical results as applying the rotations individually
- Appreciate the importance of concatenation ordering: subsequent rotations are concatenated on the left

Examine the Scene

Take a look at the `Example_8_2_QuaternionConcatenation` scene and observe three sets of variables representing the input and results of three subsequent quaternion rotations. In the following discussion, the three rotations are referred to as `q`, `q1`, and `q2`.

- Axis of rotations: `A`, `A1`, and `A2` checkered spheres. The position vectors to these objects define the axes of rotations for the three corresponding rotations `q`, `q1`, and `q2`.
- Input and results of individual rotations: `Pi` (green), `Pr` (red), `Pr1` (blue), and `Pr2` (black). The following equations summarize the relationships of these variable:

  $$P_r = q\,P_i\,q^{-1}$$

  $$P_{r1} = q_1\,P_r\,q_1^{-1} \text{ or } P_{r1} = q_1 q\,P_i\,q^{-1} q_1^{-1}$$

  $$P_{r2} = q_2\,P_{r1}\,q_2^{-1} \text{ or } P_{r2} = q_2 q_1 q\,P_i\,q^{-1} q_1^{-1} q_2^{-1}$$

  where `Pi` is the user-controlled input of the `q`-rotation. `Pr` is the result of the `q`-rotation and is the input to the `q1`-rotation with output of `Pr1` which in turn is served as the input to the `q2`-rotation with final output of `Pr2`.
- Result of the concatenated rotation: `Pc` (gray) is the result of concatenating `q`, `q1`, and `q2` rotations and applying to user input `Pi` or

  $$P_c = q_c\,P_i\,q_c^{-1}$$

where

$$q_c = q_2\, q_1\, q$$

Note that quaternion multiplication is not commutative and that the preceding concatenation order says that the order of performing rotations is q first then q1 and lastly q2.

Analyze Controller MyScript Component

The MyScript component on the Controller shows the variables with the same names as their corresponding reference game objects in the scene. Additionally, there are three floating-point variables, Theta, Theta1, and Theta2, for defining the degrees of rotations for the three rotations and corresponding toggles for showing/hiding the details of each rotation to avoid screen cluttering. The very last Boolean, DrawPc, toggles the drawing/hiding of Pc.

Interact with the Example

Click the Play button to run the example. You can see a cluttered of three independent rotations with three axes of rotations showing as vectors in red, blue, and black pointing to the three checkered spheres, A, A1, and A2. Take note that with DrawPc default to false, the gray Pc sphere is not visible.

In the following steps, your goal is to display, interact with, and examine each of the three rotations individually to verify the relationship of their inputs and results. You can begin with examining the first rotation, q, by selecting the Controller and toggling off DrawQuaternion1 and DrawQuaternion2. You are left with the details of the q-rotation defined by the axis A and Theta where the input is Pi (green) and result is Pr (red). Feel free to manipulate the positions of Pi, and A, and the value of Theta to note that as you modify the q-rotation, the positions of the other three objects (Pr, Pr1, and Pr2) follow in rigid manners maintaining constant angular displacements. This is as expected because these three objects are results of subsequent rotations. You can repeat this exercise for the other two rotations by hiding/showing the corresponding quaternions and manipulating the respective GameObjects and variables.

Verify that Pi maintains its location when you are examining the q1-rotation and that the positions of both Pi and Pr do not change when you examine the q2-rotation. These are inputs, and thus their positions are independent from the corresponding rotations.

Now, with all three quaternions showing, toggle on/off the DrawPc variable. Verify that Pc (gray) is located at exactly the same position as Pr2 (black). You can manipulate the three rotations, A (Theta), A1 (Theta1), and A2 (Theta2), and the Pi position to verify that the positions of Pr2 and Pc always overlap perfectly.

Recall that the position of Pr2 is the result of applying the three individual rotations or

$$P_{r2} = q_2\, q_1\, q\, P_i\, q^{-1}\, q_1^{-1}\, q_2^{-1}$$

while the position of Pc is the result of applying the concatenated quaternion

$$q_c = q_2\, q_1\, q$$

$$P_c = q_c\, P_i\, q_c^{-1}$$

You have verified that rotation quaternions can indeed be concatenated to capture the results of the combined rotations.

Details of MyScript

Open MyScript and examine the source code in the IDE. The instance variables and the Start() function are as follows:

```
public GameObject Pi = null;  // user control input position

public GameObject Pr = null;  // q-rotated position
public GameObject A = null;   // Axis of q-rotation
public float Theta = 30.0f;   // Angle of q-rotation
public bool DrawQuaternion = true;

public GameObject Pr1 = null;  // q1-rotated position
public GameObject A1 = null;   // Axis of q1-rotation
public float Theta1 = 40f;     // Angle of q1-rotation
public bool DrawQuaternion1 = true;

public GameObject Pr2 = null;  // q2-rotated position
public GameObject A2 = null;   // Axis of q2-rotation
public float Theta2 = 50f;     // Angle of q2-rotation
public bool DrawQuaternion2 = true;
```

```
public GameObject Pc = null;   // qc-rotated position
public bool DrawPc = false;

#region For visualizing the vectors
#endregion

void Start() {
    Debug.Assert(Pi != null);   // Verify proper setting
    Debug.Assert(Pr != null);
    Debug.Assert(A != null);
    Debug.Assert(Pr1 != null);
    Debug.Assert(A1 != null);
    Debug.Assert(Pr2 != null);
    Debug.Assert(A2 != null);
    Debug.Assert(Pc != null);

    #region For visualizing the vectors
    #endregion
}
```

All the public variables for `MyScript` have been discussed when analyzing the `Controller`'s `MyScript` component, and as in all previous examples, the `Debug.Assert()` calls in the `Start()` function ensure proper setup regarding referencing the appropriate game objects via the Inspector Window.

This example utilize the exact same three quaternion utility functions as the previous example to define `QFromAngleAxis()`, multiply `QMultiplication()`, and rotate `QRotation()` quaternions. Please refer to the previous section for the details of these functions. The details of `Update()` are as follows:

```
void Update() {
    Vector4 q  = QFromAngleAxis(Theta,
                                A.transform.localPosition);
    Vector4 q1 = QFromAngleAxis(Theta1,
                                A1.transform.localPosition);
    Vector4 q2 = QFromAngleAxis(Theta2,
                                A2.transform.localPosition);
```

```
    Pr.transform.localPosition =  QRotation(q,
                                  Pi.transform.localPosition);
    Pr1.transform.localPosition = QRotation(q1,
                                  Pr.transform.localPosition);
    Pr2.transform.localPosition = QRotation(q2,
                                  Pr1.transform.localPosition);

    Vector4 qc = QMultiplication(q1, q);
    qc = QMultiplication(q2, qc);
    Pc.transform.localPosition = QRotation(qc,
                                 Pi.transform.localPosition);

    #region  For visualizing the vectors
    #endregion
}
```

The first three lines define the three quaternion rotations q, q1, and q2 based on the user-specified angles Theta, Theta1, and Theta2 and the positions of A, A1, and A2 as position vectors for axes of rotation. The next three lines compute the three individual rotations: Pi by q to compute Pr, Pr by q1 to compute Pr1, and Pr1 by q2 to compute Pr2.

The last three lines compute the concatenated qc

$$q_c = q_2\, q_1\, q$$

and rotate Pi by qc to compute Pc.

Note The observed concatenated result being identical to applying individual quaternions is valid for any number of quaternions in the concatenation.

Takeaway from This Example

Through this example you have examined and verified that applying a sequence of quaternion rotations to a position is the same as concatenating the rotations and applying the resulting quaternion.

Relevant mathematical concepts covered include

- Multiplying multiple quaternions into a single quaternion is referred to as concatenating the quaternions.
- The inverse of a concatenated quaternion is the concatenation of the inverse of individual quaternions in the reversed order, that is, for n number of quaternions if

  $$q_c = q_n \dots q_2\, q_1$$

  then

  $$q_c^{-1} = \left(q_n \dots q_2\, q_1\right)^{-1} = q_1^{-1}\, q_2^{-1} \dots\ q_n^{-1}$$

- The rotation order of a concatenated quaternion is from the rightmost toward the left. That is, given

  $$q_c = q_n \dots q_2\, q_1$$

  The rotation q_c is the equivalent of applying q_1 first, followed by q_2 and so on, where q_n would be the last to be applied.
- Rotating a position by a sequence of quaternion rotations is identical to concatenating the rotations and rotating the position with the resulting concatenated quaternion

EXERCISES

Inverse of Concatenated Rotation Quaternion

Show the inverse of q_2q_1, or $(q_2q_1)^{-1}$, is $q_1^{-1}q_2^{-1}$. Note that q_2q_1 is applying rotation q_1 followed by q_2. Intuitively, to undo these two rotations, you would first undo the second rotation, thus applying q_2^{-1} first, and then undo the first rotation. Thus, intuitively the inverse of q_2q_1 would be $q_1^{-1}q_2^{-1}$ (apply q_2^{-1} before q_1^{-1}). Algebraically, since you know the definition of quaternion multiplication, you can simply compute and expand

$$q_2q_1 = \left(x_c, y_c, z_c, w_c\right)$$

and

$$q_1^{-1} q_2^{-1} = (x_r, y_r, z_r, w_r)$$

And verify that $x_c = -x_r$, $y_c = -y_r$, $z_c = -z_r$, and $w_c = w_r$.

The Number of Quaternions Concatenated

Verify the validity of concatenating two and four rotations. For two rotations, $q_1 q$, you can modify `MyScript` to verify

$$P_c = q_1\, q\, P_i\, q^{-1}\, q_1^{-1}$$

is identical to P_r. For four rotations, you can include support for an additional axis and theta accordingly.

The Importance of Order of Concatenation

Verify the importance of order of concatenation by modifying `MyScript` to compute

$$q_c = q\, q_1\, q_2$$

and show that the resulting location of P_c is in general very different from that of P_{r2}.

Aligning Vector Directions

Given two normalized vectors, $\hat{V}_1$ and $\hat{V}_2$,

$$\hat{V}_1 = (x_1, y_1, z_1)$$

$$\hat{V}_2 = (x_2, y_2, z_2)$$

You have learned that the cosine of the angle, θ, between these two vectors is

$$\cos\theta = \hat{V}_1 \cdot \hat{V}_2$$

or

$$\theta = \cos^{-1}\left(\hat{V}_1 \cdot \hat{V}_2\right)$$

You have also learned that when the two vectors are not parallel, if θ is not equal to 0^0 or 180^0, a plane with a normal vector, $\vec{V}_n$, can always be defined where

$$\vec{V}_n = \vec{V}_1 \times \vec{V}_2$$

Remember that vectors are independent from locations, and when depicted at the origin, $\hat{V}_1$ and $\hat{V}_2$ can be interpreted as the position vectors of positions, $P_1 = (x_1, y_1, z_1)$ and $P_2 = (x_2, y_2, z_2)$.

This fact, combined with the knowledge of quaternion rotation representation, can make the following derivation. Given any two vector directions, $\hat{V}_1$ and $\hat{V}_2$, you can compute

$$\theta = \cos^{-1}\left(\hat{V}_1 \cdot \hat{V}_2\right) \text{ and}$$

$$\vec{V}_n = \vec{V}_1 \times \vec{V}_2$$

and define the rotation, $q\left(\theta, \hat{V}_n\right)$, with rotation angle of θ and axis of $\hat{V}_n$. This rotation will rotate position P_1 to P_2 and thus is a rotation that aligns vector $\hat{V}_1$ to point to the direction of $\hat{V}_2$.

The key observations are that the angle of rotation can be derived by the dot product and that the axis of rotation is the cross product between the vectors. Since $\hat{V}_1$ and $\hat{V}_2$ are two arbitrary vectors, you have just derived a rotation that aligns the directions of any two given vectors.

The Align Vector Directions Example

This example demonstrates the derivation of angle and axis of rotation to define a quaternion rotation for aligning any two position vectors. Figure 8-5 shows a screenshot of running the `EX_8_3_AlignVectorDirections` scene from the `Chapter-8-Quaternions` project.

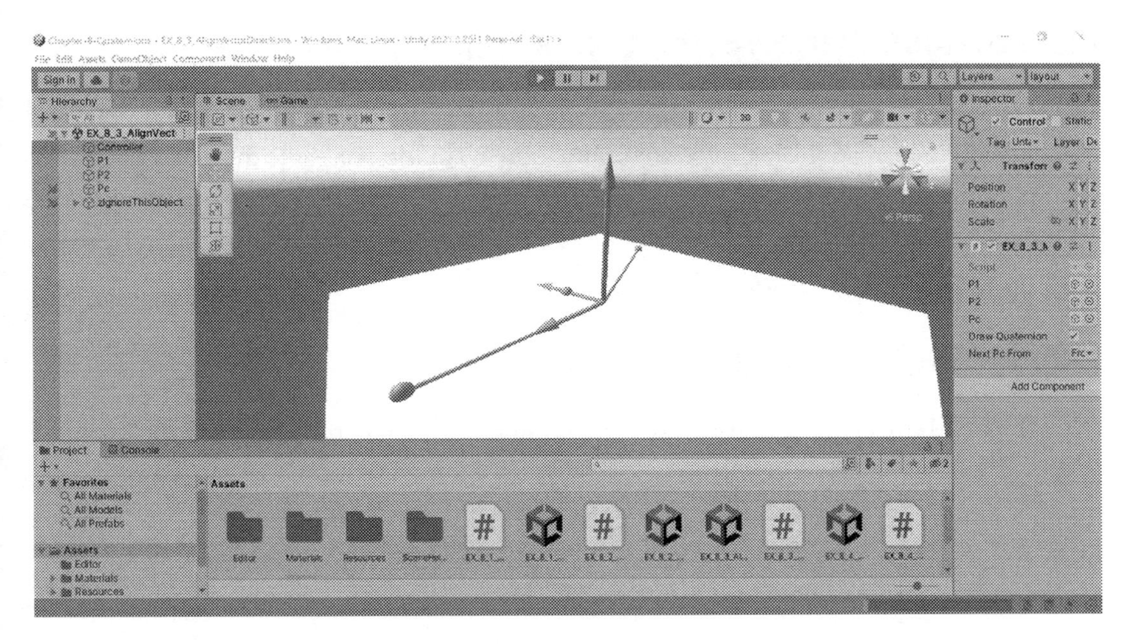

Figure 8-5. *Running the Align Vectors example*

The goals of this example are for you to

- Verify the vector direction aligning quaternion rotation
- Define and manipulate two arbitrary vectors to derive and examine the required rotation for aligning their directions
- Experience implementing the direction aligning quaternion rotation
- Appreciate that the alignment is specific to directions

Examine the Scene

Take a look at the `Example_8_3_AlignVectorDirections` scene and observe the green `P1`, red `P2`, and blue `Pc` spheres. The positions of these objects represent the position vectors where `P1` and `P2` are positions under user control while `Pc` will be in continuous motion showing the process of rotating from the directions of `P1` position vector to that of `P2`.

Analyze Controller MyScript Component

The MyScript component on the Controller shows the three variables with the same names as their corresponding reference game objects in the scene. As in previous examples, the DrawQuaternion toggles the showing/hiding of the axis and plane of rotation. The NextPcFrom option, as will be detailed, specifies one of three different ways to compute the next Pc position.

Interact with the Example

Click the Play button to run the example. You can see a red rotation axis with P1, P2, and Pc lying on the corresponding white rotation plane where Pc (blue) is in continuous motion rotating from the directions of P1 (green) to P2 (red) position vectors. You are observing the rotation that aligns the directions of $\vec{V}_1$ and $\vec{V}_2$ for position vectors of P1 and P2.

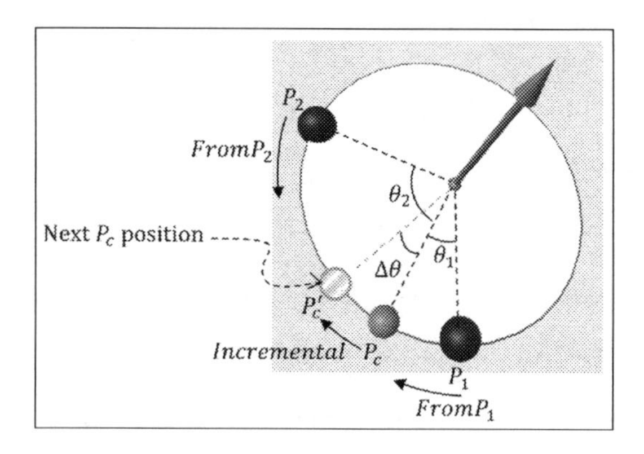

Figure 8-6. *The three rotations to compute Pc', the next position of Pc*

Note that in the following manipulations you will not affect the Pc rotation being from $\vec{V}_1$ toward $\vec{V}_2$. In other words, throughout the manipulations you will always observe Pc traveling from P1 toward P2. Your manipulation will change how Pc', the next Pc position, is computed. The interesting observation is that the same continuous rotation can be accomplished in at least three different ways.

Now, select the Controller object and iterate through each of the three options for NextPcFrom: FromPc, FromP1, and FromP2. Notice that while the color of Pc changes the rotation motion is completely unaffected. As illustrated in Figure 8-6, the angular movement of Pc is constantly from P1 toward P2 where the next Pc position, Pc', is always $\Delta\theta$ in the direction of `P2`. However, the actual `Pc'` position can be derived in three different ways according to `NextPcFrom` option:

- `FromPc`: Computes `Pc'` by rotating $\Delta\theta$ from current `Pc` and sets the color to blue
- `FromP1`: Computes `Pc'` by rotating $\theta_1 + \Delta\theta$ from `P1` and sets the color to green to match the color of `P1`
- `FromP2`: Computes `Pc'` by rotating $\theta_2 - \Delta\theta$ from `P2` and sets the color to red to match the color of `P2`

Through these options you have verified that there are multiple ways to implement a rotation and that the quaternion rotation can indeed be inversed, or reversed: the next `Pc` position, `Pc'`, can be calculated based on rotations from either `P1` or `P2`.

In the next manipulation, you will verify that the quaternion rotation aligns direction. Now, select and manipulate `P1` position to observe the red rotation axis updating to maintain the perpendicular plane of rotation that contains all three spheres; `P1`, `P2`, and `Pc`. Note the continuous motion of `Pc` rotating from the directions of $\vec{V_1}$ to $\vec{V_2}$ is independent from the length or magnitude of the $\vec{V_1}$ vector. You can further verify this by selecting and setting the position of `P1` to be located along the X-axis, for example, $(4, 0, 0)$. Now, increase and decrease the x-component value and note that the change does not affect the axis of rotation or the `Pc` motion of continuously rotating from `P1` to `P2`. In this case, changing the x-component value does not affect the direction of $\vec{V_1}$ and thus has no effect on the quaternion rotation. Feel free to repeat the manipulation with `P2`. In these interactions you have verified that the derived rotation is indeed aligning directions or unit vectors.

Note When manipulating the x-component value of the `P1` position, if you change the sign of from positive to negative, you are effectively reversing the direction of the $\vec{V_1}$ vector, and thus, you will observe a change in the rotation motion.

Lastly, you can observe the subtle and important difference of computing the next result from the current value in the `FromPc` computation vs. computing the next result from the actual initial or final value in the `FromP1` and `FromP2` options. With `NextPcFrom` set to `FromPc`, select and manipulate `P1` position away from the current plane of rotation, for example, by drastically increasing the y-component value of `P1` from the previous manipulation. Notice the blue vector to Pc, while rotating toward P2, does not reside on the plane of rotation anymore. This is not surprising, since in FromPc mode, the next Pc position is derived from the current Pc position, which in this case does not lie on the updated plane of rotation. Note that in FromP1 or FromP2 modes, since the next Pc position is derived from the actual initial or final values, the next Pc position will always be on the plane of rotation. While the behaviors are different, there is no correct, wrong, or better solution.

Different approaches to computing a solution have different characteristics. As a developer, your job is to understand these options and choose the best desired behavior.

Details of MyScript

Open MyScript and examine the source code in the IDE. The instance variables and the Start() function are as follows:

```
public enum PcPositionMode {
    FromPc,
    FromP1,
    FromP2
};
public GameObject P1 = null;    // The first position
public GameObject P2 = null;    // The second position
public GameObject Pc = null;

public bool DrawQuaternion = true;
public PcPositionMode NextPcFrom = PcPositionMode.FromPc;

private const float kDeltaTheta = 30f; // rotation speed
private const float kSmallAngle = 1f;  //

#region For visualizing the vectors
#endregion

void Start() {
    Debug.Assert(P1 != null);   // Verify proper setting
    Debug.Assert(P2 != null);
    Debug.Assert(Pc != null);
    Pc.transform.localPosition = P1.transform.localPosition;
```

```
    #region For visualizing the vectors
    #endregion
}
```

All the public variables for MyScript have been discussed when analyzing the Controller's MyScript component. The two private constants define the rate to rotate Pc and when Pc is sufficiently close to P2 for re-initializing the rotation. As in all previous examples, the Debug.Assert() calls in the Start() function ensure proper setup regarding referencing the appropriate game objects via the Inspector Window. The very last line initializes the position of Pc such that the rotation will begin from the position of P1.

As in the case of the previous examples in this chapter, this example utilizes the exact same three quaternion utility functions as the previous examples to define QFromAngleAxis(), multiply QMultiplication(), and rotate QRotation() quaternions. Please refer to the previous section for the details of these functions.

The details of Update() function are as follows:

```
void Update() {
    Vector3 V1n = (P1.transform.localPosition).normalized;
    Vector3 V2n = (P2.transform.localPosition).normalized;
    Vector3 Vcn = (Pc.transform.localPosition).normalized;

    float cosTheta = Vector3.Dot(V1n, V2n);
    if (Mathf.Abs(cosTheta) >= (1.0f-float.Epsilon)) {
        Debug.Log("V1 and V2 are almost parallel:
                                    cannot rotate to align");
        return; // V1 V2: almost parallel
    }

    float theta1 = Mathf.Acos(Vector3.Dot(Vcn, V1n)) *
                                                Mathf.Rad2Deg;
    float theta2 = Mathf.Acos(Vector3.Dot(Vcn, V2n)) *
                                                Mathf.Rad2Deg;
    float alpha = 0f;
    Vector3 axis = Vector3.zero;
    Vector3 Pf = Vector3.zero;
```

```
    if (theta2 > kSmallAngle) {
        switch (NextPcFrom) {
            case PcPositionMode.FromPc:
                  alpha = kDeltaTheta * Time.deltaTime;
                  axis = Vector3.Cross(Vcn, V2n);
                  Pf = Vcn;
            break;
            case PcPositionMode.FromP1:
                  alpha = theta1 + (kDeltaTheta * Time.deltaTime);
                  axis = Vector3.Cross(V1n, V2n);
                  Pf = V1n;
            break;
            case PcPositionMode.FromP2:
                  alpha = theta2 - (kDeltaTheta * Time.deltaTime);
                  axis = Vector3.Cross(V2n, V1n);
                  Pf = V2n;
             break;
        }
        Vector4 q = QFromAngleAxis(alpha, axis);
        Pc.transform.localPosition = QRotation(q, Pf);
    } else {
        Pc.transform.localPosition = P1.transform.localPosition;
    }

    #region  For visualizing the vectors
    #endregion
}
```

The first three lines of the `Update()` function compute the normalized position vectors to positions P1 ($\hat{V}_1$), P2 ($\hat{V}_2$), and Pc ($\hat{V}_c$). The dot product and if condition that follow check for the condition when P1 and P2 are collinear and a rotation cannot be defined. The following two lines, as illustrated in Figure 8-6, compute the angles between $\hat{V}_1$ and $\hat{V}_c$, theta1 (θ_1), and $\hat{V}_2$ and $\hat{V}_c$, theta2 (θ_2).

The if statement that follows ensures that θ_2 is sufficiently large, where $\hat{V}_2$ and $\hat{V}_c$ are not already aligned. Otherwise, the else condition re-initializes the rotation to begin from the direction of position vector to P1.

When θ_2 is sufficiently large or when the directions $\hat{V}_2$ and $\hat{V}_c$ are not already aligned, the three cases in the switch statement implement three rotations based on the value of `NextPcFrom`. The next `Pc` position, or `Pc'` in Figure 8-6, is computed by rotating a variable position, P_f, with the $q(\alpha, \hat{A})$ rotation, where depending on `NextPcFrom`

- `FromPc:` $\alpha = \Delta\theta$, $\vec{A} = \hat{V}_c \times \hat{V}_2$, and $P_f = \hat{V}_c$
- `FromP1:` $\alpha = \theta_1 + \Delta\theta$, $\vec{A} = \hat{V}_1 \times \hat{V}_2$, and $P_f = \hat{V}_1$
- `FromP2:` $\alpha = \theta_2 - \Delta\theta$, $\vec{A} = \hat{V}_2 \times \hat{V}_1$, and $P_f = \hat{V}_2$

Note that since $\Delta\theta$ is a constant positive number, although the next position of `Pc`, `Pc'` in Figure 8-6, is derived in different ways, the resulting rotation motion is always from `P1` toward `P2` position. The modulation by `deltaTime`, the wall-clock time, is to ensure the rotation speed is based on real-world time instead of the frame rate of your machine.

Takeaway from This Example

This example led you through defining two position vectors, deriving three different rotations in opposite directions to align these vectors, and examining the results of applying those rotations. It is important to remember that in this example all positions represent position vectors and that you have observed the rotation and aligning of directions.

Relevant mathematical concepts covered include

- The rotation that aligns two directions can be derived based on the angle between the directions and the axis that is defined by their cross product.
- The derived alignment rotation is specific to aligning directions.
- There are variations to the implementation of the alignment rotation where the rotation can be carried out from either of the directions.

EXERCISES

Concatenation of Quaternions

When `NextPcFrom` is `FromP1`, compute, concatenate, and apply the following two quaternions to `P1`: first, q_1 to rotate `P1` to current `Pc`, and second, q_2 to rotate `Pc` toward `P2` by $\Delta\theta$. Verify that the angular motion of `Pc` remains unchanged.

The rotation q_1 rotates from `P1` to `Pc`, and thus the angle of rotation is θ_1 and axis of rotation is $\hat{V}_1 \times \hat{V}_c$. The rotation q_2 continues the rotation toward `P2` and thus the angle of rotation is $\Delta\theta$ based on the same axis of rotation.

The concatenated result will be applied to rotate `P1` and thus the first rotation to be applied must be q_1 and followed by q_2. For this reason, the concatenated rotation is $q_c = q_2 q_1$. You can now verify that applying q_c to `P1` results in identical `Pc` motion.

Aligning P2 to P1

Modify the `Update()` function to compute the rotation that aligns the directions from `P2` to `P1`. In order words, flip the direction of the angular movement such that `Pc` always rotates from the `P2` and ends in the `P1` direction.

Interpolation and Chasing Behavior

Recall that you were able to launch an agent to travel toward a moving target in the Velocity and Aiming example, `EX_4_3_VelocityAndAiming` scene, from the `Chapter-4-Vectors` project. While interesting, you may have found the instantaneous and rigid updates of the agent's traveling direction to be unrealistic. In practice, when a target moves, it takes time for you to react and the adjustment you make should be continuous, changing gradually from your current direction to the target's new direction. This gradual change is more profound in the case of mechanical systems. For example, consider updating the aiming direction of a projectile launching turret, you would expect the device to rotate steadily from its current aim direction to the new direction.

This section first introduces the concept of interpolation as a solution to support gradual value changes over time. The interpolation of angles of rotation is then discussed to integrate interpolation into direction aligning quaternions to simulate the chasing or home-in behavior.

Interpolation: Gradual Changes

In the physical world, it takes time to react and respond. In the case of aiming at or traveling toward a target in motion, the change of direction should be gradual over time. In other words, the change of direction should be interpolated.

Figure 8-7 uses the change of an arbitrary parameter as an example to explain interpolation, where at time t_1 a parameter with an old value is to be assigned a new one. In this case, instead of updating the value abruptly, interpolation will change the value gradually over time. It will compute the intermediate results with decreasing values and complete the change to the new value at a later time t_2.

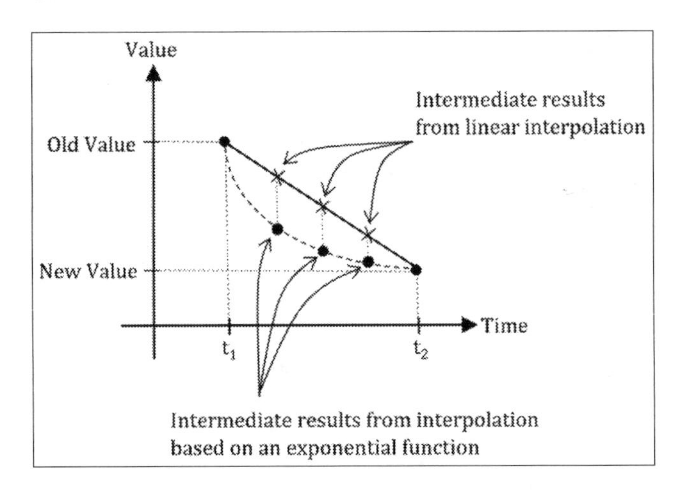

***Figure 8-7.** Interpolating values based on linear and exponential functions*

Figure 8-7 shows that there is more than one way to interpolate values over time. For example, linear interpolation computes intermediate results according to the slope of the line connecting the old and new values. In contrast, an exponential function may compute intermediate results based on percentages from previous values. With linear interpolation, the change of aiming direction would occur with a constant rotation. In comparison, interpolation based on a given exponential function would update the aim direction rapidly at first, then slow down quickly over time giving a sensation of reacting and re-aiming at the new target position.

Human motions and movements typically follow exponential interpolation functions. For example, try turning your head from facing the front to facing the right or moving your hand to pick up an object on your desk. Notice that in both cases, you began with a relatively quick motion and slowed down significantly when the destination is in close proximity. That is, you probably started by turning your head quickly and slowed down rapidly as your view approaches your right side, and it is likely your hand started moving quickly toward the object and slowed down significantly when the hand is almost reaching the object. In both of these examples, your displacements followed the exponential interpolation function as depicted in Figure 8-7—quick changes followed by a rapid slowdown as the destination approaches. This is the function you will integrate later in this section into quaternion rotations to align vector directions because it mimics organic movements.

Note Linear interpolation is often referred to as *LERP* or *lerp*. The result of lerp is the linear combination of an initial and a final value. In almost all cases, the exponential interpolation depicted in Figure 8-7 is approximated by repeatedly applying the lerp function where in each invocation, the initial value is the result of the previous lerp invocation—in effect, approximating the exponential function with a piecewise linear function. For this reason, lerp is also used to refer to the depicted exponential interpolation.

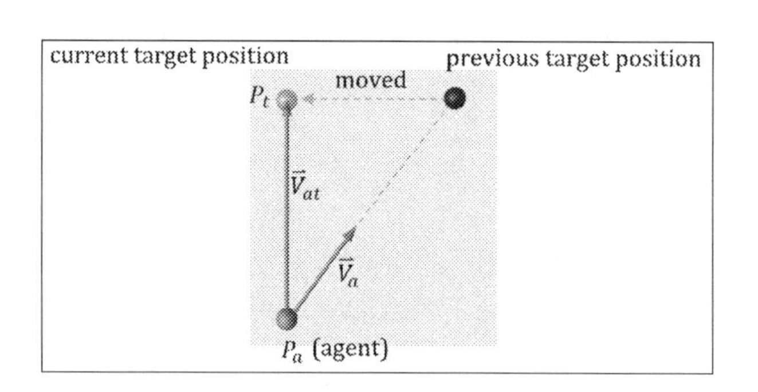

***Figure 8-8.** Current and new directions of a chasing behavior*

The Chasing or Home-In Behavior

Figure 8-8 illustrates an agent at location P_a moving toward a target at P_t, where P_t is in motion. The chasing of P_a toward the in-motion P_t can be simulated by interpolating the angle of the direction aligning quaternion rotations. In Figure 8-8, P_a and $\hat{V}_a$ are the existing agent position and traveling direction. As the target position, P_t, changes over time, the traveling direction of the agent can be gradually adjusted as follows.

The new traveling direction of the agent should be from P_a toward the current P_t, $\hat{V}_{at}$,

$$\hat{V}_{at} = (P_t - P_a).Normalized$$

Since the existing traveling direction of the agent is $\hat{V}_a$, a rotation, $q(\theta, \hat{V}_n)$ is required to align $\hat{V}_a$ to $\hat{V}_{at}$, where

$$\theta = \cos^{-1}(\hat{V}_a \cdot \hat{V}_{at}) \text{ and}$$

$$\vec{V}_n = \vec{V}_a \times \vec{V}_{at}$$

In order to support gradual rotation of $\hat{V}_a$ toward $\hat{V}_{at}$, the values of θ should be interpolated over time. Following the exponential function depicted in Figure 8-7, the direction realignment can be accomplished via a series of rotations, each with a fraction of the actual angle required

$$\theta' = Rate \times \cos^{-1}(\hat{V}_a \cdot \hat{V}_{at})$$

where

$$0.0 < Rate < 1.0$$

When traveling with a constant speed and a direction that is constantly adjusted by the rotation $q(\theta', \hat{V}_n)$, the agent would result in gradually approaching homing into or chasing after the target position.

Note Linearly interpolating the angle of a quaternion rotates the head of a vector following the circumference of a sphere and is referred to as spherical linear interpolation, or SLERP.

The Chasing Behavior Example

This example demonstrates how chasing behavior can be improved by using gradual instead of instantaneous direction changes. This example allows you to interactively manipulate a target and an observer position, examine gradual direction changes, and launch an agent from the observer position to home in to or chase after the target position. Figure 8-9 shows a screenshot of running the `EX_8_4_ChasingBehavior` scene from the `Chapter-8-Quaternions` project.

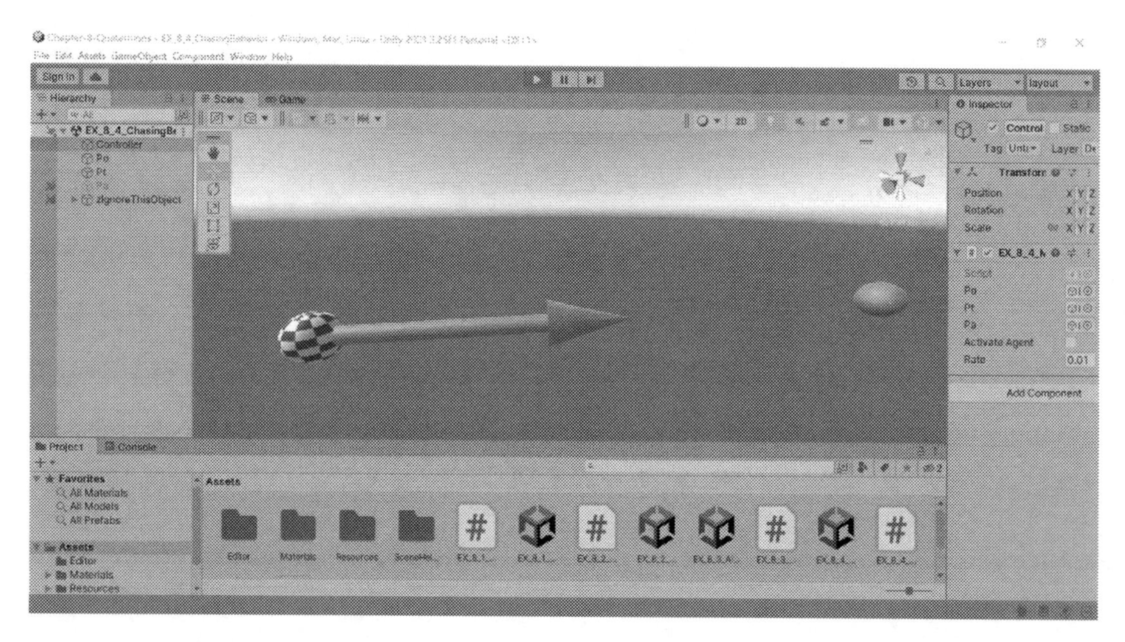

Figure 8-9. Running the Chasing Behavior example

The goals of this example are for you to

- Examine the implementation of interpolating directions
- Interact and gain experience with the results of linearly interpolating the angle for rotation, or SLERP
- Observe the results of direction interpolation
- Verify the home-in or chasing behavior

Examine the Scene

Take a look at the `Example_8_4_ChasingBehavior` scene and examine the three spheres: checkered observer, `Po`; red target, `Pt`; and green agent, `Pa`. In this example, the user can interactively manipulate the positions of `Po` and `Pt` and activate the agent to chase after the target position.

Analyze Controller MyScript Component

The `MyScript` component on the `Controller` shows the three variables with the same names as their corresponding reference game objects in the scene. The `ActivateAgent` toggle launches the green agent to chase after the target position, and the `Rate` variable controls the rate of interpolation where values of zero would mean ignoring the target and a value of around `60` would change agent traveling direction instantaneously.

Note To maintain consistency in performance, as you will observe when analyzing the source code, the `Rate` variable is modulated by the wall-clock elapsed time. The value 60 corresponds to an approximate frame refresh rate of your machine. Your actual frame refresh rate may be higher or lower than 60, but a value of 60 will approximately give you an instantaneous update.

Interact with the Example

Click the Play button to run the example. You can see a green vector attached to the checkered observer, `Po`. The green vector represents the direction from the checkered observer to the red target, `Pt`. On start, the green vector begins by pointing toward the positive x-direction and rotates gradually to align with the direction from the checkered observer to the red target sphere.

Select and manipulate the positions of the checkered observer or the red target to verify that the green vector always follows and gradually matches the actual direction from the observer to the target. You can compare and contrast this behavior to that of `EX_4_3_VelocityAndAiming`, where without interpolation, the aiming at the target is instantaneous and rigid and lacks the realism of organic reaction time.

Select the Controller and set the Rate to 0. You can verify that the green vector will not update as the positions of the observer and target change. Recall that a rate of zero means ignoring the final value and to not change the current value. Set the Rate to a larger value, for example, 10, to observe that the interpolation occurs too quickly for you to observe any gradual changes. In this implementation, the values of the Rate variable convey a sense of stiffness, or how quickly and rigidly the green vector follows the actual direction.

Now, set the Rate value to 0.8 and enable the ActivateAgent toggle. The green vector on the green agent is the direction of its velocity. Observe that the green agent initially travels toward the x-direction and then adjusts gradually to the direction toward the target. Upon reaching the target position, since there is no support for collision, the agent continuously moves beyond the target and attempts to adjust its traveling direction resulting in orbiting the target. You can manipulate the red target position to observe the green agent always chases after and attempts to home in on the target. You can toggle ActivateAgent to relaunch the agent.

Details of MyScript

Open MyScript and examine the source code in the IDE. The instance variables and the Start() function are as follows:

```
public GameObject Po = null;    // Observer position
public GameObject Pt = null;    // Target position

public GameObject Pa = null;    // Agent position
public bool ActivateAgent = false;
public  float Rate = 0.8f;

private Vector3 Vot = Vector3.right;  // (1,0, 0)
private Vector3 Vat =  Vector3.right; // (1, 0, 0)

private const float kAgentSpeed = 0.01f;
private const float kSmallAngle = 1f;

#region For visualizing the vectors
#endregion
```

```
void Start() {
    Debug.Assert(Po != null);    // Verify proper setting
    Debug.Assert(Pt != null);
    Debug.Assert(Pa != null);

    #region For visualizing the vectors
    #endregion
}
```

All the public variables for `MyScript` have been discussed when analyzing the `Controller's` `MyScript` component. The private variables, `Vot` and `Vat`, are the vectors representing the directions from the observer to the target, $\vec{V}_{ot}$, and from the agent to the target, $\vec{V}_{at}$. Note that these two vectors are initialized to point in the positive x-direction. The two constants define the speed of the traveling agent and the condition when directions are aligned. As in all previous examples, the `Debug.Assert()` calls in the `Start()` function ensure proper setup regarding referencing the appropriate game objects via the Inspector Window.

In this example, in addition to the three previously defined quaternion utility functions, `QFromAngleAxis()`, `QMultiplication()`, and `QRotation()`, an additional function `AlignVectors()` is introduced to compute and interpolate vectors with details as follows:

```
Vector3 AlignVectors(Vector3 from, Vector3 to, float rate) {
    from.Normalize();
    to.Normalize();
    float theta = Mathf.Acos(Vector3.Dot(from, to))
                                                * Mathf.Rad2Deg;
    Vector4 q = new Vector4(0, 0, 0, 1); // Quaternion identity
    if (theta > kSmallAngle) {
        Vector3 axis = Vector3.Cross(from, to);
        q = QFromAngleAxis(rate * Time.smoothDeltaTime * theta,
                           axis);
    }
    return QRotation(q, from);
}
```

The first three lines of the function normalize the input `from` and `to` vectors and perform a dot product to compute the angle, θ, between the two input vectors. When θ is sufficiently large, the vector aligning quaternion is defined to rotate the `from` vector by an angle that is `rate`×θ toward the `to` vector. The `Time.smoothDeltaTime` modulation is to ensure that the rate of rotation is independent from the performance of your machine. In this way, the value of `rate` scales the angle for rotation and is spherically linearly interpolated; thus, the returned vector is a SLERP between the `from` and `to` vectors. The details of `Update()` are as follows:

```
void Update() {
    Vector3 o2t = Pt.transform.localPosition -
                  Po.transform.localPosition;
    Vot = AlignVectors(Vot, o2t, Rate);

    if (ActivateAgent) {
        Vector3 a2t = Pt.transform.localPosition -
                      Pa.transform.localPosition;
        Vat = AlignVectors(Vat, a2t, Rate);
        Pa.transform.localPosition += kAgentSpeed * Vat;
    } else {
        Pa.transform.localPosition = Po.transform.localPosition
        Vat = Vector3.right;
    }

    #region  For visualizing the vectors
    #endregion
}
```

The first two lines compute the vector, `o2t`, from the observer to target and call `AlignVectors()` to compute the SLERP result `Vot`. The `Vot` vector is the one shown on the checkered observer. When `ActivateAgent` is enabled, a similar computation is performed for the agent position to derive `a2t` and `Vat`, where the `Vat` direction is used as the velocity direction for updating the position of the agent, `Pa`. Since the agent's velocity direction, `Vat`, is constantly updated and gradually points toward the target position, the agent's motion showcases that it is chasing the target position.

Takeaway from This Example

Through this example you have observed the importance of gradual changing based on interpolation and gained experienced with the chasing behavior, a common application of the vector aligning quaternion rotation.

Relevant mathematical concepts covered include

- Interpolation computes a result that is in between the inputted initial and final values.
- Linear interpolation (LERP) computes the results based on a constant change factor.
- Spherical linear interpolation (SLERP) linearly interpolates the angle of a rotation.

EXERCISES

Chasing with Constant Rotation

Instead of SLERP with a constant rate, you can experience rotating directions based on a constant angular speed. In the `AlignVectors()` function, instead of computing the rotation

```
q = QFromAngleAxis(rate * Time.smoothDeltaTime * theta, axis);
```

try defining the rotation with a constant angular speed, for example,

```
q = QFromAngleAxis(1.0f, axis);
```

Now run the example to observe that a constant angular rotation speed seems mechanical and lacks the organic realism of SLERP.

Aligning Axis Frames

With the knowledge of quaternion rotation, concatenation, and alignment of vector directions, you can now derive the solution to align axis frames. The problem is straightforward: after a user manipulates an object, for example, a spaceship, how can you align objects with the rotated axis frame, that is, the navigated spaceship. This is an important issue to resolve because you may want to supply the spaceship with emergency equipment where it is crucial that the container boxes land on the spaceship appropriately.

Recall that an axis frame is defined by three perpendicular axes or vectors. It is always the case that the direction of the third vector is defined by the cross product of the first two. This means, the orientation of an axis frame can be completely specified by the directions of two of the vectors. For this reason, when aligning axis frames, you only need to ensure two of the vectors are aligned. In other words, when given two axis frames, if the directions of two of the vectors are aligned, then it is guaranteed that the directions of the third vector must also be aligned.

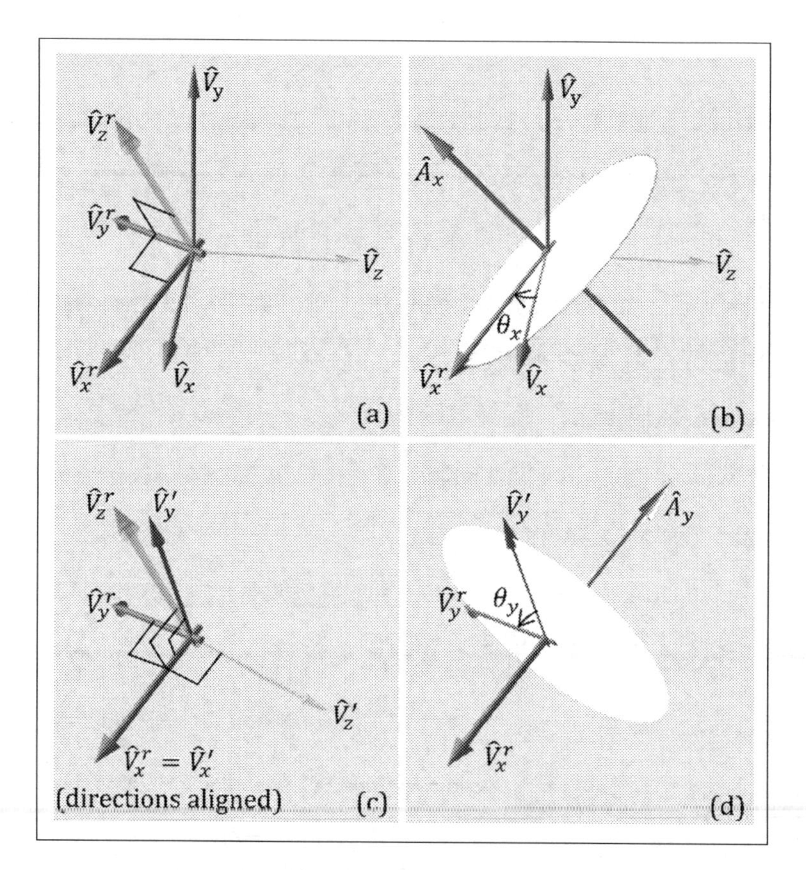

Figure 8-10. *Rotations to align the default to a rotated axis frame. (a) The two axis frames, (b) the first rotation to align* $\hat{V}_x$ *to* $\hat{V}_x^r$*, (c) the resulting axis frames after the first rotation, (d) the second rotation along* $\hat{V}_x^r$ *to align* $\hat{V}_y'$ *to* $\hat{V}_y^r$

For clarity, instead of depicting alignment from a rotated axis frame, Figure 8-10 illustrates the rotations required based on the default axis frame to an arbitrarily rotated axis frame. It is important to recognize that in the following derivation there are no assumptions made on the actual directions of any of the vectors. For this reason, the derived results are applicable to align any two arbitrarily rotated axis frames.

Figure 8-10 (a) shows two sets of axis frame drawn at the origin: the first thinner set on the right defined by $\hat{V}_x$, $\hat{V}_y$, and $\hat{V}_z$ and the rotated thicker set to the left defined by $\hat{V}_x^r$, $\hat{V}_y^r$, and $\hat{V}_z^r$. The goal is to derive an operator to align any two of the three vectors, for example, align $\hat{V}_x$ to $\hat{V}_x^r$ and $\hat{V}_y$ to $\hat{V}_y^r$.

The actual choice of directions for alignment does not affect the result. In Unity the Y- and Z-axes are used as the upward and forward directions and thus are the choice of directions for alignment. In the following derivation, x- and y-directions are used. In the exercise at the end of this section, you will verify that the alignment results are independent from the directions of choice.

Figure 8-10 (b) illustrates the rotation, $q\left(\theta_x, \hat{A}_x\right)$, required to align $\hat{V}_x$ to $\hat{V}_x^r$ direction. Vectors $\hat{V}_y^r$ and $\hat{V}_z^r$ are not shown to avoid cluttering the figure and because they do not contribute in the derived rotation. For the rotation, $q\left(\theta_x, \hat{A}_x\right)$, you know

$$\theta_x = \cos^{-1}\left(\hat{V}_x \cdot \hat{V}_x^r\right) \text{ and}$$

$$\vec{A}_x = \vec{V}_x \times \vec{V}_x^r$$

Figure 8-10 (c) shows the results of applying $q\left(\theta_x, \hat{A}_x\right)$ to the axis frame, $\hat{V}_x$, $\hat{V}_y$, and $\hat{V}_z$. The rotation aligns the thinner $\hat{V}_x$ with the thicker $\hat{V}_x^r$; thus the rotated $\hat{V}_x$, or $\hat{V}_x'$, is occluded by $\hat{V}_x^r$ and not visible in the figure. It is crucial to recognize that the rotation is applied to all three vectors where the resulting axis frame is now $\hat{V}_x'$, $\hat{V}_y'$, and $\hat{V}_z'$. Take note that at this point, $\hat{V}_x' = \hat{V}_x^r$, and that this vector is the x-direction of both axis frames. This is to say $\hat{V}_x'$ is perpendicular to all four vectors, $\hat{V}_y'$, $\hat{V}_z'$, $\hat{V}_y^r$, and $\hat{V}_z^r$. For this reason, in the following rotation to align $\hat{V}_y'$ with $\hat{V}_y^r$, the axis of rotation is along the positive or negative $\hat{V}_x'$ direction.

Lastly, Figure 8-10 (d) illustrates the rotation, $q\left(\theta_y, \hat{A}_y\right)$, required to align $\hat{V}_y'$ to $\hat{V}_y^r$ direction. There are two key points to this rotation. First, as discussed, $\hat{A}_y$, the axis of rotation will be along the positive or negative $\hat{V}_x^r$ direction. Second, the rotation is defined to be applied to the results of the $q\left(\theta_x, \hat{A}_x\right)$ rotation, or $\hat{V}_y'$ and $\hat{V}_z'$, and not the original $\hat{V}_x$, $\hat{V}_y$, and $\hat{V}_z$. Once again, to avoid cluttering, $\hat{V}_z'$ and $\hat{V}_z^r$ are not shown in Figure 8-10 (d). In this case, you know

$$\theta_y = \cos^{-1}\left(\hat{V}_y^{\prime} \cdot \hat{V}_y^{r}\right) \text{ and}$$

$$\vec{A}_y = \hat{V}_y^{\prime} \times \hat{V}_y^{r}$$

The final rotation operator that aligns the two given axis frames, q_c, is,

$$q_c = q\left(\theta_y, \hat{A}_y\right)\, q\left(\theta_x, \hat{A}_x\right)$$

Once again, the importance of concatenation ordering cannot be overstressed. In this derivation, it is important that the x-alignment rotation, $q\left(\theta_x, \hat{A}_x\right)$, is applied before the y-alignment rotation, $q\left(\theta_y, \hat{A}_y\right)$, and thus $q\left(\theta_x, \hat{A}_x\right)$ must be on the right-hand side of the concatenation.

The Unity Quaternion Class

In the next example, the results from the derived axis frame alignment formulation will be compared to the solutions defined by the Unity `Quaternion` class. This is an excellent opportunity to relate and contrast relevant concepts learned. Unity API documents the Quaternion class (`https://docs.unity3d.com/ScriptReference/Quaternion.html`) as follows:

> *Quaternions are used to represent rotations.*

If you browse through their utility methods, you will notice the following similarities:

- `AngleAxis`: This is the `QFromAngleAxis()` utility function.
- `FromToRotation`: This is similar to the `QAlignVectors()` utility function.
- `Slerp`: This is covered in the example scene `Example_8_4_ChasingBehavior`.

Additionally, you have also learned about the `Inverse()` function and the *-operator (concatenation operator). Pay attention to the `LookRotation()` function:

> *Creates a rotation with the specified forward and upward directions*

Note that this is precisely the subject of coverage in this section and you will work with this function in the next example.

Finally, notice the absence of an actual rotation function. That is, there is no correspondence of the `QRotation()` function defined in the Unity `Quaternion` class. Recall that a significant limitation of the quaternion representation for rotation is its inability to describe rotations when the axis of rotation does not pass through the origin. As pointed out when first introduced, this is not an issue because quaternions are typically integrated with matrices in representing coordinate transformation. Together, the tools can address the off-origin rotation limitation. In the case of Unity, the integration of quaternions with matrices occurs in the `Transform` class (`https://docs.unity3d.com/ScriptReference/Transform.html`), where rotations are represented by quaternions and the transformation functionality is encoded as matrices. It is the `Transform` class that defines the relevant position and vector rotation functions.

The details of the `Transform` class, the subject of coordinate transformation, are an advanced topic that is out of the scope of this book. However, you have been working with the `Transform` class in all of the examples where you have set the `transform.localPosition` to control the location of objects. In the example that follows, you will compute and set the `transform.localRotation` to control the orientation of objects to verify the axis frame alignment formulation.

Note The Unity `Transform` class explicitly maintains the axis frame of an object. The `x`-, `y`-, and `z`-directions of a transformed axis frame are accessible via the `transform.right`, `transform.up`, and `transform.forward` properties on a `Transform` object.

The Align Frames Example

This example demonstrates the results of applying the derived rotation to align with a user-manipulated axis frame. To assist in gaining insights into the alignment, this example also shows the results of applying only the first axis alignment rotation. Additionally, to assist in verifying the solution, the results from the Unity quaternion utility are also displayed. Figure 8-11 shows a screenshot of running the `EX_8_5_AlignFrames` scene from the `Chapter-8-Quaternions` project.

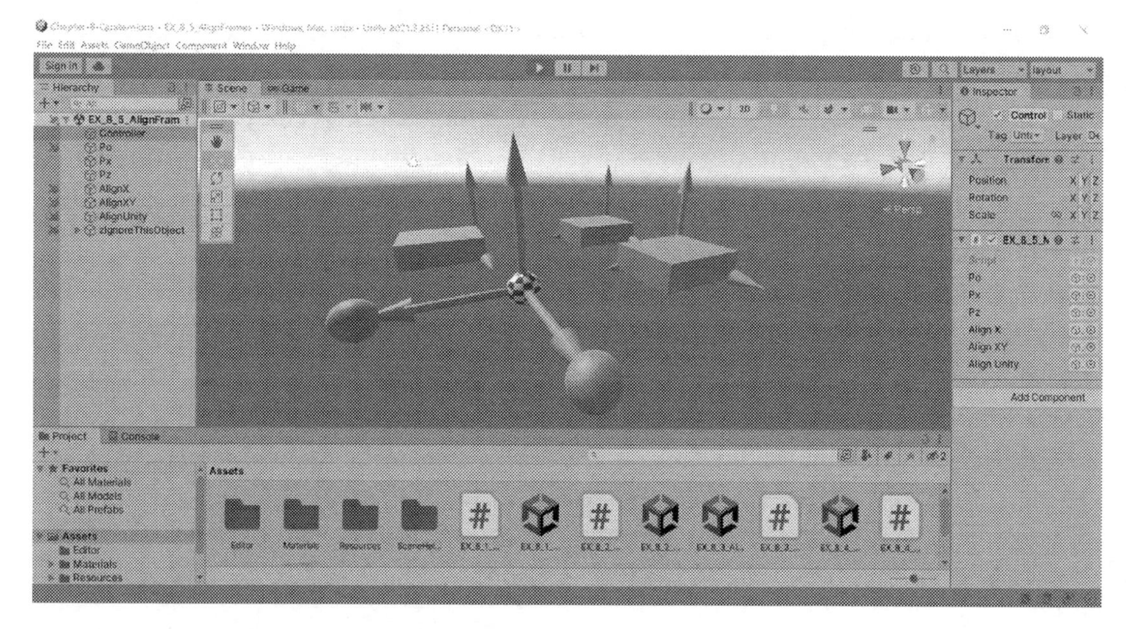

Figure 8-11. *Running the Align Frames example*

The goals of this example are for you to

- Interact with the smooth manipulation of positions that define an axis frame
- Verify the results of aligning the first of the directions in axis frames
- Observe that the concatenation of the two axis aligning rotations can indeed define an axis frame alignment rotation operator
- Examine the implementation of the axis frame alignment formulation
- Validate the alignment results by comparing with the results from the Unity quaternion utility

Examine the Scene

Take a look at the `Example_8_5_AlignFrames` scene and observe the three spheres and three flattened rectangular cubes. Similar to examples from the previous chapter, the spheres `Po`, `Px`, and `Pz` are the three non-collinear positions that you can manipulate to define an arbitrary axis frame. The orientations of the three flattened rectangular cubes represent the results of aligning with the user-defined axis frame: the red,

AlignX, with only the first X-axis alignment rotation applied; the green, AlignXY, with the concatenated xy-axis rotations applied; and the blue, AlignUnity, with alignment performed based on the quaternion utility from Unity.

Analyze Controller MyScript Component

The MyScript component on the Controller shows the six variables with the same names as their corresponding reference game objects in the scene.

Interact with the Example

Click the Play button to run the example. You can see four sets of three vectors representing axis frames wrapping around each of the four axis frames: the three flattened rectangular cubes and the spheres. In all cases, the red vector is the x-direction, green is the y-direction, and blue is the z-direction. In this context, alignment refers to the matching of the vector directions with the same colors. For example, the X-axis is aligned when the red vectors are pointing in the same direction. Two axis frames are aligned when all three colored vectors are pointing in the same directions.

Select and adjust the y-component of the blue sphere, Pz. This manipulation results in rotating the axis frame around the x-direction where the red vector, or the X-axis direction, does not changed. Observe that the green, AlignXY, and blue, AlignUnity, cubes always align exactly with the manipulated frame. This is in contrast to the red cube, AlignX, where it is only rotated by the x-direction alignment rotation, and in the absence of x-direction changes, the red cube stays stationary.

Select and manipulate either Px or Pz freely to observe that the green and blue cubes continue to always align exactly with the user-defined axis frame while the orientation of the red cube only guarantees that the red X-axis is aligned. Now compare the red and the green cubes and observe that the orientations of these two cubes are always different by one rotation about their red vector. In other words, the alignment can be achieved by rotating either the red or the green cube about the red vector. A straightforward way to establish this observation is by analyzing the green vectors on these two cubes when viewing the red vector straight down. You will see that the green vectors are a simple rotation apart.

In these manipulations, you have observed and interacted with the two-step axis frame alignment rotation. You have also verified that the derived alignment formulation matches the results from the Unity quaternion utility.

Lastly and very importantly, take note that in this example all three cubes are located at positions other than the origin where they can be moved to any position and yet you were able to flawlessly manipulate their rotations. In other words, you have worked with but did not encounter the quaternion limitation that the axis of rotation must pass through the origin. As pointed out earlier, the Unity Transform class strategically integrates quaternions with matrices and avoids that limitation completely.

Details of MyScript

Open MyScript and examine the source code in the IDE. The instance variables and the Start() function are as follows:

```
public GameObject Po = null; // Origin of the reference frame
public GameObject Px = null; // X-position defining the x-axis
public GameObject Pz = null; // Z-position defining the z-axis
public GameObject AlignX = null;  // X-axis aligned
public GameObject AlignXY = null; // X,Y-both aligned
public GameObject AlignUnity = null;  // Unity aligned
private const float kSmallAngle = 1f;

#region For visualizing the vectors
#endregion

void Start()  {
    Debug.Assert(Po != null);   // Verify proper setting
    Debug.Assert(Px != null);
    Debug.Assert(Pz != null);
    Debug.Assert(AlignX != null);
    Debug.Assert(AlignXY != null);
    Debug.Assert(AlignUnity != null);

    #region For visualizing the vectors
    #endregion
}
```

All the public variables for MyScript have been discussed when analyzing the Controller's MyScript component. The private kSmallAngle defines when two vectors are in the same direction and that the alignment rotation is not necessary.

This example defines the same quaternion utility functions: QFromAngleAxis(), QMultiplication(), and QRotation(). The previous AlignVectors() function is replaced by a similar QAlignVectors() function with details as follows:

```
Vector4 QAlignVectors(Vector3 from, Vector3 to) {
    from.Normalize();
    to.Normalize();
    float theta = Mathf.Acos(Vector3.Dot(from, to))
                                                * Mathf.Rad2Deg;
     Vector4 q = new Vector4(0, 0, 0, 1); // Quaternion identity
     if (theta > kSmallAngle) {
         Vector3 axis = Vector3.Cross(from, to);
         q = QFromAngleAxis(theta, axis);
     }
     return q;
}
```

This new function removed the SLERP functionality and returned a quaternion rotation instead of a rotated vector. The last additional utility function, V4ToQ(), is defined for type conversion to be compatible with the Unity Quaternion class. The details are as follow:

```
Quaternion V4ToQ(Vector4 q) {
    return new Quaternion(q.x, q.y, q.z, q.w);
}
```

With these utilities, the details of Update() are as follows:

```
void Update() {
    Vector3 vxr = (Px.transform.position -
                  Po.transform.position).normalized;
    Vector3 vzr = (Pz.transform.position -
                  Po.transform.position).normalized;
    Vector3 vyr = Vector3.Cross(vzr, vxr);

    Quaternion qUnity = Quaternion.LookRotation(vzr, vyr);
    AlignUnity.transform.localRotation = qUnity;
```

```
    Vector4 qx = QAlignVectors(Vector3.right, vxr);
    AlignX.transform.localRotation = V4ToQ(qx);

    Vector4 qy = QAlignVectors(AlignX.transform.up, vyr);
    Vector4 qc = QMultiplication(qy, qx);
    AlignXY.transform.localRotation = V4ToQ(qc);

    #region  For visualizing the vectors
    #endregion
}
```

The first three lines compute the user-defined axis frame, the $\hat{V}_x^r$, $\hat{V}_y^r$, and $\hat{V}_z^r$ in Figure 8-10. The next two lines call the Unity `Quaternion.LookRotation()` utility with $\hat{V}_z^r$ as the forward and $\hat{V}_y^r$ as the upward directions to compute and set the rotation to the `transform.localRotation` of the `AlignUnity` object. Recall that `AlignUnity` is a reference to the blue cube. The matching alignment of the blue cube axis frame verifies that `Quaternion.LookRotation()` indeed computes an axis frame alignment rotation.

In the line that follows, the variable `qx` represents $q\left(\theta_x,\hat{A}_x\right)$, rotating `Vector3.right`, or $(1,0,0)$ or $\hat{V}_x$ in Figure 8-10, to $\hat{V}_x^r$. This rotation is set to `AlignX`, or the red cube. Note that when the x-direction is not changed, $\hat{V}_x^r$ would remain $(1,0,0)$ and `qx` would be a quaternion identity. This is why in the previous interaction the red cube would stay stationary when the axis frame is rotated about the x-direction.

The variable `qy` represents $q\left(\theta_y,\hat{A}_y\right)$, rotating `AlignX.transform.up` to $\hat{V}_y^r$. In this case, `AlignX.transform.up` is the result of $\hat{V}_y$ rotated by $q\left(\theta_x,\hat{A}_x\right)$, or $\hat{V}_y^{'}$ in Figure 8-10(c). The last two lines concatenate `qx` with `qy` to compute the actual axis frame aligning operator `qc` and set the rotation to `AlignXY`, or the green cube. The fact that the blue and green cubes, or `AlignUnity` and `AlignXY`, align identically verifies that the computed `qc` is indeed the same as the results from the Unity `Quaternion.LookRotation()` function.

Note The Unity `GameObjects`, `AlignX`, `AlignXY`, and `AlignUnity`, are located at positions other than the origin and with axes of rotations that do not pass through the origin. The Unity `Transform` class, where the computed quaternion rotations are set via transform.localRotation, integrates matrix transformation functionality and seamlessly overcomes the quaternion rotation limitation.

Takeaway from This Example

Through this example you have examined and interacted with each of the two rotations involved in aligning axis frames. You have also verified that strategically concatenating two rotations can indeed result in an axis frame aligning operator.

Relevant mathematical concepts covered include

- To align two axis frames, you only need to ensure two of the three axes are aligned.
- You can choose to align any of the two axes to accomplish axis frame alignment.
- The second rotation of axis frame alignment aligns the results from the first rotation and not the original axis directions.
- The limitation of quaternion rotation that the axis of rotation must pass through the origin can be avoided with strategic integration with matrices.

Unity tools

- `Quaternion.LookRotation()`: Aligns the default to a given axis frame based on forward, z-directions, and up, y-directions
- `Transform.localRotation`: Encodes rotation with a quaternion
- `Transform.right/up/forward`: The major axes' directions of a rotated `GameObject`

EXERCISES

Replace Our Functions with Unity Quaternion

Replace `QAlignVectors()`, `QFromAngleAxis()`, and `QMultiplication()` with the corresponding Unity `Quaternion` class utility functions and verify that the exact same results can be observed.

Align Based on Two Other Axes

Replace the X- and Y-axes with z- and y-directions to verify that the choice of axes for alignment indeed does not affect the results. You can repeat this exercise with any other two axes, for example, X and Z, if desired.

Align a Rotated Axis Frame to the Default Axis Frame

Derive and display the rotations required to align the user-defined axis frame to the default axis frame.

In this case, the first rotation required is to align vxr to the default x-direction, $\hat{V}_x = (1,0,0)$. In other words, $q(\theta_x, \hat{A}_x)$ has

$$\theta_x = \cos^{-1}\left(\hat{V}_x^r \cdot \hat{V}_x\right) \text{ and}$$

$$\vec{A}_x = \vec{V}_x^r \times \vec{V}_x$$

Not surprisingly, the direction of the axis of rotation is reversed from that in this example. The rotation $q(\theta_x, \hat{A}_x)$ would be applied to the user-defined axis frame: $\hat{V}_x^r$, $\hat{V}_y^r$, or $\hat{V}_z^r$. The second rotation should align the rotated y-direction, $\hat{V}_y'$, the $q(\theta_x, \hat{A}_x)$ rotated $\hat{V}_y^r$, to align with the default Y-axis, $\vec{V}_y = (0,1,0)$, where $q(\theta_y, \hat{A}_y)$ has

$$\theta_y = \cos^{-1}\left(\hat{V}_y' \cdot \hat{V}_y\right) \text{ and}$$

$$\vec{A}_y = \vec{V}_y' \times \vec{V}_y$$

You can now edit `MyScript` to implement the preceding formulation. With this exercise, you have verified that not only can you rotate the default to a user-defined axis frame, you can indeed reverse the alignment from a user-defined axis frame to the default axis frame. Since you can align an axis frame, A, with the default and then align the default with another axis frame, B, you can indeed align any two given axis frames A and B.

An alternative and much more straightforward approach is to recognize that quaternion rotations are reversible. The inverse of the computed qc in the existing code will accomplish the specified axis frame alignment.

Integrate SLERP to Axis Frame Alignment

Integrate the SLERP functionality of `AlignVectors()` from the previous example to the `QAlignVectors()` function and experience with gradual and smooth axis frame alignment that more resembles the steering of a spaceship.

Navigation with Axis Frame

As discussed, navigating a spaceship is simply aligning the ship with an axis frame and moving along the front direction. If the `AlignXY` object represents a spaceship with `transform.forward` as the front direction, then you can navigate the `AlignXY` object by including the following line at the end of the `Update()` function:

```
AlignXY.transform.localPosition =
        0.5f * Time.deltaTime * AlignXY.transform.forward
```

Now, if you run the game, you will observe the green cube moving toward the positive z-direction. Try manipulating the positions of `Po` and `Pz` to verify that you can indeed steer the traveling of the `AlignXY` object.

Summary

This chapter introduces the four-tuple quaternion to represent a rotation. You have learned that three of the numbers describe the axis of rotation where the forth number encodes the angle to be rotated. The mathematical rules for working with quaternion, or quaternion algebra, are well established for supporting rotation operations. You have learned the inverse of a quaternion reverses a rotation and the concatenation of quaternions aggregates and captures the results of multiple rotations. The limitation of the compact four-number representation of a rotation is that there is no way to encode the location of the axis of rotation: quaternion representation and the involved algebra implicitly assume that the axis of rotation passes through the origin of the Cartesian coordinate.

You have examined quaternion rotation as a tool for aligning directions. Chapter 5 has taught you that the angle between two normalized vectors is the arccosine of the dot product. From Chapter 6, you know that the axis of rotation for aligning two vectors is simply the cross product of the vectors. Based on this knowledge, you have derived the formulation for aligning the directions of any two vectors. By analyzing how you would turn your head when changing viewing directions, you recognized that real-world organic and mechanical movements are gradual and continuous. You have learned to emulate such movements by continuously applying quaternion rotations based on repeatedly linear interpolated angle of rotation, or SLERP. Lastly, you learned that by strategically computing and concatenating two rotations, you can align any two given axis frames. Through working with the Unity `Transform` class, you have witnessed that the quaternion rotation limitation of requiring the axis of rotation to pass through the origin can be avoided completely. The steering and navigation of a spaceship will be further explored in the next chapter via the motion of a traveling agent.

It is important to recognize that this chapter has led you to investigate quaternions as being used as a tool for rotation. Thus, the focus of this chapter has been on the characteristics of quaternions in effectively rotating vectors. This is very different from learning quaternions as a field of mathematical study. You may have noticed some of the missing details, such as the derivation or justification of quaternion multiplication definition. Though important, such details are outside of the scope of using quaternions as a tool for rotations. The limited coverage of quaternion fundamentals means that while you are able to use quaternion as a tool to align vectors and axis frames, it may be challenging for you to use it as a general mathematical tool for solving other problems.

Lastly, you may have noticed a slight deviation of topic coverage in this chapter. While the other chapters in the book analyzed and studied the application of points and vectors, this chapter examined how to manipulate and change them. For example, instead of applying vectors in representing axis frames, this chapter examined how to manipulate a defined axis frame. This subtle shift serves as the introduction to the next topic area in mathematics for supporting video game development: matrices and transformation. A more involved topic for a more advanced book.

CHAPTER 9

Conclusion

With your background in basic algebra and trigonometry, this book took you on the journey from the review of the Cartesian Coordinate System to the application of vector algebra to solve frequently encountered problems in video game development. In Chapter 1, you reviewed and familiarized yourself with the Unity system as a learning tool. Then, in Chapter 2, you learned about bounding boxes, one of the most used tools in game engines, by revising and generalizing number intervals. Along the way, you also examined issues related to bounding volumes.

In Chapters 3 and 4, you studied the relationships between positions. You began studying these relationships in Chapter 3 through exploring bounding volumes by examining another important tool: bounding spheres. From here, you were led into Chapter 4 where you were introduced to the concept of vectors. That chapter provided you with a comprehensive and formal foundation for discussing relationships between positions in the form of directions and distances. It was also in that chapter that you gained experience in applying vector concepts to model and implement object velocity manipulation and how to calculate object motions under external factors such as wind or current flow conditions.

In Chapters 5 and 6, you learned to relate vectors to each other and to the space that defines them. The vector dot product introduced in Chapter 5 demonstrated that two vectors are related by the angle they subtend and their mutual projected sizes. You applied this knowledge to describe and analyze line segments and then connected these vector line segments back to the simple number intervals reviewed in Chapter 2. You then applied these concepts to solve the problem of high-speed objects' missing collisions. Then, in Chapter 6, you learned about the vector cross product and used it to analyze 2D planes. This analysis included exploring 2D planes from additional perspectives including the ability to define general axis frames and to create your own line intervals to define 2D regions on 2D planes.

K. Sung and G. Smith, *Basic Math for Game Development with Unity 3D*,
https://doi.org/10.1007/978-1-4842-9885-5_9

In Chapter 7, you analyzed axis frames and began to appreciate complex situations with independent movements of elements that are geometrically related or connected. You generalized axis frames and learned that they can be located at any position with any orientation. You then applied that knowledge to define multiple overlapping coordinate systems and learned about the conversion between these systems so that you can describe and control character motion in a navigating spaceship. The attempt to navigate the spaceship brought up the next topic: an operator for manipulating orientation, specifically, the quaternion. In Chapter 8, you learned and represented rotations with quaternions. Building on your knowledge of dot and cross products, you derive solutions for aligning vectors and axis frames. You have also observed and emulated organic movements with gradual changes through repeated linear interpolation, LERP and SLERP.

The insights gained from learning these basic math concepts have enabled you to analyze and solve some of the most encountered problems in video game development. This chapter summarizes the book, continuing with the philosophy that interactive exploration is an important and integral part of learning, by presenting the concepts learned throughout this book in a straightforward and comprehensive example. Though not a video game, this example highlights solutions that are implemented in many modern video games.

The Final Comprehensive Example

This example integrates and demonstrates the concepts learned in this book in a comprehensive and coherent application. This example allows you to interactively manipulate the speed and direction of a traveling agent. On the agent and within its bounds, you can control the movement of a hero. You will also be able to manipulate a 2D plane that represents a wall that the traveling agent can reflect off of and cast a shadow upon. Finally, you will also be able to manipulate the radius of a treasure bounding sphere that the agent can collide against. During the interaction, you can suspend all movements and examine the computed projection and collision results, the paths of the agent and the hero on it, and the results of the treasure collision. Figure 9-1 shows a screenshot of running the `EX_9_1_FinalComprehensiveExample` scene from the `Chapter-9-Conclusion` project.

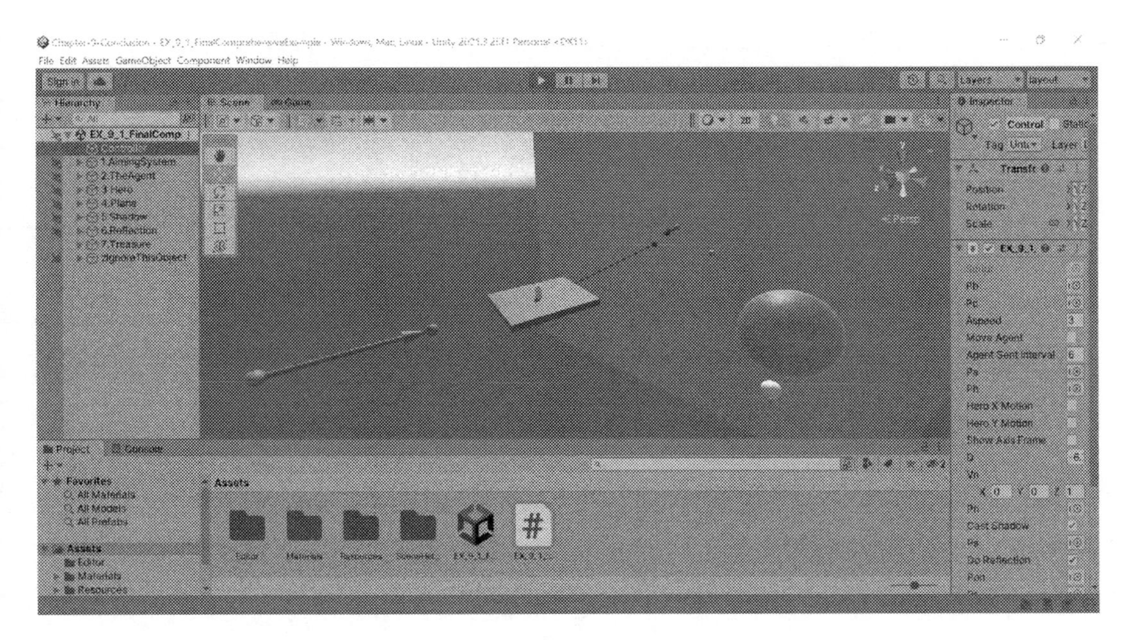

Figure 9-1. Running the Final Comprehensive Example

The goals of this example are for you to

- Experience an interaction session based on a coherent collection of vector-based solutions
- Examine solutions studied in the context of a simple yet comprehensive application
- Examine the implementation source code of a non-trivial system

Examine the Scene

Take a look at the `Example_9_1_FinalComprehensiveExample` scene and observe the predefined game objects in the Hierarchy Window. Due to the slight complexity of this scene, the game objects are categorized into seven groups according to their roles. Each group is an empty game object that serves as the parent or, in this case, a holder, for all the relevant objects that you will actually manipulate. Please pay attention to and only manipulate the relevant game objects when interacting with this example. Additionally, make sure to avoid changing the transforms of the empty grouping game objects during your interactions as it will also change the transforms of the game objects within them. Figure 9-2 depicts the grouping and object names in this scene.

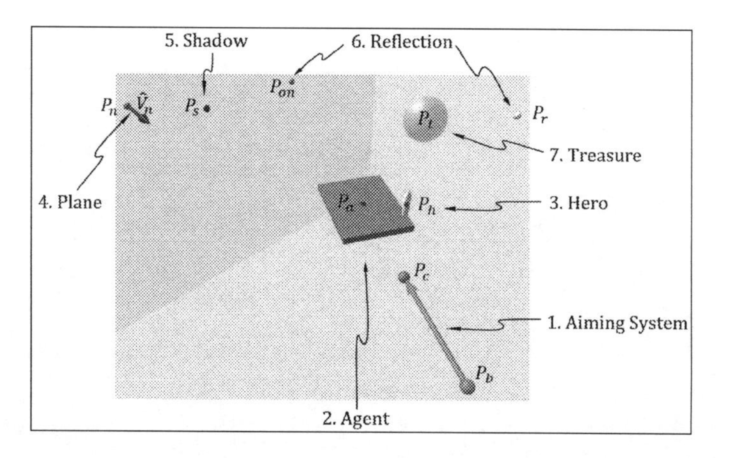

***Figure 9-2.** The groups and game objects in the Final Comprehensive Example*

The six groups of objects are as follows. You can click the small triangle icon beside each object's name in the Hierarchy Window to expand the group.

- `1. Aiming System`: The two spheres in this group are the base, `Pb` in blue, and the control, `Pc` in green. The positions of these two spheres and the distance between them serve to define the direction and speed of the traveling agent.
- `2. Agent`: The only object in this group is the red flattened rectangle, the agent `Pa`. This rectangular object represents the position and orientation of the traveling agent.
- `3. Hero`: The only object in this group is a white capsule, `Ph`, representing the hero in motion referencing the axis frame of the agent object.
- `4. Plane`: The only object in this group is the position on the reflecting wall or the checkered sphere `Pn`. This object exists to assist with visualization. As with all 2D plane examples in Chapter 6, `Pn` is the intersection of the plane normal position vector with the plane. In other words, if the vector plane equation of the wall is

$$p \cdot \hat{V}_n = D$$

Then,

$$P_n = D\hat{V}_n$$

where P_n is the position on the plane along the $\hat{V}_n$ direction from the origin.

- `5. Shadow`: The only object in this group is a semi-transparent black sphere, `Ps`, indicating the shadow of the agent object or the projection of the position `Pa` on the plane that represents the wall.
- `6. Reflection`: The two objects in this group are `Pon`, the striped sphere, and `Pr`, the white sphere. `Pon` is the predicted intersection position of the agent with the wall, and `Pr` is the agent position, `Pa`, reflected across the wall.
- `7. Treasure`: The only object in this group is the semi-transparent red sphere, `Pt`, representing the bounding sphere of a treasure located at this position.

In all cases, the objects' `transform.localPosition` will be referenced as the positions for performing the necessary vector computations and the orientation of the agent will be updated via `transform.localRotation`. Additionally, since `Pt` represents a bounding sphere, its `transform.localScale` property represents the radius and is also referenced.

Analyze Controller MyScript Component

The `MyScript` component on the `Controller` shows variables that can be categorized into the same groups as those of the scene hierarchy. These groups and their accompanying `MyScript` variables are listed as follows:

- Aiming System
    - `Pb`: A reference to the `Pb` game object
    - `Pc`: A reference to the `Pc` game object
    - `Aspeed`: The speed of the traveling agent and also the distance between `Pb` and `Pc`

- Agent
    - `MoveAgent`: A toggle controlling the agent's motion
    - `AgentSentInterval`: The time period before a traveling agent will have its position reset to the control position, `Pc`, and repeat the entire traveling path
    - `Pa`: A reference to the `Pa` game object
- Hero
    - `Ph`: A reference to the `Ph` game object
    - `HeroXMotion`: A toggle controlling the x-direction motion of the hero
    - `HeroYMotion`: A toggle controlling the y-direction motion of the hero
- Plane
    - `ShowAxisFrame`: A toggle to show or hide the Cartesian Coordinate axis frame for verifying the vector plane equation
    - `D`: The plane distance from the origin of the vector plane equation, $p \cdot \hat{V}_n = D$
    - `Vn`: The plane normal vector of the vector plane equation, $p \cdot \hat{V}_n = D$
    - `Pn`: A reference to the `Pn` game object
- Shadow
    - `CastShadow`: A toggle to show or hide the shadow computation results
    - `Ps`: A reference to the `Ps` game object
- Reflection
    - `DoReflection`: A toggle to show or hide the reflection computation
    - `Pon`: A reference to the `Pon` game object
    - `Pr`: A reference to the `Pr` game object

- Treasure
  - `CollideTreasure`: A toggle to show or hide the collision computation
  - `Pt`: A reference to the `Pt` game object
  - `Tr`: The radius of the treasure bounding sphere

The very last variable in the `MyScript` component of `Controller` is the `ShowDebugLines` toggle which is used for showing or hiding all the debug lines in the scene.

Interact with the Example

Click the Play Button to run the example. Notice that initially the red rectangle, or the agent, `Pa`, is stationary. This is by design. You will analyze and understand the scene before setting the agent in motion.

The aiming system, the blue and green spheres, `Pb` and `Pc`, is connected by a red vector representing the direction and speed of the velocity of the agent. The red agent is in front of the aiming system with the white capsule hero pacing back and forth on the agent. A thin black line extending from the center of the agent toward the plane visualizes the location of the projected shadow on the plane, `Ps`. The two thin red lines connecting the agent to `Pon` on the plane and `Pr` in the mirrored reflection direction show, when in motion, the intersection position with the 2D plane and the reflection of the agent across the plane. The transparent bounding sphere at `Pt` is red because it intersects the reflection ray.

During your interaction, be careful to avoid adjusting the transforms of the empty container parent or holder objects. Additionally, pay attention to the Console Window printout. If you accidentally set the application to an ill-defined state, for example, by overlapping `Pb` and `Pc` positions, warning messages will be printed to the Console Window and the script will reset its state to ensure that the application does not crash.

Now toggle off `CastShadow`, `DoReflection`, and `CollideTreasure` such that you can focus on and examine each of the seven functionalities separately.

Interact with the Aiming System

Figure 9-3 focuses on the aiming system and the orientation of the agent. Details of the hero object, the white capsule, will be discussed later. The objects are annotated with their corresponding variable names in the implementation such that you can observe their behaviors to examine the mathematics of the vector solution.

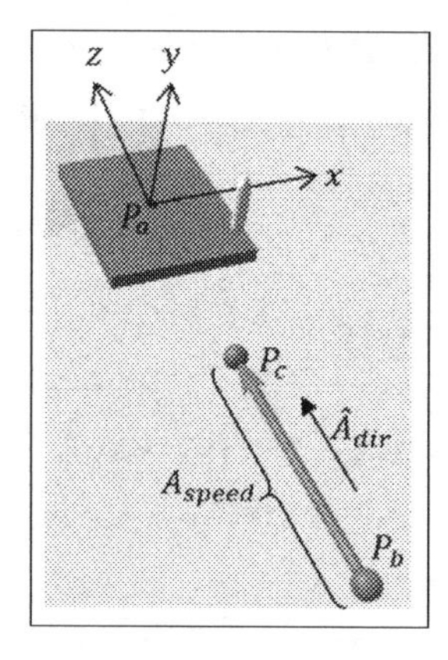

Figure 9-3. *The aiming system and the agent orientation*

As illustrated in Figure 9-3, the vector from P_b to P_c defines the direction of the agent velocity, $\hat{A}_{dir}$, and the distance between these two positions is A_{speed}, which is under the user control via the variable Aspeed. The agent, Pa, always aligns its forward and up directions with that of $\hat{A}_{dir}$ and the vertical direction $\hat{V}_y = (0,1,0)$ of the Cartesian Coordinate System. In this way, as illustrated in Figure 9-3, the agent defines a separate and independent axis frame with its center, Pa, being the origin of this axis frame.

In the Hierarchy Window, expand the `1.AimingSystem` game object by clicking the triangle icon beside it. Select Pb and manipulate its position. You will observe that changes to Pb always result in corresponding changes in Pc and the agent, Pa, ensuring that the agent is always located in front of and aligned with the velocity direction, $\hat{A}_{dir}$. The center of the agent is located at a constant distance of 2xAspeed away from Pb. You can change the Aspeed magnitude to observe the in-between space adjusting accordingly.

Select Pc and adjust its position to observe that by maintaining a constant distance from Pb, Pc can only orbit Pb. That is, the position Pc can only change along the circumference of the circle centered at Pb with radius Aspeed. Note that as the velocity direction, $\hat{A}_{dir}$, changes, so does the position and orientation of Pa. This is because the distance and direction between Pb and Pc is same as the distance between Pc and the center of Pa, and the agent's front or z-direction is always aligned with that of $\hat{A}_{dir}$.

As described, the velocity direction, $\hat{A}_{dir}$, is simply the vector between Pb and Pc. The behaviors you just walked through identify Pb as the base, or tail, of the aiming system, controlling both the Pc and the agent, Pa, positions. The aiming direction and Pc position can be computed as follows:

$$\hat{A}_{dir} = (P_c - P_b).Normalize \quad \text{direction from } P_b \text{ to } P_c$$

$$P_c = P_b + A_{speed}\hat{A}_{dir} \quad A_{speed} \text{ from } P_b$$

where the agent's position and orientation can be determined by

$$P_a = P_b + 2A_{speed}\hat{A}_{dir} \quad 2\times \text{ the distance}$$

$$P_a.localRotation = Quaternion.LookRotation\left(\hat{A}_{dir}, \hat{V}_y\right)$$

Interact with the Agent

Enable the agent motion by switching on the MoveAgent toggle. For now, continue to ignore the pacing hero on the agent. Notice that Pa orientates along and moves in the $\hat{A}_{dir}$ direction, and at about every six-second interval, the position of Pa is reset to that of Pc and the motion repeats. This interval period is the time period controlled by AgentSentInterval, which uses seconds as its unit of time. You can adjust this variable to observe its effect. Notice that when AgentSentInterval is a negative number or zero, Pa's position is being reset at every update, and as a result, it becomes stationary at position Pc. You can verify the direction of the agent velocity by adjusting Pc's position and the speed of the agent by manipulating the Aspeed value. The agent's orientation and traveling direction only update at the beginning of each interval period. These observations suggest that when MoveAgent is true and AgentSentInterval time limit is reached, the position of Pa is reset to that of Pc with orientation updated to align with $\hat{A}_{dir}$ or

$$P_a = P_c$$

$$P_a.rotation = Quaternion.LookRotation\left(\hat{A}_{dir}, \hat{V}_y\right)$$

And during motion, Pa position is updated according to

$$P_a = P_a + \left(ElapsedTime \times A_{speed}\right)\hat{A}_{dir}$$

where the new position is the old position plus *time* × *speed*. Note that the "×" symbol in this case is a floating-point multiplication and not a vector cross product. You know this because the cross product between floating-point numbers is undefined; therefore, it must be multiplication.

Lastly, tumble the Scene View camera to observe that while traveling in space, it is actually rather challenging to resolve the relative distance and position between the agent and the plane. To assist with distance determination, the ShowDebugLines is switched on by default where you can observe a thin red line in the direction of $\hat{A}_{dir}$ in front of Pa indicating the pathway of Pa. This thin red line is informative because it assists in resolving relative positions. However, it is also distracting because in real life such indicating lines do not exist. As you will verify soon, dropping a shadow can also be an effective way of addressing the challenge of resolving relative distance.

Interact with the Hero Motion

Please restart the game to ensure a proper initial setting. In the following, before enabling the agent to travel, you will first focus on analyzing and understanding the pacing motion of the hero, Ph, the white elongated capsule on the agent.

Now, observe the back and forth pacing of the hero along the direction defined by the aiming system, Pb to Pc, or $\hat{A}_{dir}$. Select and adjust the position of Pc to manipulate $\hat{A}_{dir}$ and verify that the pacing direction indeed followed. Now, enable the HeroYMotion toggle and observe the hero hopping vertically on the agent with respect to and along the $\hat{A}_{dir}$ direction. You can adjust the y-value of Pc to aim $\hat{A}_{dir}$ up- or downward and verify that the hero's hopping direction is indeed aligned perpendicular to the flat surface of the agent. Now, disable the HeroYMotion and enable the HeroXMotion toggle. Notice that in this case the hero is sweeping along a sinusoidal pathway on the surface of the agent. Once again, manipulate Pc position to alter the agent's orientation and verify that the hero movement pathway remains. Feel free to enable both motions of the hero and manipulate Pc position to examine and admire the hero's constant sinusoidal hopping that follows the changing orientation of the agent.

You have interacted with and observed the movement of the hero being defined with respect to the axis frame of the agent. Recall that the axis frame of the agent has its origin located the agent's center position, P_a, and is defined by $\hat{A}_{dir}$ being the forward or z-direction and $\hat{V}_y = (0,1,0)$ being the y-direction. This means that the back and forth pacing of the hero is z-direction, the hopping is y-direction, and sinusoidal sweeping is x-direction movements. In this way, the hero position, Ph, is a vector, $\vec{V}_h = (\Delta x, \Delta y, \Delta z)$, offset from the origin, P_a, of the agent axis frame or

$$P_h = P_a + \Delta x\ \hat{x} + \Delta y\ \hat{y} + \Delta z\ \hat{z}$$

where $\hat{x}$, $\hat{y}$, and $\hat{z}$ are the directions of the major axes of the agent axis frame. In this case, let the constant pacing speed be HeroSpeed, the y-direction hopping is implemented as an absolute cosine, and x-direction sweeping is a simple sine function:

$$\Delta z = ElapsedTime \times HeroSpeed$$

$$\Delta y = abs(\cos(\pi \Delta z))$$

$$\Delta x = \sin(\pi \Delta z)$$

Interact with the Plane

With the agent in motion (ensure MoveAgent is toggled on), please switch on the ShowAxisFrame toggle, and begin to investigate the plane and its spatial relationship with the agent. First, note the white line extending from position Pn to the origin overlapping with the plane normal vector. This shows that Pn is indeed a position vector in the direction of the plane normal vector.

Adjust the parameter D to change the distance between the plane and the axis frame as well as components of Vn to see the plane rotating about the axis frame. Because of the large size of the plane, you may have to zoom out the camera to observe the effects of adjusting Vn. Notice that Pn is always located at the intersection of the plane normal vector extending from the origin. You have verified that this plane is indeed defined by the vector plane equation

$$p \cdot \hat{V}_n = D$$

and that

$$P_n = D\hat{V}_n$$

is a position on the plane along the $\hat{V}_n$ direction from the origin. When examining the relative position of the agent, its motion, and the normal direction of the plane, along with the anticipation for later shadow and reflection computations, there are few concerns. Please refer to Figure 9-4 for the details.

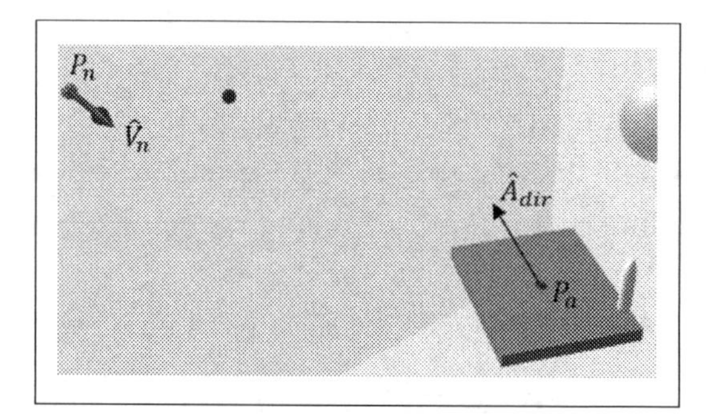

Figure 9-4. *The plane and its relationship to the position and motion of the agent*

This example specifies that shadow casting and reflection can only occur when P_a is on the side pointed toward by the plane normal vector or along the direction of $\hat{V}_n$. Additionally, you have already verified that reflection computation should not occur if the agent's velocity, $\hat{A}_{dir}$, is parallel to the plane or perpendicular to the normal vector, $\hat{V}_n$. Lastly, note that a reflection cannot occur if P_a is moving away from the plane. These discussions identify three geometric conditions of interests:

- In front of condition: This is when the position of the agent is on the side of the 2D plane that is pointed to by the plane normal vector, $\hat{V}_n$. To determine if this is true, you can simply verify that the projected size of position vector P_a in the plane normal direction, $\hat{V}_n$, is greater than the plane distance, D, or

$$In\ front: \left(P_a \cdot \hat{V}_n\right) > D$$

- Perpendicular or not parallel condition: When a velocity vector is perpendicular to a plane normal vector, the velocity is parallel to and will never intersect with the plane. This condition can be determined by one of the following tests:

- $Perpendicualr\ to\ normal\ vector: \left(\hat{A}_{dir} \cdot \hat{V}_n\right) \approx 0$

  subtended angle≈90°

- $Not\ parallel\ to\ plane: \left(\hat{A}_{dir} \cdot \hat{V}_n\right) \neq 0$

  subtended angle≠90°

- Approaching condition: When an object is in front of and moving toward a plane, its velocity will be pointing in the direction opposite to the plane normal vector or

$$is\ approaching\ from\ front: \left(\hat{A}_{dir} \cdot \hat{V}_n\right) < 0$$

$$90° \leq \text{subtended angle} \leq 180°$$

Interact with the Shadow

Please restart the game to ensure a proper initial setting and then toggle off `DoReflection` and `CollideTreasure`, switch on `MoveAgent`, and increase `Aspeed` to 8. Now, you can toggle the `ShowDebugLines` on and off to experience the full effect of the shadow object, `Ps`, in conveying the relative spatial relationship.

Notice that, as defined by the application, shadow casting does not occur once the agent moves past the plane. You can verify this as follows. First, set the plane normal, `Vn`, to (0, 1, 0) to observe the shadow when the agent velocity is parallel and in front of the plane. Then, if you flip the plane around, by setting `Vn` to (0, −1, 0) and `D` to positive 6, you can now notice that the agent is not on the side pointing to by the plane normal and thus shadow casting does not occur. Figure 9-5 illustrates the solution for computing `Ps` when `Pa` is in front of the plane.

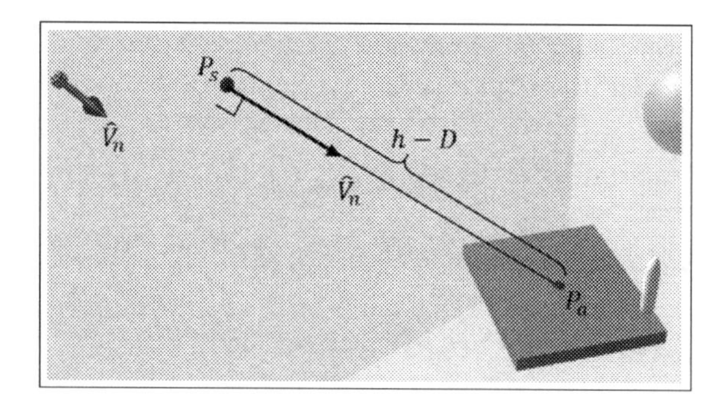

Figure 9-5. *The shadow* `Ps` *computation*

A quick review of "Projections onto 2D Planes" discussion from Chapter 6 says that the projected length of the position vector P_a onto the plane normal, $\hat{V}_n$, is

$$h = P_a \cdot \hat{V}_n \qquad P_a \text{ length along } \hat{V}_n$$

Position P_s is simply $h - D$ distance from the position P_a in the negative $\hat{V}_n$ direction

$$P_s = P_a - (h - D)\hat{V}_n$$

Interact with the Reflection

Once again, please restart the game to ensure a proper initial setting, toggle off `CollideTreasure`, switch on `MoveAgent`, and set the `Aspeed` to 5. Feel free to switch `CastShadow` toggle off if you find the shadow distracting.

Observe how the red agent and the white `Pr` sphere approach the `Pon` intersection position in perfect synchronization. When the distance between `Pa` and `Pon` is very small, the bounding spheres around these two objects will collide. After the collision, since the agent is moving away from the plane and its velocity does not reflect with the 2D plane anymore, the white sphere representing the agent's reflection, `Pr`, disappears leaving the red agent to continue with its motion in the mirrored reflection direction. You can adjust the plane by manipulating the `Vn` and `D` parameters and observe that the reflected motion adjusts correctly.

If you flip the 2D plane from its initial orientation by setting `Vn` to $(0, 0, -1)$ and `D` to 6 you, will notice that the reflection computation does not occur. This example only computes reflection when the agent travels into the plane from the front. Note that this is not a limitation of the solution; rather, this is a design choice for showcasing the in front of test with a 2D plane. Now, if you set Vn to (0, 1, 0), the plane will be parallel to the agent velocity direction, $\hat{A}_{dir}$. When this occurs, notice that both Pr and Pon disappear. In this case, the reflection is not defined and therefore the computation for these positions is not invoked.

Restart the game again, switch on MoveAgent, and this time, set the Aspeed to a large number, for example, 15. Notice now that the agent sometimes fails to collide with the plane and instead simply crosses the plane. What you are observing is the exact same problem as the one described in Figure 5-13 of the "Line to Point Distance" section in Chapter 5 or the problem of failed collision for fast-moving objects. You will resolve this issue in an exercise. It is interesting that the collision detection only fails some of the time depending on the actual rate that the Update() function is called. Unfortunately, these types of uncertainty are rather common in typical video game development and must be predicted and resolved.

Figure 9-6 depicts the reflection computation that supports the behaviors you just observed.

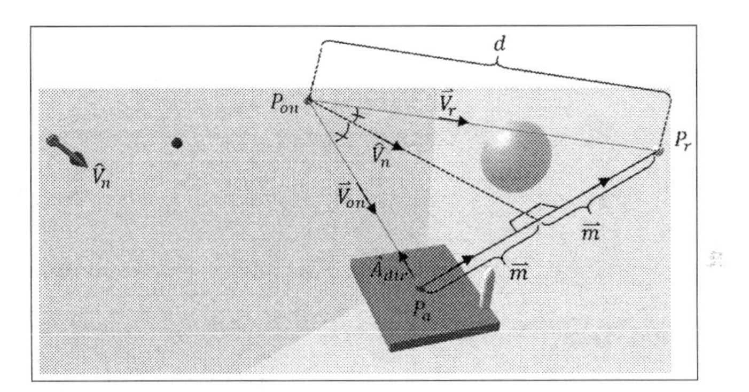

***Figure 9-6.** Reflecting Pa across the wall*

As seen in Figure 9-6, reflection computation will only proceed when the agent is in front of the plane and has a velocity direction, $\hat{A}_{dir}$, that is not parallel to and is headed toward the plane. In this case, the reflection direction can be derived by first computing the position, P_{on}, where the line segment that begins at P_a with a direction of $\hat{A}_{dir}$ intersects the plane, $p \cdot \hat{V}_n = D$,

$$P_{on} = P_a + d\hat{A}_{dir} \qquad d \text{ along } \hat{A}_{dir} \text{ from } P_a$$

in this case, d, which as shown in the discussion of "Line to Plane Intersection" in Chapter 6 as illustrated in Figure 6-16, can be derived as

$$d = \frac{D - \left(P_a \cdot \hat{V}_n\right)}{\left(\hat{A}_{dir} \cdot \hat{V}_n\right)} \quad \text{from } P_a \text{ to plane along } \hat{A}_{dir}$$

and

$$\vec{V}_{on} = P_a - P_{on} \quad \text{from } P_{on} \text{ to } P_a$$

The "Mirrored Reflection Across a Plane" discussion, as illustrated in Figure 6-18, showed that

$$\vec{m} = \left(\vec{V}_{on} \cdot \hat{V}_n\right)\hat{V}_n - \vec{V}_{on} \quad \text{perpendicular to } \hat{V}_n \text{ at } P_a$$

and the reflection direction is

$$\vec{V}_r = \vec{V}_{on} + 2\vec{m} \quad \text{reflection of } \vec{V}_{on} \text{ across } \hat{V}_n$$

where

$$P_r = P_{on} + \vec{V}_r \quad \text{mirrored reflection of } P_a$$

Interact with the Colliding Treasure

For the last time, please restart the game to ensure a proper initial setting. For now, please do not enable `MoveAgent`. Feel free to switch the `CastShadow` toggle off if you find the shadow distracting.

Notice that the `Pt` sphere is highlighted in red because the reflection vector, $\vec{V}_r$, passes through this sphere. Now, select position `Pc` in `1.AimingSystem` and adjust its x-component value. This will change the velocity direction of the agent, $\hat{A}_{dir}$, and thus affect the reflection vector, $\vec{V}_r$. Notice the `Pt` sphere changing to white when the reflection vector is outside of the sphere. This application is designed to detect the condition when the reflection vector is sufficiently close to the `Pt` sphere.

You can adjust the `Pt` position and the sphere's radius via `Tr` to modify the reflection vector and the bounding sphere respectively to verify the correctness of the vector inside sphere results. Now if you enable the `MoveAgent` toggle and increase the agent speed or its interval time so that collision can occur before the agent motion is reset, you can verify the correctness of the results for a changing reflection vector. Notice that after the collision at Pon, the $\vec{V}_r$ vector is not defined anymore and thus the Pt sphere becomes white in color.

As illustrated in Figure 5-13 and discussed in the "Line to Point Distance" section of Chapter 5, the vector cutting through a bounding sphere functionality can be implemented as a point to line distance computation. The details of this computation are illustrated in Figure 9-7.

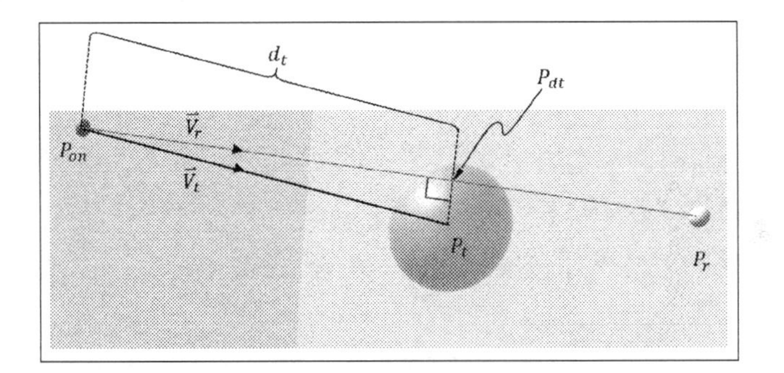

***Figure 9-7.** Point to line distance for bounding sphere collision detection*

Refer to Figure 9-7 and note that $\vec{V}_t$ is defined to be the vector from P_{on} to the center of the treasure bounding sphere, P_t,

$$\vec{V}_t = P_t - P_{on} \qquad \text{from } P_{on} \text{ to } P_t$$

Then note that the projected distance of $\vec{V}_t$ along $\vec{V}_r$ is d_t,

$$d_t = \vec{V}_t \cdot \hat{V}_r \qquad \vec{V}_t \text{ length in } \hat{V}_r \text{ direction}$$

And that when d_t is larger than zero and less than the magnitude of $\vec{V}_r$, then the closest point, P_{dt}, between P_t and the line segment is

$$P_{dt} = P_{on} + d_t \hat{V}_r \qquad d_t \text{ along } \hat{V}_r \text{ from } P_{on}$$

And finally, the line segment intersects the given bounding sphere when

$$\left\| P_{dt} - P_t \right\| \le \textit{Bounding Sphere Radius}$$

Summary of Interaction

Now that you have a comprehensive understanding of this example and insights into the solutions, please feel free to adjust any and all parameters to examine the consistency of the results.

Details of MyScript

Open MyScript and examine the source code in the IDE. The instance variables and the Start() function are as follows:

```
// Aim System
public GameObject Pb = null;
public GameObject Pc = null;
public float Aspeed = 2.0f;              // Agent Speed

// Agent Support
public bool MoveAgent = false;
public float AgentSentInterval = 4f;   // Re-send Interval
public GameObject Pa = null;
private Vector3 Adir = Vector3.zero;
private float AgentSinceTime = 100f;   // Since resent

// Hero
public GameObject Ph = null;
public bool HeroXMotion = true;
public bool HeroYMotion = true;
private Vector3 Vh = Vector3.zero;
private float HeroSpeed = 0.5f;
private const float kHeroZMotionRange = 1f;

//  Plane
public bool ShowAxisFrame = false;
public float D = -6.7f; // The distance to the plane
public Vector3 Vn;      // Normal vector of reflection plane
public GameObject Pn;   // Location where the plane center is

// Shadow
```

```
public bool CastShadow = true;
public GameObject Ps;  // Location of Shadow of Agent

// Reflection
public bool DoReflection = true;
public GameObject Pon; // Collision point of Agent
public GameObject Pr;  // Reflection of current Agent position

// Treasure Collision
public bool CollideTreasure = true;
public GameObject Pt;    // Treasure position
public float Tr = 2f;    // Treasure radius
public bool ShowDebugLines = true;

#region For visualizing
#endregion

void Start() {
    Debug.Assert(Pa != null);      // Verify proper setting
    Debug.Assert(Pb != null);
    Debug.Assert(Pc != null);
    Debug.Assert(Pn != null);
    Debug.Assert(Ps != null);
    Debug.Assert(Pon != null);
    Debug.Assert(Pr != null);
    Debug.Assert(Pt != null);
    Debug.Assert(Ph != null);
    #region For visualization
    #endregion
}
```

All public variables for MyScript have been discussed when analyzing the Controller's MyScript component. The only internal states or private variables maintained are for supporting the hero movement, reflection of the agent's velocity, Adir ($\hat{A}_{dir}$), and for keeping track of the elapsed time since the previous agent position and velocity were reset, AgentSinceTime.

As in all previous examples, the Debug.Assert() calls in the Start() function ensure proper setup regarding referencing the appropriate game objects via the Inspector Window. The Update() function is organized into the following regions where the details will be examined accordingly:

```
void Update() {
    Step 0: Initial Error Checking

    Step 1: The Aiming System

    Step 2: The Agent

    Step 3: The Hero motion

    Step 4: The Plane and infront/parallel checks

    Step 5: The Shadow

    Step 6: The Reflection

    Step 7: The collision with treasure
    #region  For visualization
    #endregion
}
```

Step 0: Initial Error Checking

Expand this region and examine the following:

```
# #region Step 0: Initial error checking
Debug.Assert((Pc.transform.localPosition -
      Pb.transform.localPosition).magnitude > float.Epsilon);
Debug.Assert(Vn.magnitude > float.Epsilon);
Debug.Assert(Aspeed > float.Epsilon);
Debug.Assert(Tr > float.Epsilon);
// recoveries from the errors
if ((Pc.transform.localPosition -
     Pb.transform.localPosition).magnitude < float.Epsilon)
        Pc.transform.localPosition
        = Pb.transform.localPosition - Vector3.forward;
```

```
if (Vn.magnitude < float.Epsilon)
    Vn = Vector3.forward;
if (Aspeed < float.Epsilon)
    Aspeed = 0.01f;
if (Tr < float.Epsilon)
    Tr = 0.01f;
#endregion
```

These lines of code are simple edge case error checking before any computation begins. The first three nonzero assertions are to avoid working with zero vectors and the last assertion ensures that the treasure bounding sphere has a valid radius. The four `if` statements are attempts to recover from ill-defined states. Notice the error recoveries are rather ad hoc, where the application state is simply set to a defined situation. In a real application, it is the responsibility of the game designers to ensure that inputs from the users are not capable of setting or creating such ill-defined states. For example, in this scenario, the game designer is responsible for defining limitations such that during the aiming process, the user will not accidentally set the agent speed to zero or a negative value.

Step 1: The Aiming System

Expand this region and examine the following:

```
#region Step 1: The Aiming System
Vector3 aDir = Pc.transform.localPosition -
               Pb.transform.localPosition;
aDir.Normalize(); // assuming not located at the same point
Pc.transform.localPosition =
               Pb.transform.localPosition + Aspeed * aDir;
if (!MoveAgent) { // controls only when agent is not moving
    Pa.transform.localPosition =
                 Pb.transform.localPosition + 2 * Aspeed * aDir;
    Pa.transform.localRotation =
                 Quaternion.LookRotation(aDir, Vector3.up);
    Adir = aDir;
}
#endregion
```

This code computes

$$\hat{A}_{dir} = (P_c - P_d).Normalize$$

$$P_c = P_b + A_{speed}\hat{A}_{dir}$$

and when the agent is not in motion, the code also computes

$$P_a = P_b + 2A_{speed}\hat{A}_{dir}$$

Step 2: The Agent

Expand this region and examine the following:

```
#region Step 2: The Agent
if (MoveAgent) {
    Pa.transform.localPosition += Aspeed * Time.deltaTime * Adir
    AgentSinceTime += Time.deltaTime;
    if (AgentSinceTime > AgentSentInterval) {  // Should re-send
        Pa.transform.localPosition = Pc.transform.localPosition
        Adir = aDir;
        Pa.transform.localRotation =
                 Quaternion.LookRotation(aDir, Vector3.up);
        AgentSinceTime = 0f;
    }
}
if (ShowVelocity && ShowDebugLines)
    Debug.DrawLine(Pa.transform.localPosition,
           Pa.transform.localPosition + 20f * Adir, Color.red);
#endregion
```

This code shows that actual computations are required for the agent object only when MoveAgent toggle is enabled. When this toggle is enabled, the agent's new position is updated via its current velocity

$$P_a = P_a + \left(ElapsedTime \times A_{speed}\right)\hat{A}_{dir}$$

Then, when the wall-clock elapsed time is more than the user-specified AgentSentInterval, the agent position is reset to P_c and its velocity is set to the current $(P_c - P_d)$. *Normalize*. The last line of code in this region draws a red line with length of 20 units from the agent position in its velocity direction when the user settings are favorable.

Step 3: The Hero Motion

Expand this region and examine the following:

```
#region Step 3: The Hero motion
// Hero's follows Agent (Pa) axis frame
Vector3 po = Pa.transform.localPosition;
Vector3 vx = Pa.transform.right;
Vector3 vy = Pa.transform.up;
Vector3 vz = Pa.transform.forward;

Vh.z += HeroSpeed * Time.deltaTime;  // moved
if (Mathf.Abs(Vh.z) > kHeroZMotionRange) {
    Vh.z = (Vh.z>0f) ? 1f : -1f;
    HeroSpeed = -HeroSpeed;
}

if (HeroYMotion)
    Vh.y= Mathf.Abs(Mathf.Cos(Mathf.PI * Vh.z));

if (HeroXMotion)
    Vh.x= Mathf.Sin(Mathf.PI * Vh.z);

Vector3 vhc = Vh.x * vx + Vh.y * vy + Vh.z * vz;
Ph.transform.localPosition = po + vhc;
Ph.transform.localRotation = Pa.transform.localRotation;
#endregion
```

The first four lines extract the agent axis frame: po being the origin and vx, vy, and vz are the directions of the major axes. The last two lines set the position and orientation of the hero

$$P_h = P_a + Vh.x\ \hat{x} + Vh.y\ \hat{y} + Vh.z\ \hat{z}$$

$$P_h.locationRotation = P_a.locationRotation$$

The lines in between compute and set the hero movement vector, Vh,

$$Vh.z = HeroSpeed \times ElapsedTime$$

$$Vh.y = abs(\cos(\pi\, Vh.z))$$

$$Vh.x = \sin(\pi\, Vh.z)$$

Step 4: The Plane

Expand this region and examine the following:

```
#region Step 4: The Plane and infront/parallel checks
Vn.Normalize();
Pn.transform.localPosition = D * Vn;

// agent position checks
float paDotVn = Vector3.Dot(Pa.transform.localPosition, Vn);
bool infrontOfPlane = (paDotVn > D);

// Agent motion direction checks
float aDirDotVn = Vector3.Dot(Adir, Vn);
bool isApproaching = (aDirDotVn < 0f);
bool notParallel = (Mathf.Abs(aDirDotVn) > float.Epsilon);
#endregion
```

This region ensures a proper vector plane equation and computes object and velocity to plane relationships. The first two lines compute

$$\hat{V}_n = \frac{\vec{V}_n}{\|\vec{V}_n\|} \quad \text{normalization after user manipulations}$$

$$P_n = D\hat{V}_n$$

Next, the in front of, approaching, and not parallel conditions are computed as follows:

$$In\,frontOfPlane = \left(P_a \cdot \hat{V}_n\right) > D$$

$$isApproaching = \left(\hat{A}_{dir} \cdot \hat{V}_n\right) < 0$$

$$notParallel = \left\|\hat{A}_{dir} \cdot \hat{V}_n\right\| > 0$$

These conditions will assist in determining if shadow casting, reflection, and collision with the treasure bounding sphere should occur.

Step 5: The Shadow

Expand this region and examine the following:

```
#region Step 5: The Shadow
Ps.SetActive(CastShadow && infrontOfPlane);
if (CastShadow && infrontOfPlane) {
    float h = Vector3.Dot(Pa.transform.localPosition, Vn);
    Ps.transform.localPosition =
                Pa.transform.localPosition - (h-D) * Vn;
    if (ShowDebugLines)
        Debug.DrawLine(Pa.transform.localPosition,
                    Ps.transform.localPosition, Color.black);
}
#endregion
```

The first line shows or hides the Ps game object depending on user command. The next conditional statement determines if shadow computation should occur. This computation will occur only if the user wants to examine shadow casting and if the agent is in front of the plane. Shadow is computed by

$$h = P_a \cdot \hat{V}_n$$

$$P_s = P_a - (h - D)\hat{V}_n$$

Lastly, when users specify, a black line is drawn from Pa to Ps to assist in visualizing the projection.

Step 6: The Reflection

Expand this region and examine the following:

```
#region Step 6: The Reflection
Pon.SetActive(DoReflection && notParallel
                            && infrontOfPlane && isApproaching);
Pr.SetActive(DoReflection && notParallel
                           && infrontOfPlane && isApproaching);
Vector3 vr = Vector3.up;  // Reflection vector
bool vrIsValid = false;
if (DoReflection && notParallel && isApproaching) {
    if (infrontOfPlane) {
        float d = (D -
                  Vector3.Dot(Pa.transform.localPosition, Vn))
                 / aDirDotVn;
        Pon.transform.localPosition =
                 Pa.transform.localPosition + d * Adir;
        Vector3 von = Pa.transform.localPosition -
                     Pon.transform.localPosition;
        Vector3 m = (Vector3.Dot(von, Vn) * Vn) - von;
        vr = 2 * m + von;
        Pr.transform.localPosition =
                     Pon.transform.localPosition + vr;
        vrIsValid = true;
        if (ShowDebugLines) {
            Debug.DrawLine(Pa.transform.localPosition,
                  Pon.transform.localPosition, Color.red);
            Debug.DrawLine(Pon.transform.localPosition,
                  Pr.transform.localPosition, Color.red);
        }
        // What will happen if you do this?
        // if (von.magnitude < float.Epsilon)
        if (von.magnitude < 0.1f) {
            // collision with "virtual" bounding sphere
            Adir = vr.normalized;
```

```
            Pa.transform.localRotation =
                Quaternion.LookRotation(Adir, Vector3.up);
        }
    } else {
        Debug.Log("Potential problem!: high speed Agent,
                    missing the plane collision?");
        // What can you do?
    }
}
#endregion
```

The first two lines show or hide the Pon and Pr game objects based on user command and the relationship between the agent and the plane. Reflection computation will occur only if the user wants to examine the reflection, when the agent is in front of the plane, has a velocity that is not parallel to the plane, and the velocity is moving toward the plane. The in front of condition is a design choice; the parallel condition is required to avoid undefined solutions; and the last condition is required because when an object is in front of and moving away from the plane, no collision will occur and thus no reflection computation is necessary.

Note that the outer if condition checks for user command, "not parallel", and "is approaching" conditions, whereas the "in front of" condition is checked in an inner if statement. When all conditions are satisfied, the reflection position, Pr, is computed as

$d = \frac{D - (P_a \cdot \hat{V}_n)}{(\hat{A}_{dir} \cdot \hat{V}_n)}$	agent to plane distance
$P_{on} = P_a + d\hat{A}_{dir}$	agent intersects plane at P_{on}
$\vec{V}_{on} = P_a - P_{\mathrm{on}}$	plane to agent $(-d\hat{A}_{dir})$
$\vec{m} = (\vec{V}_{on} \cdot \hat{V}_n)\hat{V}_n - \vec{V}_{on}$	perpendicular to plane
$\vec{V}_r = \vec{V}_{on} + 2\vec{m}$	reflection direction
$P_r = P_{on} + \vec{V}_r$	mirrored reflection of agent

The two red lines from Pa to Pon and from Pon to Pr are then drawn according to user's command. The last if statement compares $\|\vec{V}_{on}\|$ to a small number, $0.1f$. This is essentially checking for the intersection of the bounding spheres around the agent and the Pon position. When these two positions are close to each other or when $\|\vec{V}_{on}\|$ is very small, a collision is detected and $\hat{A}_{dir}$ becomes the reflected direction, $\hat{A}_{dir} = \hat{V}_r$. The vrIsValid flag informs the next step, collision with the treasure bounding sphere, when there is a valid reflection vector. Recall from Chapter 3 that bounding spheres are less than ideal for detecting collisions for the rectangular agent, and yet, as in this case, when rectangular objects are not aligned with the major axes, it is often the default solution.

You can now analyze the reason for checking the "in front of" condition in the inner if statement. Recall that in the initial setup, the AimingSystem sends the agent toward the plane. If a condition should occur where the agent's velocity indicates that it is approaching the plane and yet its current position is not in front of the plane, then there are two possible cases. First, the agent's initial position is behind the plane, and it continues to move away from the plane. In this situation, there is no cause for concern as everything is functioning as it should. However, if it is the second case, then something should be done. Recall that the agent's position was already updated in Step 2; it therefore may be the case that, in one update, the agent has moved from a position that is in front of the plane to a position that is behind the plane. As you have observed, this situation can occur for an agent traveling at high speeds. In this implementation, such a situation is detected, and a warning message is printed to the Console Window. In an exercise, you will be led to develop a solution for this missing collision problem.

Step 7: The Collision with Treasure

Expand this region and examine the following:

```
#region Step 7: The collision with treasure
Pt.SetActive(DoReflection && CollideTreasure);
Pt.transform.localScale = new Vector3(2 * Tr, 2 * Tr, 2 * Tr);
                                  // this is the diameter
Pt.GetComponent<Renderer>().material.color =
                                 MyDrawObject.NoCollisionColor;
if (DoReflection && CollideTreasure && vrIsValid) {
    Vector3 vt = Pt.transform.localPosition -
                 Pon.transform.localPosition;
```

```
    float dt = Vector3.Dot(vt, vr.normalized);
    if ((dt >= 0) && (dt <= vr.magnitude)) {
        Vector3 pdt = Pon.transform.localPosition +
                      dt * vr.normalized;
        if ((pdt - Pt.transform.localPosition).magnitude <= Tr)
            Pt.GetComponent<Renderer>().material.color =
                                      MyDrawObject.CollisionColor;
    }
}
#endregion
```

The first two lines of code show or hide the Pt sphere and set its radius according to the user commands. The third line initializes the sphere to the no-collision color, white. The treasure bounding sphere collision computation is performed only when the user demands it and when reflection was successful in the previous step. The two lines of code in the if condition compute

$$\vec{V}_t = P_t - P_{on} \qquad \text{from } P_{on} \text{ on the plane to } P_t$$

$$d_t = \vec{V}_t \cdot \hat{V}_r \qquad \text{project } \vec{V}_t \text{ along } \hat{V}_r$$

The inner if condition checks for $0 \le d_t \le \|\vec{V}_r\|$, or the condition when the projected length is within the bounds of the reflected vector, and computes the $\vec{V}_t$ projection on $\vec{V}_r$, P_{dt},

$$P_{dt} = P_{on} + d_t\hat{V}_r \qquad \text{treasure position } \hat{V}_r$$

Since the position, P_{dt}, on the reflection vector is closest to the treasure position, P_t, the reflection vector will intersect the treasure bounding sphere when the distance between these two positions is less than the radius of the sphere; in other words, a collision occurs when this condition is true:

$$\|P_{dt} - P\|_t \le T_r$$

closest distance is less than the treasure bounding sphere radius.

Takeaway from This Example

This has been the most complex example in this book. This example demonstrates many of the concepts discussed throughout this book in a straightforward and coherent application. Though the exact form and details involved can vary, all of the interactions you have gone through in this example can be found in popular video games. Notice how you approached the analysis and examination of both the scene and the implementation. You first understood the entire narration: the aiming system, agent traveling, hero movement, casting shadow, reflecting, and colliding. After that, you categorized the scene and the implementation into distinct steps. This is a top-down, divide, and conquer approach to problem analysis and solution derivation. The lesson here is to understand the problem space, subdivide into smaller tasks, solve each individually, and then combine the results as the final solution to the original problem. Video games and the vast majority of software applications, graphical or otherwise, can be intimidating when you first examine their requirements. The key is to avoid being overwhelmed by the complicated problem narrative and to break the narrative down into pieces that you can understand and accomplish, just like you did for this example.

Relevant mathematical concepts covered include most of the concepts learned in this book. The important lesson here is that when combining concepts in solving a series of related problems, it is critical to subdivide the problems into individual tasks and then to apply the concepts to accomplish each task independently.

Relevant observations on implementation include what to avoid when building software solutions. It is important to recognize that all example implementations in this book were designed to serve a narrow purpose—to best showcase the math concepts. This single goal overrides all other vital software development guidelines, including the very important concepts of information hiding and abstraction. A significant strategic effort was made to ensure that all solutions can be presented in a single execution unit, `MyScript`, with most of variables being publicly accessible. Though the code in the `MyScript` files is straightforward to comprehend and interact with, they can be challenging to expand, generalize, and build upon. In the case of the last example, you may have noticed the important and yet messy relationships between the individual steps in the implementation. For example, the agent velocity is computed and updated conditionally in Steps 1, 2, and 5. While the implementation of this last example served well as a demonstration of vector operations, it does not serve to demonstrate how to structure a video game. Properly designed software should hide essential information and define abstract interfaces.

EXERCISES

Line to Plane Intersection Solution to the Missed Collision Problem

You have witnessed the agent traveling right through the wall at high speed. This condition is even detected in Step 5 of `MyScript`. In general, a straightforward solution for an object traveling toward the wall is to define a line segment representing the current motion of the object, in this case, the line segment

$$l(d) = P_a + d\hat{A}_{dir}$$

and to compute the intersection of this line segment with the 2D plane that represent the wall

$$p \cdot \hat{V}_n = D$$

If the computed *d* value is less than zero, then the intersection position is behind the object and the object has overshot. Please refer to Figure 9-6 and observe that this computation is already performed. Now, modify `MyScript` to avoid the overshooting situation by reflecting the agent accordingly.

Your Own Quaternion Rotations

Replace the Unity `Quaternion.LookRotation()` function with your own quaternion rotation functions.

Location of Hero on the Agent

Unity capsules are defined with respect to their center. This is why only half of the pacing hero is above the agent. In this example, the height of the agent is exactly 1.0; you can place the hero above the agent by offsetting its position by 0.5 when computing the offset vector, `vhc`, in the `Update()` function

```
Vector3 vhc = Vh.x * vx + (Vh.y + 0.5f) * vy + Vh.z * vz;
```

With this fix, you can see the hero pacing on top of instead of "in" the agent.

Before-After Position Solution to the Missed Collision Problem

Examine the line to plane intersection solution to the missed collision problem and observe that the computation result is the actual amount of overshooting. This is invaluable information if precision is important. For example, you can always backtrack the object by the overshot amount and then perform the reflection. In other cases, such as in this example, where the precise position of the agent is of less consequence, there is a simpler solution. You can compute the in front of status for both the current and the next agent positions. If the status of these two positions is different, you know during this update, the agent will overshoot the wall. Notice that this solution only provides a binary answer, yes or no, and does not provide the information on the amount overshot. Now, modify `MyScript` to support this solution.

Proper Treasure Collision Support

It is somewhat annoying that the treasure bounding sphere interacts with the reflection vector and not the actual agent. For example, after the reflection, the treasure bounding sphere does not detect when the agent actually passes through it! Please modify `MyScript` to support the highlight of the treasure bounding sphere after the reflected agent collides with it instead of just its reflection vector.

What's Next

This book approached introductory mathematical concepts from the perspective of video game development. The relevant concepts in vectors are introduced, examined, and applied in solving problems related to this one application area. Through this book you have learned one of a large variety of flavors of vector applications. Though you haven't learned everything about vectors and their applications, what you have learned is a powerful tool set for solving some very important problems, both in and out of video games and other interactive graphical applications.

You have learned that quaternion rotations only work when the rotation axis passes through the origin. However, you have also witnessed and experienced that the integration with matrix math can resolve this limitation but no details were provided. It is hoped that the awareness of available, yet inaccessible information can serve as a motivation to continue this fun and rewarding journey of learning.

In the meantime, you can begin practicing and experimenting with your newly acquired powerful knowledge in vector applications. As a first step, you can tweak and enhance the example from this chapter in the following ways:

- Project shadows onto either side of the wall.
- Compute shadow size as a function of object distance or projection angle.
- Reflect the agent when it approaches from either side of the wall.
- Replace the wall definition to be based on three positions and support the definition of a 2D region for shadow casting and reflection.
- Include an external wind factor to affect the agent's motion.

Next, you can consider supporting "gaming features" in the form of challenges, accomplishments, and rewards. For example, include hazardous barriers that must be avoided, treasures that can be collected when passed in close proximity, and power ups in the form of speed increments when sufficient treasures are acquired. During this process, you should constantly apply object-oriented design principles and design separate classes to support and hide the behaviors of each element in the interaction.

As you can see, you are on your way to building your first agent exploration game! The key is to describe what you want, depict the solution with careful drawings and consistent symbol labels, and then implement and verify your solution, just as you have followed in this book. It is fun, and practice really does make perfect.

Index

A

K. Sung and G. Smith, *Basic Math for Game Development with Unity 3D*,
https://doi.org/10.1007/978-1-4842-9885-5

E, F

G

H

I, J, K

L

M, N, O

P

Q, R

S

T

U

V, W, X, Y, Z

Printed in the United States
by Baker & Taylor Publisher Services